OPTIMIZATION ALGORITHMS

FOR

NETWORKS

AND

GRAPHS

SECOND EDITION,   REVISED AND EXPANDED

JAMES R. EVANS

University of Cincinnati
Cincinnati, Ohio

EDWARD MINIEKA

The University of Illinois at Chicago
Chicago, Illinois

CRC Press
Taylor & Francis Group
Boca Raton London New York

CRC Press is an imprint of the
Taylor & Francis Group, an **informa** business

CRC Press
Taylor & Francis Group
6000 Broken Sound Parkway NW, Suite 300
Boca Raton, FL 33487-2742

First issued in paperback 2019

© 1992 by Taylor & Francis Group, LLC
CRC Press is an imprint of Taylor & Francis Group, an Informa business

No claim to original U.S. Government works

ISBN-13: 978-0-8247-8602-1 (hbk)
ISBN-13: 978-0-367-40280-8 (pbk)

Library of Congress Cataloging-in-Publication Data

Evans James R. (James Robert).
    Optimization algorithms for networks and graphs. – 2nd ed., rev.
and expanded / James R. Evans, Edward Minieka.
       p.   cm.
    Rev. ed. of: Optimization algorithms for networks and graphs /
Edward Minieka. c1978.
    Includes bibliographical references and index.
    ISBN 0-8247-8602-5
    1. Graph theory.   2. Network analysis (Planning)   3. Algorithms.
L Evans, James R.   II. Minieka, Edward.   Optimization algorithms
for networks and graphs.   III. Title.
QA166.E92   1992
658.4'032--dc20   92-3323
                                  CIP

Visit the Taylor & Francis Web site at
http://www.taylorandfrancis.com

and the CRC Press Web site at
http://www.crcpress.com

PREFACE TO THE SECOND EDITION

The 1970s ushered in an exciting era of research and applications of networks and graphs in operations research, industrial engineering, and related disciplines. Network optimization has been an important area of research and application in its own right and, in addition, is increasingly important as a component of broader and more powerful decision support systems.

This is a text about optimization algorithms for problems that can be formulated on graphs and networks. The first edition of this text was unique in providing a comprehensive, cohesive, and clear treatment of this body of knowledge. Many new results and applications have appeared in the literature since that time. This edition provides many new applications and algorithms while maintaining the classic foundations on which contemporary algorithms have been developed.

The major changes in this edition are described below.

1. The original chapters have been expanded and updated and include new material that originated over the past decade and a half. The latter includes such topics as partitioning algorithms for shortest paths, preflow-push algorithms for maximum flow, extensions of the postman problem, the network simplex method, and heuristic procedures for the traveling salesman problem. In addition, the introductory section of each chapter now includes many new and realistic application examples.

2. A new chapter on computer representation, algorithms, heuristics, and computational complexity has been added. The chapter on flow algorithms has been divided into two separate chapters: one on minimum-cost flow algorithms, and one dealing exclusively with maximum-flow algorithms.

3. A computer software package, NETSOLVE, developed by James Jarvis and Douglas Shier, has been integrated with the text to provide students with the ability to solve larger applications than one can expect to by hand.

Every effort has been made to retain the style and flavor of the first edition. The audience consists of advanced undergraduate or beginning graduate students in a variety of disciplines; an operations research or mathematics background is, of course, helpful but not essential. Many sections that rely on prior knowledge of linear programming can be skipped easily. The focus is on an intuitive approach to the inner workings, interdependencies, and applications of the algorithms. Their place in the hierarchy of advanced mathematics and the details of their computer coding are not stressed. The text can be treated comprehensively in a one-semester course or, less thoroughly but without significant omissions, in a one-quarter course.

We gratefully acknowledge the helpful comments provided by James P. Jarvis and Douglas R. Shier, as well as their permission to use NETSOLVE and examples from *Practical Aspects of Network Analysis*, copyright ©1986, 1989 by the authors.

James R. Evans
Edward Minieka

PREFACE TO THE FIRST EDITION

This is not another graph theory text; it is a text about algorithms—optimization algorithms for problems that can be formulated in a network or graph setting.

As a thesis student studying these algorithms, I became aware of their variety, elegance, and interconnections and also of the acute need to collect and integrate them between two covers from their obscure hiding places in scattered journals and monographs. I hope that I have been able in the confines of this text to make a stab at a comprehensive, cohesive, and clear treatment of this body of knowledge.

This text is self-contained at the level of an advanced undergraduate or beginning graduate student in any discipline. An operations research or mathematics background is, of course, helpful but hardly essential.

The text aims at an intuitive approach to the inner workings, interdependencies, and applications of the algorithms. Their place in the hierarchy of advanced mathematics and the details of their computer coding are not stressed.

Chapter 1 contains background information and definitions. Aside from Chapter 1, I have tried to make all chapters as independent of one another as possible, except where one algorithm uses another as a subroutine. Even in this situation, the reader can continue if he is willing to accept the subroutine algorithm on faith. This text can be treated comprehensively in a one-semester course and less thoroughly, but without significant omissions, in a one-quarter course.

It takes a lot of ink to go from source to sink. I wish to thank Randy Brown at Kent State University, Ellis Johnson at IBM, George Nemhauser at Cornell University, and Douglas Shier at the National Bureau of Standards for their careful readings and suggestions. Also, emphatic thanks to my colleague Leonard Kent for his confidence in this project and for years of encouragement. Lastly, thanks to my students who graciously endured three years of classroom testing of the various manuscript stages. Most of all, this book is for you and for your successors, everywhere.

Edward Minieka

CONTENTS

Preface to the Second Edition   iii
Preface to the First Edition   v

1 INTRODUCTION TO GRAPHS AND NETWORKS   1
  1.1 Introduction and Examples   1
  1.2 Some Essential Concepts and Definitions   5
  1.3 Modeling With Graphs and Networks   11
      Appendix: Linear Programming   18
      Exercises   22
      References   25

2 COMPUTER REPRESENTATION AND SOLUTION   27
  2.1 Introduction and Examples   27
  2.2 Data Structures for Networks and Graphs   27
  2.3 Algorithms   33
  2.4 Computational Complexity   35
  2.5 Heuristics   38
      Appendix: NETSOLVE—An Interactive Software Package
      for Network Analysis   39
      Exercises   43
      References   44

3   TREE ALGORITHMS   **46**
  3.1   Introduction and Examples   46
  3.2   Spanning Tree Algorithms   49
  3.3   Variations of the Minimum Spanning Tree Problem   54
  3.4   Branchings and Arborescences   59
        Appendix: Using NETSOLVE to Find Minimum Spanning Trees   67
        Exercises   72
        References   76

4   SHORTEST-PATH ALGORITHMS   **77**
  4.1   Introduction and Examples   77
  4.2   Types of Shortest-Path Problems and Algorithms   81
  4.3   Shortest Paths from a Single Source   82
  4.4   All Shortest-Path Algorithms   93
  4.5   The k-Shortest-Path Algorithm   102
  4.6   Other Shortest Paths   112
        Appendix: Using NETSOLVE to Solve Shortest-Path Problems   114
        Exercises   116
        References   121

5   MINIMUM-COST FLOW ALGORITHMS   **123**
  5.1   Introduction and Examples   123
  5.2   Basic Properties of Minimum-Cost Network Flow Problems   130
  5.3   Combinatorial Algorithms for Minimum-Cost Network Flows   138
  5.4   The Simplex Method for Minimum-Cost Network Flows   145
  5.5   The Transportation Problem   148
  5.6   Flows with Gains   151
        Appendix: Using NETSOLVE to Find Minimum-Cost Flows   167
        Exercises   171
        References   177

6   MAXIMUM-FLOW ALGORITHMS   **178**
  6.1   Introduction and Examples   178
  6.2   Flow-Augmenting Paths   184
  6.3   Maximum-Flow Algorithm   189
  6.4   Extensions and Modifications   198
  6.5   Preflow-Push Algorithms for Maximum Flow   201
  6.6   Dynamic Flow Algorithms   206
        Appendix: Using NETSOLVE to Find Maximum Flows   227
        Exercises   229
        References   233

7 MATCHING AND ASSIGNMENT ALGORITHMS   **234**
   7.1 Introduction and Examples   234
   7.2 Maximum-Cardinality Matching in a Bipartite Graph   239
   7.3 Maximum-Cardinality Matching in a General Graph   241
   7.4 Maximum-Weight Matching in a Bipartite Graph: The
       Assignment Problem   250
   7.5 Maximum-Weight Matching Algorithm   257
       Appendix: Using NETSOLVE to Find Optimal Matchings
       and Assignments   267
       Exercises   273
       References   276

8 THE POSTMAN AND RELATED ARC ROUTING PROBLEMS   **278**
   8.1 Introduction and Examples   278
   8.2 Euler Tours   279
   8.3 The Postman Problem for Undirected Graphs   283
   8.4 The Postman Problem for Directed Graphs   287
   8.5 The Postman Problem for Mixed Graphs   296
   8.6 Extensions to the Postman Problem   303
       Exercises   310
       References   315

**9 THE TRAVELING SALESMAN AND RELATED VERTEX
   ROUTING PROBLEMS**   **317**
   9.1 Introduction and Examples   317
   9.2 Basic Properties of the Traveling Salesman Problem   320
   9.3 Lower Bounds   328
   9.4 Optimal Solution Techniques   336
   9.5 Heuristic Algorithms for the TSP   341
   9.6 Vehicle Routing Problems   350
       Appendix: Using NETSOLVE to Find Traveling Salesman Tours   353
       Exercises   356
       References   360

10 LOCATION PROBLEMS   **362**
   10.1 Introduction and Examples   362
   10.2 Classifying Location Problems   364
   10.3 Mathematics of Location Theory   366
   10.4 Center Problems   372
   10.5 Median Problems   379
   10.6 Extensions   386
       Exercises   387
       References   388

11 PROJECT NETWORKS   **390**
    11.1 Introduction and Examples   390
    11.2 Constructing Project Networks   395
    11.3 Critical Path Method   398
    11.4 Generalized Project Networks   405
        Appendix: Using NETSOLVE to Find Longest Paths   411
        Exercises   413
        References   417

NETSOLVE USER'S MANUAL   **419**

Index   **465**

1
INTRODUCTION TO GRAPHS AND NETWORKS

1.1 INTRODUCTION AND EXAMPLES

Graph theory is a branch of mathematics that has wide practical application. Numerous problems arising in such diverse fields as psychology, chemistry, industrial and electrical engineering, transportation planning, management, marketing, and education can be posed as problems from graph theory. Because of this, graph theory is not only an area of interest in its own right but also a unifying basis from which results from other fields can be collected, shared, extended, and disseminated.

Unlike other scientific fields, graph theory has a definite birthday. The first paper on graphs was written by the Swiss mathematician Leonhard Euler (1707–1783) and was published in 1736 by the Academy of Science in St. Petersburg. Euler's study of graphs was motivated by the so-called Königsberg bridge problem. The city of Königsberg (now called Kaliningrad) in East Prussia was built at the junction of two rivers and the two islands formed by them (see Fig. 1.1). In all, there were seven bridges connecting the islands to each other and to the rest of the city. Euler was posed the following question: Could a Königsberger start from home and cross each bridge exactly once and return home? The answer, which Euler proved, is no. We shall see why later (Chapter 8) when we study a generalized version of this problem called the postman problem.

The growth of graph theory continued in the late nineteenth and early twentieth centuries with advances motivated by molecular theory and electrical theory. By the 1950s, the field had taken two essentially different directions: the *algebraic* aspects of graph theory and the *optimization* aspects of graph theory. The latter were greatly advanced by the advent of the computer and the discovery of linear programming techniques. This text is concerned almost exclusively with the optimization aspects of graph theory.

1

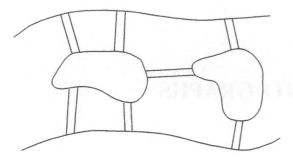

Fig. 1.1   The Königsberg bridge problem.

What is a graph? A graph consists of two parts, points and lines joining these points. The points can be depicted as points in a plane or, if you prefer, points without any specific physical location. For example, in the Königsberg bridge problem of Fig. 1.1, let us label each of the regions of land by the points 1, 2, 3, and 4 and join a line between two points whenever a bridge connects them. The graph corresponding to the Königsberg bridge problem is shown in Fig. 1.2. In keeping with standard terminology, we shall refer

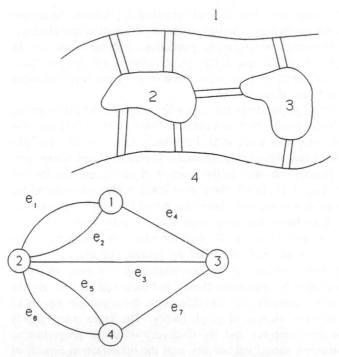

Fig. 1.2   Graph of the Königsberg bridge problem.

to the points of a graph as *vertices* and the lines of a graph as *edges*. Each edge often is given an identifier such as $e_1, e_2, \ldots$ or a letter of the alphabet $(a, b, \ldots)$.

The same graph could be specified without using a picture simply by listing the vertices 1, 2, 3, 4 and listing the edges $e_1 = (1, 2)$, $e_2 = (1, 2)$, $e_3 = (2, 3)$, $e_4 = (1, 3)$, $e_5 = (2, 4)$, $e_6 = (2, 4)$, $e_7 = (3, 4)$ as pairs of points. From any such listing, we can always draw a picture of the graph. For instance, suppose we provide a list of vertices 1, 2, 3, 4, 5 and edges $a = (1, 3)$, $b = (1, 4)$, $c = (2,3)$, $d = (3, 5)$, $e = (5, 1)$, $f = (5, 3)$. The graph corresponding to these sets is given in Fig. 1.3.

With this motivation, we can now state a formal definition for a graph:

A *graph G* is a set X whose elements are called *vertices* and a set E of pairs of vertices, called *edges*. The graph G is denoted by (X, E).

Thus, in the previous example, $X = \{1, 2, 3, 4, 5\}$ and $E = \{(1, 3), (1, 4), (2, 3), (3, 5), (5, 1), (5, 3)\}$. Throughout, we shall assume that both the set X and the set E contain only a finite number of elements.

Any situation in which you can list a set of elements and a relationship between *pairs* of elements can be depicted as a graph. For example, let X denote the set of street intersections in a city. Let E denote the set of all street segments that join the intersections. Clearly, (X, E) is a graph. Edges need not correspond to physical objects such as bridges or highways, nor must their lengths be proportional to actual physical distances. As another example, let X denote the set of all airports in Illinois. Let E denote the set of all pairs of airports (x, y) such that there is a nonstop commercial flight between airport x and airport y. Then (X, E) is a graph. Throughout this book we shall see many more examples of problems that can be modeled as graphs.

Whenever the set E consists of *unordered* pairs of vertices, we have an *undirected graph*. In an undirected graph, an edge (x, y) and an edge (y, x) are indistinguishable; we are only concerned with the endpoints of each edge. Both the graphs in Figs. 1.2 and 1.3 are undirected.

Some graphs have more than one edge between a given pair of vertices. In Fig. 1.3, for example, we see that there are two edges between vertices 3 and 5. Since these edges

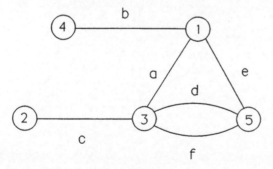

Fig. 1.3   An example of a graph.

are undirected, either of them can be written as (5, 3) or as (3, 5). Whenever we have multiple edges between the same two vertices, we can denote them by a subscripted pair of vertices if we wish to distinguish between them. Thus, in Fig. 1.3, we could denote edge d by $(3, 5)_1$ and edge f by $(3, 5)_2$. When no confusion will develop, we shall omit the subscripts.

In many practical situations, however, such as when we have one-way streets, the *direction* of an edge must be specified. We specify direction in a graph by drawing arrows instead of lines between the vertices. Directed edges are often called *arcs*, and the graph is called a *directed graph*. Vertices in directed graphs are more commonly called *nodes*. The set of arcs is denoted by A. An example of a directed graph is shown in Fig. 1.4. The set $A = \{(1, 2), (2, 1), (2, 3), (2, 4), (4, 3), (3, 1)\}$. Notice that in this case the arcs (1, 2) and (2, 1) are not identical, but represent two different ordered pairs of vertices. If (x, y) is a directed arc, then x is called the *tail* of the arc and y is called the *head* of the arc. As an example of a directed graph, let X denote the set of all passengers aboard a certain transatlantic flight. Let A denote the set of all pairs (x, y) of passengers such that passenger x is older than passenger y and both speak a common language. Clearly, (X, A) is a directed graph with node set X and arc set A. Is it possible for this graph to have both an arc (x, y) and an arc (y, x)?

In this book, we shall normally use the terms *vertex* and *edge* when referring to an undirected graph and the terms *node* and *arc* when referring to a directed graph. We will use the notation (X, E) to denote an undirected graph with vertex set X and edge set E, and we will use the notation (X, A) to denote a directed graph with node set X and arc set A. Unless otherwise stated, we will normally use m to denote the number of vertices or nodes, and n to denote the number of edges or arcs. Occasionally, however, an algorithm or concept might apply to either a directed or an undirected graph. In these cases, either pair of terms can be used, but no confusion should result in context. We caution you that many of the definitions used throughout this book are not uniform in the literature on graphs and networks. Thus, you must be careful when reading other references on the subject.

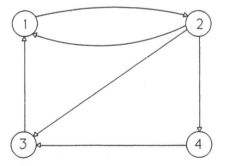

Fig. 1.4 An example of a directed graph.

A *network* is a graph with one or more numbers associated with each edge or arc. These numbers might represent distances, costs, reliabilities, or other relevant parameters.

Graph and network optimization represents one of the most important areas of operations research for several reasons. Graphs and networks can be used to model many diverse applications such as transportation systems, communication systems, vehicle routing, production planning, and cash flow analysis. They are more readily accepted by nontechnical people since they are often related to the physical system, which enhances model understanding. Finally, graphs and networks have special properties that allow the solution of much larger problems than can be handled by any other optimization technique and are much more computationally efficient than other types of solution procedures.

Each remaining chapter of this text is devoted to a single type of practical optimization problem on a graph or network. The procedures that are given for seeking optimum solutions to these problems are called *algorithms*; hence, the title of this text. Algorithms in general will be discussed in more detail in Chapter 2. As some of these optimization algorithms build upon others, the order of presentation is restricted, and these considerations have dictated the sequencing of the chapters of this text.

1.2 SOME ESSENTIAL CONCEPTS AND DEFINITIONS

To decrease the dependence between chapters, some basic concepts and definitions that are needed throughout are presented here. The motivation for these definitions will be reserved to later chapters where applications are discussed in greater depth. Unless specified otherwise, these definitions apply to both directed and undirected graphs.

We assume that the reader is familiar with basic set notation such as set membership ($\in$), union ($\cup$), intersection ($\cap$), and subset ($\subseteq$). If S is a set, then $S \cup \{e\}$ will be written as $S + e$. If $e \in S$, then $S - \{e\}$ will be written as $S - e$. $|S|$ denotes the *cardinality* of the set S, that is, the number of elements in S. Finally, $S \oplus T = S \cup T - S \cap T$ is the *symmetric difference* of sets S and T.

An edge that has both of its endpoints as the same vertex is called a *loop*. In Fig. 1.5, edge e_2 is a loop. (Most graphs that we shall consider in this text will contain neither loops nor multiple edges. This will be assumed unless otherwise stated.) If a graph is loopless, does not contain multiple edges, and $|X| = m$ and $|E| = n$ (that is, the graph has m

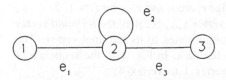

Fig. 1.5 A graph containing a loop.

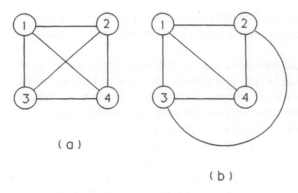

(a)

(b)

Fig. 1.6 Example of a planar graph.

vertices and *n* edges), then $n \leq m \ (m - 1)/2$. Why? A graph in which every pair of vertices are connected by an edge is called a *complete graph*.

A graph is *planar* if it *can* be drawn with no two edges crossing each other. The graph in Fig. 1.6a is planar even though it is drawn with two edges crossing each other. Figure 1.6b shows how it can be redrawn in a different fashion.

Every planar graph *G* has a *geometric dual*, which we denote by G^*. The dual graph is constructed by placing a vertex in every region of the original graph and connecting the new vertices by drawing an edge across every original edge. Figure 1.7 shows the construction of a dual graph. The dual of a dual graph always results in the original graph.

A vertex and an edge are said to be *incident to one another* if the vertex is an endpoint of the edge. Thus in Fig. 1.8 edge e_1 is incident with vertex 1 and with vertex 2. Two edges are said to be *adjacent* if they are both incident to the same vertex. In Fig. 1.8, edges e_1 and e_2 are adjacent because they are both incident with vertex 2. Two vertices are said to be *adjacent* to one another if there is an edge joining them. In Fig. 1.8, vertices 2 and 3 are adjacent to one another because there is an edge (e_2) that joins them.

The *degree* of a vertex is the number of edges incident with it. Thus, in Fig. 1.8 the degree of vertex 2 is three. For directed graphs, we could also define the *in-degree* of a node as the number of arcs directed toward it and the *out-degree* of a node as the number of arcs directed from it. Thus, in Fig. 1.9 the degree of node 2 is 3, the in-degree is 2, and the out-degree is 1.

Consider any sequence $x_1, x_2, \ldots, x_n, x_{n+1}$ of vertices. A *path* is any sequence of edges $e_1, e_2, \ldots, e_n$ such that the endpoints of edge e_i are x_i and x_{i+1} for $i = 1, 2, \ldots, n$. Vertex x_1 is called the *initial vertex* of the path; vertex x_{n+1} is called the *terminal vertex* of the path. The path is said to extend from its initial vertex to its terminal vertex. The *length* of the path equals the number of edges in the path. In Fig. 1.8, the sequence of edges e_1, e_2, e_3 forms a path of length 3 from vertex 1 to vertex 4.

The concept of a path in a directed graph is the same as in an undirected graph. That is, it does not matter what the directions of the arcs are in a path in a directed graph; either

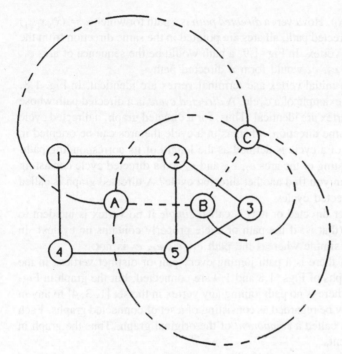

Fig. 1.7 Construction of the geometric dual graph. (———) G; (– – –) G^*.

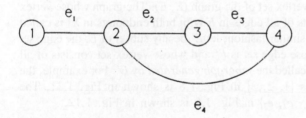

Fig. 1.8 An undirected graph.

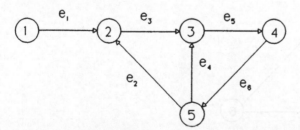

Fig. 1.9 A directed graph.

$e_i = (x_i, x_{i+1})$ or $e_i = (x_{i+1}, x_i)$. However, a *directed path* is a path for which $e_i = (x_i, x_{i+1})$ for $i = 1, 2, \ldots, n$. In a directed path, all arcs are pointed in the same direction from the initial vertex to the terminal vertex. In Fig. 1.9, a path would be the sequence of arcs e_1, e_2, e_4, e_5, whereas arcs e_1, e_3, e_5 would form a directed path.

A *cycle* is a path whose initial vertex and terminal vertex are identical. In Fig. 1.8, edges e_2, e_3, e_4 would be an example of a cycle. A *directed cycle* is a directed path whose initial vertex and terminal vertex are identical. Thus, for a directed graph, a directed cycle has all arcs pointed in the same direction, whereas in a cycle the arcs can be oriented in either direction. The length of a cycle is defined as the length of its corresponding path. Figure 1.9 has a cycle consisting of the arcs e_2, e_3, and e_4 and a directed cycle consisting of the arcs e_4, e_5, and e_6. Can you find another directed cycle? A directed graph is called *acyclic* if it contains no directed cycles.

A path or cycle (whether directed or not) is called *simple* if no vertex is incident to more than two of its edges (that is, if the path or cycle properly contains no cycles). In Fig. 1.9, the path e_1, e_2 is simple whereas the path e_1, e_3, e_4, e_2 is not.

A graph is *connected* if there is a path joining every pair of distinct vertices in the graph. For example, the graphs of Figs. 1.8 and 1.9 are connected, but the graph in Fig. 1.10 is not connected since there is no path joining any vertex in the set $\{1, 3, 4\}$ to any in the set $\{2, 5, 6\}$. A graph may be regarded as consisting of a set of connected graphs. Each of these connected graphs is called a *component* of the original graph. Thus the graph in Fig. 1.10 has two components.

A graph g is a *subgraph* of a graph G if all the vertices and edges of g are in G and each edge of g has the same two end vertices that it has in G.

Let X' be any subset of X, the vertex set of the graph (X, E). The graph whose vertex set is X' and whose edge set consists of all edges in E with both endpoints in X' is called the *subgraph generated by X'*. In a similar fashion, let E' be any subset of E, the edge set of the graph (X, E). The graph whose edge set is E' and whose vertex set consists of all vertices incident to an edge in E' is called the *subgraph generated by E'*. For example, the subgraph generated by the vertices $\{1, 2, 3\}$ in Fig. 1.8 is shown in Fig. 1.11. The subgraph generated by the arcs $\{e_1, e_2, e_3\}$ in Fig. 1.9 is shown in Fig. 1.12.

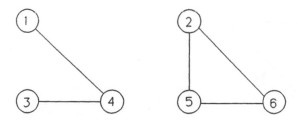

Fig. 1.10   A disconnected graph.

Fig. 1.11 Subgraph generated by a set of vertices in Fig. 1.8.

A graph is called a *tree* if it satisfies two conditions:

1. The graph is connected.
2. The graph contains no cycles.

In Fig. 1.8, the subgraphs generated by the following sets of edges each form a tree: $\{e_1, e_3, e_4\}$, $\{e_3, e_4\}$, $\{e_1, e_2, e_4\}$, $\{e_3\}$. The edges $\{e_1, e_2, e_3, e_4\}$ and their incident vertices do not form a tree since the subgraph generated contains a cycle.

A *forest* is any graph that contains no cycles. Thus, a forest consists of one or more trees. A *spanning tree* of a graph is any tree that includes every vertex in the graph. In Fig. 1.9 the subgraph generated by arcs $\{e_1, e_2, e_3, e_5\}$ forms a spanning tree since it includes each vertex 1, 2, 3, 4, and 5. Clearly, no spanning tree can exist for a graph with more than one component, and every connected graph possesses a spanning tree. An edge of a spanning tree is called a *branch*; an edge in the graph that is not in the spanning tree is called a *chord*.

A tree with one edge contains two vertices; a tree with two edges contains three vertices; a tree with three edges contains four vertices; and, in general, a tree with $m - 1$ edges must contain m vertices. Hence, each spanning tree of a (connected) graph with m vertices consists of $m - 1$ edges.

Since a tree is necessarily connected, there is at least one path between any two vertices. Suppose there were more than one path between a pair of vertices. Taking these together, we would create a cycle. But a tree has no cycles. Therefore, there must be a *unique path* between any two vertices in a tree.

A set of edges in a connected graph whose removal from the graph increases the number of components in the graph is called a *disconnecting set*. If the addition of any edge of a disconnecting set necessarily decreases the number of components in the remaining graph, the disconnecting set is called a *cutset*. A cutset that contains as a proper

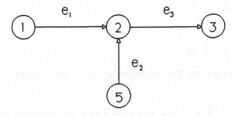

Fig. 1.12 Subgraph generated by a set of arcs in Fig. 1.9.

subset no other cutset is called a *simple cutset* (also called a *proper cutset* or *minimal disconnecting set*). In Fig. 1.3, the set of edges {*a, b, c*} form a disconnecting set since the number of components in the remaining graph is increased to three. However, this is not a cutset since if edge *a* is readded to the graph, the number of components remains three. The set {*b, c*} is a cutset of the graph, but is not a simple cutset since {*b*} and {*c*} alone are also cutsets.

A *flow* is a way of sending objects from one place to another in a network. For example, the shipment of finished goods from a manufacturer to a distributor, the movement of people from their homes to places of employment, and the delivery of letters from their point of posting to their destinations can all be regarded as flows. Many important problems involve flows in networks. For example, one might wish to maximize the amount transported from one place to another, or might wish to determine the least-cost way to send a given number of objects from their sources of supply to their destinations.

The objects that travel or flow through a network are called *flow units* or *units*. Flow units can be finished goods, people, letters, information, or almost anything. Nodes from which units begin their journey through a network are called *source nodes*. Nodes to which units are routed are called *sink nodes*. Source nodes usually have a *supply*, which represents the number of units available at the node. Sink nodes usually have a *demand*, representing the number of units which must be routed to them.

In a flow problem, each arc often has a *capacity* associated with it. The capacity of an arc is the maximum allowable flow that may travel over it. Arc capacities are simply upper bounds on the flow units.

The fundamental equation governing flows in networks is known as *conservation of flow*. Simply stated, conservation of flow states that at every node

Flow out − flow in = 0

The supply at a source node is considered as "flow in"; the demand at a sink node is considered as "flow out." To illustrate this, consider the network in Fig. 1.13. Node 1 is a source node with a supply of 10; nodes 2 and 4 are sink nodes with demands of 5 each. Node 3 has neither supply nor demand; it is called a *transshipment node*. Let $f(x,y)$ denote the flow from node x to node y. The conservation of flow equations are written as

Node 1: $f(1, 2) + f(1, 3) - 10 = 0$

Node 2: $5 + f(2, 4) - f(1, 2) - f(4, 2) = 0$

Node 3: $f(3, 4) - f(1, 3) = 0$

Node 4: $5 + f(4, 2) - f(2, 4) - f(3, 4) = 0$

If we rewrite these equations, putting all constants on the right-hand side, we obtain

$f(1, 2) + f(1, 3) = 10$
$f(2, 4) - f(1, 2) - f(4, 2) = -5$
$f(3, 4) - f(1, 3) = 0$
$f(4, 2) - f(2, 4) - f(3, 4) = -5$

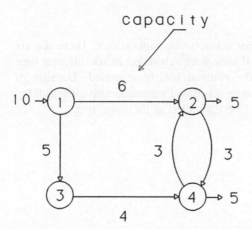

Fig. 1.13 Illustration of conservation of flow.

Notice that the supplies will be positive numbers whereas demands will be negative numbers. Also, note that each variable appears in exactly two equations, one with a $+1$ coefficient and one with a -1 coefficient. The node (equation) corresponding to the $+1$ corresponds to the tail of the arc, while the node corresponding to the -1 corresponds to the head of the arc.

The arc capacities in Fig. 1.13 can be written as the constraints

$$f(1, 2) \leq 6$$
$$f(1, 3) \leq 5$$
$$f(3, 4) \leq 4$$
$$f(2, 4) \leq 3$$
$$f(4, 2) \leq 3$$

A flow is called *feasible* if it satisfies the conservation of flow and capacity constraints. Since the conservation of flow equations and capacity constraints are linear, you might suspect that flow problems involve linear programming. Indeed, they are one of the most important classes of linear programming problems, as we shall see in a later chapter.

1.3 MODELING WITH GRAPHS AND NETWORKS

As we have noted, graphs and networks can be used to model many important problems in engineering, business, and the physical and social sciences. In this section we shall present several examples of how graphs and networks can be useful modeling tools.

A Scheduling Problem

A university is sponsoring a half-day seminar on network applications. There are six speakers, each planning to speak one hour. If they were scheduled in six different time slots, the seminar would extend beyond the planned four-hour period. Because of nonoverlapping interests, certain speakers can be scheduled simultaneously. The following matrix shows which speakers *should not* be scheduled at the same time.

	1	2	3	4	5	6
1					x	
2						
3				x	x	x
4			x		x	x
5	x		x	x		x
6			x	x	x	

We can model this problem as a graph by letting the speakers correspond to vertices. An edge is drawn between two vertices if the speakers should not be scheduled at the same time. This is shown in Fig. 1.14. A solution exists if the vertices can be labeled A, B, C, and D (corresponding to the four time slots) such that no two adjacent vertices have the same label. A related problem would be to find the minimum number of time slots necessary to schedule all speakers.

Offshore Pipeline Design

An oil company owns several oil drilling platforms in the Gulf of Mexico. Oil that is recovered from each platform must be shipped to refineries in Louisiana. A network of pipelines can be constructed between the platforms and the Louisiana shore. How should the pipeline network be designed to minimize the construction cost?

Each drilling platform can correspond to a vertex of a graph. The edges in the graph can be assigned weights representing the distance between platforms. One vertex also corresponds to the refinery. The problem then becomes one of finding a connected subgraph having the minimum total distance. The graph in Fig. 1.15 illustrates a small example.

Clearly, the pipeline network must be connected. We claim that the optimal solution must be a spanning tree. If it were not, then it would contain a cycle. This means that there would be two different ways to transport the oil from one vertex to another on the cycle. Clearly, one could remove some edge from the cycle and reduce the total cost while still retaining a connected graph. Can you find the minimum-cost spanning tree in Fig. 1.15?

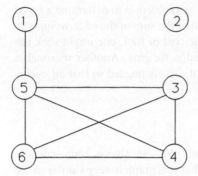

Fig. 1.14 Graph for the speaker scheduling problem.

Delivering "Meals on Wheels"

"Meals on Wheels" is a service which delivers prepared lunches to people who are unable to shop or cook for themselves. The meals are prepared early in the morning and delivered from a services center to the individual locations between 10 a.m. and 2 p.m. The program director needs to decide how many drivers are needed and in what order the meals should be delivered to ensure that all will be delivered within the specified time window.

The problem can be modeled as a graph by letting each delivery location as well as the services center correspond to a vertex on a graph. Edges connecting the vertices represent the shortest route that can be driven between the locations. Each edge can be

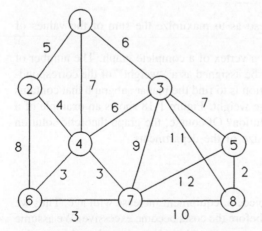

Fig. 1.15 An oil pipeline network.

assigned the travel time between the incident vertices. The problem is to determine a set of cycles that pass through the services center vertex such that the sum of the edge weights in each cycle does not exceed four hours. To minimize the cost of fuel, one might seek the set of feasible cycles having the smallest sum of the edge weights. Another reasonable problem would be to determine the minimum number of drivers needed so that all cycles are feasible.

Facility Layout

Graph theory has useful applications in facility layout. We can always represent any layout as a planar graph as shown in Fig. 1.16. Notice that this graph is very similar to the concept of a dual graph. We associate a vertex with each department in the layout and draw an edge between two vertices if the departments are adjacent. The more difficult problem is to find an appropriate graph in order to design a layout.

Suppose that we wish to lay out m facilities, for example, different departments in a library. A traffic study might be conducted, resulting in the construction of a matrix of specifying the number of trips made between each pair of facilities. This matrix is as follows:

		1	2	3	4	5
1.	Entrance	—				
2.	Catalogue	200	—			
3.	Photocopy	4	77	—		
4.	Journals	80	125	64	—	
5.	New books	32	42	19	26	—

We would like to locate the departments so as to maximize the sum of the values of *adjacent* pairs of departments.

We may represent each department by a vertex of a complete graph. The number of trips between each pair of departments will be assigned as a "weight" of the corresponding edge of the graph (Fig. 1.17). The solution is to find the planar subgraph that contains all the vertices and has maximum total edge weight. Figure 1.18 shows an example of a feasible solution. Can you find a better solution? Of course, the graph theoretic solution must be scaled to reflect the appropriate size of the departments.

Equipment Replacement

The operating and maintenance costs of a piece of equipment increase with age. Thus we want to replace the equipment periodically before the costs become excessive. We assume that we know the costs over the life of the equipment and can determine the salvage value at the end of each year. Let c_{ij} represent the total cost of purchasing a new piece of

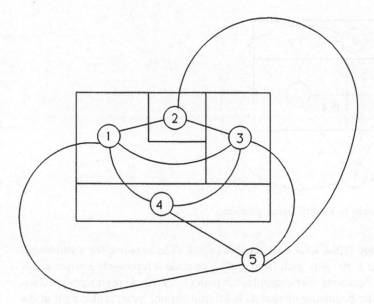

Fig. 1.16 A department layout and its planar graph.

equipment at the beginning of period i and selling it at the beginning of period j. This value would consist of the purchase cost, plus all operating and maintenance costs for periods i through $j - 1$, less the salvage value at the end of period $j - 1$.

We can find the optimal replacement plan over a time horizon of n periods by letting each period correspond to a node of a network. An arc from node i to node j represents purchasing the equipment at the beginning of period i and keeping it through period $j - 1$.

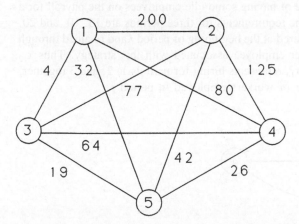

Fig. 1.17 Complete graph for facility layout problem.

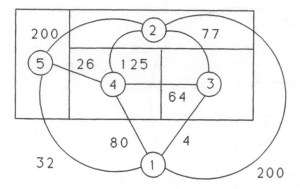

Fig. 1.18 Feasible solution to facility layout problem.

Associated with each arc is the total cost for this policy. The network for a three-year period is shown in Fig. 1.19. Any path from node 1 to node 4 represents a sequence of purchasing and selling decisions. For example, the path (1, 2), (2, 4) represents purchasing a new machine at the beginning of year 1, holding it for one year, replacing it at the beginning of year 2, and holding this machine for two years. Since each arc has a cost associated with it, the shortest path through the network will provide the minimal-cost policy.

Employment Scheduling

In many industries, the demand for services varies considerably over time. One strategy for reducing costs is to employ part-time workers for short periods of time. The minimum number of employees required in each time period is known. We seek to balance the cost of hiring and firing with the expense of having some idle employees on the payroll for a short period of time. Suppose that the requirements for three periods are 15, 10, and 20. Let x_{ij} be the number of employees hired at the beginning of period i and retained through period $j - 1$ and c_{ij} be the cost per employee associated with this strategy. Thus x_{12} represents hiring for period 1 only; x_{14} represents hiring for periods 1, 2, and 3 together. Finally, let s_j be the excess number of workers employed in period j.

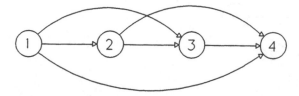

Fig. 1.19 Equipment replacement problem network.

The number of workers available in period 1 is

$$x_{12} + x_{13} + x_{14}$$

Using the relationship that the number available less the excess must equal the requirements, we have

$$x_{12} + x_{13} + x_{14} - s_1 = 15$$

Similarly, for periods 2 and 3 we have

$$x_{13} + x_{14} + x_{23} + x_{24} - s_2 = 10$$

and

$$x_{14} + x_{24} + x_{34} - s_3 = 20$$

If we subtract the first equation from the second, and the second equation from the third, and multiply the third equation by -1, we have the following

$$
\begin{array}{rcl}
x_{12} + x_{13} + x_{14} & - s_1 & = 15 \\
-x_{12} + x_{23} + x_{24} + s_1 - s_2 & & = -5 \\
- x_{13} - x_{23} + x_{34} + s_2 - s_3 & & = 10 \\
- x_{14} - x_{24} - x_{34} + s_3 & & = -20
\end{array}
$$

Each column on the left-hand side has exactly one $+1$ and one -1; this is characteristic of all network flow problems. Through algebraic manipulation we have created a set of equations that represents the node-arc incidence matrix of a network flow problem. If we associate a node with each equation, we can draw a directed arc corresponding to each column from the row with the $+1$ to the row with the -1. This leads to the network in Fig. 1.20. The sign of the right-hand side values determines whether the nodes are sources or sinks. If we assign a cost to each arc, we can find the flow that minimizes the total cost.

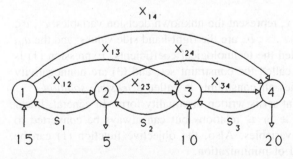

Fig. 1.20 Employment scheduling network.

These examples show only a fraction of the types of problems that can be modeled and solved using graphs and networks. Throughout this book you will see many more examples.

APPENDIX: LINEAR PROGRAMMING

Many of the network optimization problems considered in subsequent chapters can be formulated as linear programming problems. Often, insights from the linear programming formulation can assist in developing efficient solution procedures. This section gives a brief review of linear programming results that will be needed later in this book. The presentation here, however, is by no means intended to produce a profound or intuitive understanding of linear programming. A more complete treatment of linear programming can be found in the references.

Basic Concepts and Theory

The general form of a linear program is

$$\text{Minimize} \quad c_1x_1 + c_2x_2 + \cdots + c_nx_n \tag{1}$$

subject to

$$a_{11}x_1 + a_{12}x_2 + \cdots + a_{1n}x_n = b_1 \tag{2}$$
$$a_{21}x_1 + a_{22}x_2 + \cdots + a_{2n}x_n = b_2$$
$$\vdots$$
$$a_{m1}x_1 + a_{m2}x_2 + \cdots + a_{mn}x_n = b_m$$
$$x_1, x_2, \cdots, x_n \geq 0 \tag{3}$$

In this linear program, $x_1, x_2, \ldots, x_n$ represent the unknown decision variables; $c_1, c_2, \ldots, c_n$ are cost coefficients; $b_1, b_2, \ldots, b_m$ are the right-hand side values; and the a_{ij}, $i = 1$ to m and $j = 1$ to n, are called the technological coefficients. Expression (1) is called the objective function, (2) is called the constraint set, and (3) are nonnegativity constraints. Any solution that satisfies all constraints is called a *feasible solution*.

In this formulation, the constraint set is written in equality form. In general, linear programming constraints may have $\geq$ or $\leq$ relations but can always be converted to equalities by the addition of slack variables. Also, the objective function (1) can be expressed as a maximization instead of minimization.

Two other mathematical representations of the formulation given above are often used for notational convenience. One way is to express the problem in summation notation:

Minimize $\displaystyle\sum_{j=1}^{n} c_j x_j$

subject to

$$\sum_{j=1}^{n} a_{ij} x_j = b_i, \qquad i = 1, 2, \ldots m$$
$$x_j \geq 0, \qquad j = 1, 2, \ldots n$$

A second way is to use matrix notation. Let $\mathbf{c} = [c_1, c_2, \ldots, c_n]$, $\mathbf{x} = [x_1, x_2, \ldots, x_n]$, $\mathbf{b} = [b_1, b_2, \ldots, b_m]$, $\mathbf{0}$ be an n-dimensional vector of zeros, and A be the matrix of the technological coefficients a_{ij}. An equivalent formulation is (we assume that vectors are transposed appropriately for the proper multiplication)

Minimize $\mathbf{cx}$

subject to

$A\mathbf{x} = \mathbf{b}$

$\mathbf{x} \geq \mathbf{0}$

Every linear program (called the *primal problem*) has a related problem called the *dual problem*. We associate a dual variable u_i with each primal constraint. If the primal objective is to minimize, the dual objective is maximization. The sign restrictions on the dual variables are determined by the type of primal constraint. If the primal problem is minimization, then an equality constraint results in u_i being unrestricted; if the primal constraint is $\geq$, the $u_i \geq 0$; if the primal constraint is $\leq$, then $u_i \leq 0$. The dual to problem (1)–(3) is

Maximize $\quad b_1 u_1 + b_2 u_2 + \cdots + b_m u_m$ $\qquad\qquad\qquad$ (1′)

subject to

$a_{11} u_1 + a_{21} u_2 + \cdots + a_{m1} u_m \leq c_1$ $\qquad\qquad\qquad$ (2′)

$a_{12} u_1 + a_{22} u_2 + \cdots + a_{m2} u_m \leq c_2$

$\qquad\qquad\vdots$

$a_{1n} u_1 + a_{2n} u_2 + \cdots + a_{mn} u_m \leq c_n$

$u_1, u_2, \cdots, u_m$ unrestricted in sign $\qquad\qquad\qquad$ (3′)

In terms of the alternate notation, this dual problem can be expressed in summation notation:

Maximize $\displaystyle\sum_{i=1}^{m} b_i u_i$

subject to

$\displaystyle\sum_{i=1}^{m} a_{ij} \leq c_j, \qquad j = 1, \ldots n$

u_i unrestricted

or in matrix notation:

Maximize **bu**

subject to

$\mathbf{u}A \leq \mathbf{c}$

u unrestricted

Suppose we know a feasible solution for a linear program and a feasible solution for its dual problem. A fundamental result in linear programming is that these primal and dual feasible solutions are both optimal if they satisfy *complementary slackness conditions*. Using the matrix notation, we have

$\mathbf{u}(A\mathbf{x} - \mathbf{b}) = \mathbf{u}A\mathbf{x} - \mathbf{u}\mathbf{b} = 0$

and

$(\mathbf{c} - \mathbf{u}A)\mathbf{x} = \mathbf{c}\mathbf{x} - \mathbf{u}A\mathbf{x} \geq 0$

It follows that $\mathbf{u}\mathbf{b} = \mathbf{u}A\mathbf{x} \leq \mathbf{c}\mathbf{x}$; that is, the value of the dual objective function (max) is always less than or equal to the value of the primal objective function (min). If $\mathbf{u}\mathbf{b} = \mathbf{c}\mathbf{x}$ then clearly **x** and **u** must be optimal to their respective problems. This is true whenever $(\mathbf{c} - \mathbf{u}A)\mathbf{x} = 0$. The complementary slackness conditions are

$\mathbf{u}(A\mathbf{x} - \mathbf{b}) = 0 \quad \text{and} \quad (\mathbf{c} - \mathbf{u}A)\mathbf{x} = 0$

Any feasible solutions satisfying these conditions will be optimal.

The Simplex Algorithm

The simplex algorithm is a method for solving linear programs by systematically moving from one solution to another until the optimal solution is found (or the problem is determined to be unbounded or infeasible). Consider the problem

$$\min \quad z = \mathbf{cx}$$
$$A\mathbf{x} = \mathbf{b}$$
$$\mathbf{x} \geq \mathbf{0}$$

We can partition A into a set of m linearly independent columns called the basis matrix B and the remaining columns called the nonbasis matrix N. Letting $\mathbf{x}_B$ and $\mathbf{x}_N$ be the associated variables, the problem can be written as

$$\min \quad z = \mathbf{c}_B\mathbf{x}_B + \mathbf{c}_N\mathbf{x}_N$$
$$B\mathbf{x}_B + N\mathbf{x}_N = \mathbf{b}$$
$$\mathbf{x}_B, \ \mathbf{x}_N \geq 0$$

Solving the constraint set for $\mathbf{x}_B$ and substituting, we obtain

$$\min \quad z = \mathbf{c}_B B^{-1}\mathbf{b} + (\mathbf{c}_N - \mathbf{c}_B B^{-1}N)\mathbf{x}_N$$
$$\mathbf{x}_B + B^{-1}N\mathbf{x}_N = B^{-1}\mathbf{b}$$
$$\mathbf{x}_B, \ \mathbf{x}_N \geq \mathbf{0}$$

If $\mathbf{x}_N = \mathbf{0}$, the solution is called a *basic feasible solution*. By defining $\mathbf{c}_B B^{-1}\mathbf{b}$ as z_0, $B^{-1}\mathbf{b}$ as $\mathbf{b}'$, $\mathbf{c}_B B^{-1}N$ as the vector $\mathbf{z}$, the columns of $B^{-1}N$ as the vectors $\mathbf{y}_j$, and R as the index set of the variables $\mathbf{x}_N$, we have

$$\min \quad z = z_0 + \sum_{j \in R} (c_j - z_j)x_j$$
$$\mathbf{x}_B + \sum_{j \in R} \mathbf{y}_j x_j = \mathbf{b}'$$
$$\mathbf{x}_B, \ \mathbf{x}_N \geq \mathbf{0}$$

The solution $\mathbf{x}_B = \mathbf{b}'$ is optimal if $c_j - z_j \geq 0$ for all $j \in R$. If not, then choose the index k with the most negative value of $c_j - z_j$. The variable x_k will enter the basis and replace one of the current basic variables. We determine which variable to leave the basis by finding the minimum $\{b'_i/y_{ik}, y_{ik} > 0\}$, where b'_i is the ith component of $\mathbf{b}'$. This is the maximum value that x_k can be increased until a basic variable is reduced to zero. Suppose this occurs for the index r. Then the rth basic variable will leave the basis. We *pivot* to a new basic feasible solution by setting $x_k = b'_r/y_{rk}$ and $x_{Bi} = b'_i - (b'_r/y_{rk})y_{ik}$ for i not equal to r. The column corresponding to x_k replaces the column corresponding to x_{Br} in the basis B and the procedure is repeated.

To summarize, the steps of the simplex algorithm are

1. *Priceout*: Determine which nonbasic variable should enter the basis to improve the objective function.

2. *Ratio test*: Determine which current basic variable would leave the basis first as the value of the new nonbasic variable is increased.
3. *Pivot*: Exchange the entering and exiting variables in the basis and adjust the values of all remaining basic variables.

In Chapter 6 we shall see how properties of networks enable the simplex algorithm to be streamlined very efficiently.

EXERCISES

1. Consider the graph shown in Fig. 1.21. Formally list the sets X and E that define this graph.

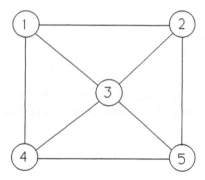

Fig. 1.21 Graph for Exercise 1.

2. Draw the graph corresponding to the following sets: $X = \{1, 2, 3, 4, 5, 6\}$ and $E = \{(1, 2), (1, 3), (1, 6), (2, 4), (2, 6), (5, 6)\}$.

3. Is a five-vertex complete graph planar? What can you say about complete graphs with more than five vertices?

4. Construct the dual graph for the graph in Exercise 1 (Fig. 1.21).

5. Show that for any graph G the number of vertices with odd degree is even.

6. Show that for any directed graph, the sum of the out-degrees must equal the sum of the in-degrees.

7. Find all simple cycles in the graph of Exercise 1 (Fig. 1.21).

8. For the graph in Fig. 1.21, construct the subgraph generated by the vertices $\{1, 2, 3, 4\}$.

9. For the graph in Fig. 1.21, construct the subgraph generated by the edges (1, 4), (3, 4), (3, 5), and (4, 5).

10. For the graph in Fig. 1.21, find five different spanning trees.

11. Is it possible for a cutset and a cycle to contain exactly one edge in common? Why?

12. Is there a relationship between cutsets in a graph and cycles in its dual graph? If so, what is it?

13. Construct a graph whose vertex set is the set of courses you are required to pass for your degree. Place an arc from vertex x to vertex y if course x is a prerequisite for course y. Give an interpretation for each of the following:
    (a) Path
    (b) Directed path
    (c) Cycle
    (d) Directed cycle
    (e) Connected component

14. Is every subset of a tree also a tree? Is every subset of a forest also a forest?

15. Show that if T is a spanning tree, then $T + e$, where e is a chord, contains exactly one cycle.

16. Suppose forest F consists of t trees and contains v vertices. How many edges are in forest F?

17. Let the subgraph generated by edges $\{e_1, e_2, e_3, e_4\}$ in Fig. 1.22 define a spanning tree. The set of cycles formed by adding each chord to the tree, one at a time, is called a *fundamental set of cycles*. Show that any cycle in the graph can be expressed as the symmetric difference of some subset of fundamental cycles. Does this have a familiar analogy with a well-known result in linear algebra?

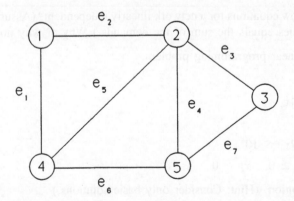

Fig. 1.22 Graph for Exercise 17.

18.   Any partition of vertices into two sets generates a cutset. For a graph of n vertices, how many such cutsets can be found? Illustrate your answer with the graph in Fig. 1.21. Does this enumerate all possible cutsets?

19.   Many of the original 13 states share common borders. Construct a graph G whose vertices represent the original 13 states. Join two vertices by an edge if the corresponding states have a common border. Is this graph connected? Find the cutset with the smallest number of edges. Is there any state which, if it did not join the Union, would have severed all land travel between the remaining states?

20.   Write down the conservation of flow and capacity constraints for the network in Fig. 1.23.

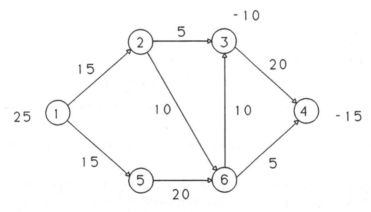

Fig. 1.23   Network for Exercise 20.

21.   Is the graph in Fig. 1.23 acyclic? Why or why not?

22.   Are the conservation of flow equations for a network linearly independent? (Assume that the sum of the supplies equals the sum of the demands.) Why or why not?

23.   Consider the following linear programming problem:

Maximize

$$3x_1 + 2x_2 + 1x_3 - 4x_4$$

such that

$$2x_1 + 1x_2 + 3x_3 + 7x_4 \leq 10$$

$$x_1 \geq 0, \quad x_2 \geq 0, \quad x_3 \geq 0, \quad x_4 \geq 0$$

(a)   Find an optimal solution. (Hint: Consider only basic solutions.)
(b)   What is the dual linear programming problem?

(c)   Use the complementary slackness conditions to find an optimal solution to the dual.

24.   A classic problem is the "knight's tour." On an $n \times n$ chessboard, starting from some square, move the knight so that it lands on every space exactly once. Show how a graph can be used to solve this problem. Illustrate on a 4×4 chessboard.

25.   Show how to use a graph to solve the following puzzle. Place the numbers 1 through 8 in the boxes so that no two consecutive numbers are horizontally, vertically, or diagonally adjacent to each other.

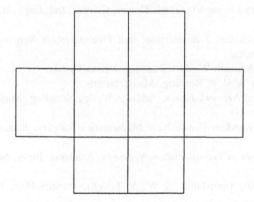

26.   A caterer must provide n_j napkins for d successive days. He can buy new napkins at c dollars each or launder the dirty napkins. The slow laundry takes 2 days and costs a dollars per napkin; the fast laundry takes 1 day and costs b dollars per napkin. Model this problem as a network flow problem to meet demands at minimum total cost.

27.   A product can be produced in each of T time periods. The product can be either sold or held until the next time period. The demand for period t is D_t. All demands must be met, the unit cost of production in period t is p_t, and the unit cost of carrying inventory from one period to the next is h_t. Show how to construct a network flow model to determine the optimal production and inventory policy.

REFERENCES

Bartholdi, J. J., L. K. Platzman, R. L. Collins, and W. H. Warden, III, 1983. A Minimal Technology Routing System for Meals on Wheels, *Interfaces*, *13*(3), pp. 1–8.

Bazaraa, M. S., J. J. Jarvis, and H. D. Sherali, 1990. *Linear Programming and Network Flows*, 2nd ed., Wiley, New York.

Bennington, G. E., 1974. Applying Network Analysis, *Ind. Eng. 6*(1), pp. 17–25.

Berge, C., 1973. *Graphs and Hypergraphs* (translated by E. Minieka), North-Holland, Amsterdam.

Busacker, R. and T. Saaty, 1965. *Finite Graphs and Networks*, McGraw-Hill, New York.

Chachra, V., P. M. Ghare, and J. M. Moore, 1979. *Applications of Graph Theory Algorithms*, North-Holland, New York.

Dantzig, G. B., 1963. *Linear Programming and Extensions*, Princeton University Press, Princeton, New Jersey.

Deo, N., 1974. *Graph Theory with Applications to Engineering and Computer Science*, Prentice-Hall, Englewood Cliffs, New Jersey.

Ford, L. R., and D. R. Fulkerson, 1962. *Flows in Networks*, Princeton University Press, Princeton, New Jersey.

Foulds, L. R., and H. V. Tran, 1986. Library Layout Via Graph Theory, *Comput. Ind. Eng. 10*(3), pp. 245–252.

Frank, H., and I. Frisch, 1971. *Communication, Transmission, and Transportation Networks*, Addison-Wesley, Reading, Massachusetts.

Hadley, G., 1962. *Linear Programming*, Addison-Wesley, Reading, Massachusetts.

Harary, F., 1969. *Graph Theory*, Addison-Wesley, Reading, Massachusetts.

Hu, T. C., 1969. *Integer Programming and Network Flows*, Addison-Wesley, Reading, Massachusetts.

Ore, O., 1963. *Graphs and Their Uses*, Random House New Mathematical Library, Random House, New York.

Potts, R. B., and R. M. Oliver, 1972. *Flows in Transportation Networks*, Academic Press, New York.

Simonnard, M., 1966. *Linear Programming* (translated by W. S. Jewell), Prentice-Hall, Englewood Cliffs, New Jersey.

Swamy, M. N. S., and K. Thulasiraman, 1981. *Graphs, Networks, and Algorithms*, Wiley, New York.

Wilson, R., 1972. *Introduction to Graph Theory*, Academic Press, New York.

Wu, N., and R. Coppins, 1981. *Linear Programming and Extensions*, McGraw-Hill, New York.

2
COMPUTER REPRESENTATION AND SOLUTION

2.1 INTRODUCTION AND EXAMPLES

Because graph and network problems have such a unique structure, fast and efficient computer programs can be developed for solving them. For example, special network codes for flow problems that can be modeled as linear programs have been developed that are hundreds of times faster than general-purpose linear programming codes. Such codes take advantage of new data structures developed by computer scientists. Researchers began investigating the role of computer science techniques in network algorithms in the early 1970s. Their result led to vast improvements in the ability to solve large practical problems. This book has the major objective of developing a fundamental understanding of optimization techniques for networks and graphs and is not focused on computer implementation. Nevertheless, any serious student of network optimization should have a fundamental understanding of computer science, particularly data structures and algorithms. The purpose of this chapter is to introduce these concepts as well as to discuss issues of computational complexity, that is, the efficiency of algorithms.

2.2 DATA STRUCTURES FOR NETWORKS AND GRAPHS

Suppose that $G = (X, E)$ is an undirected graph with m vertices and n edges. There exist several ways to represent G for computer processing. These include:

1. Adjacency matrix
2. Incidence matrix
3. Edge list

4. Adjacency list
5. Star

To illustrate these concepts, we shall use the graphs shown in Figs. 2.1 and 2.2.

Vertex (Node) Adjacency Matrix

Matrix representation provides a convenient way to describe a graph without listing vertices or edges or drawing pictures. We define the vertex adjacency matrix as follows. Let A be an $m \times m$ matrix in which $a_{ij} = 1$ if vertices i and j are adjacent, that is, connected by an edge, and 0 otherwise. The vertex adjacency matrix corresponding to the graph in Fig. 2.1 is

$$
\begin{array}{c}
 \\
1 \\
2 \\
3 \\
4
\end{array}
\begin{array}{cccc}
1 & 2 & 3 & 4 \\
\end{array}
\left[
\begin{array}{cccc}
0 & 1 & 1 & 0 \\
1 & 0 & 1 & 1 \\
1 & 1 & 0 & 1 \\
0 & 1 & 1 & 0
\end{array}
\right]
$$

Notice that this matrix is symmetric (for an undirected graph) and that the number of ones in each row gives the number of edges incident to that vertex.

For a directed graph, we define $a_{ij} = 1$ if there exists an arc (i, j) from node i to node j. Thus, for Fig. 2.2, the node adjacency matrix is

$$
\begin{array}{c}
 \\
1 \\
2 \\
3 \\
4
\end{array}
\begin{array}{cccc}
1 & 2 & 3 & 4 \\
\end{array}
\left[
\begin{array}{cccc}
0 & 1 & 1 & 0 \\
0 & 0 & 1 & 1 \\
0 & 0 & 0 & 1 \\
0 & 0 & 0 & 0
\end{array}
\right]
$$

For a directed graph, a computer would require m^2 storage locations for a node adjacency matrix. For an undirected graph, only $m(m - 1)/2$ storage locations are necessary. (Why?) If the graph is *sparse*, that is, has relatively few edges, then the adjacency matrix will have many 0s and few 1s; this is wasteful of space, and other data structures may be more appropriate.

Vertex-Edge (Node-Arc) Incidence Matrix

Let N be a matrix with m rows, one for each vertex, and n columns, one for each edge. Let n_{ij} denote the element in the ith row and jth column. Then $n_{ij} = 1$ if edge j is incident with

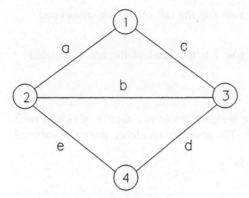

Fig. 2.1 An undirected graph.

vertex i, and zero otherwise. It is clear that each column of N contains exactly 2 ones. The matrix N is called the *vertex-edge incidence matrix of G*. The vertex-edge incidence matrix associated with the graph in Fig. 2.1 is

$$
\begin{array}{c}
\begin{array}{ccccc} a & b & c & d & e \end{array} \\
\begin{array}{c} 1 \\ 2 \\ 3 \\ 4 \end{array}
\left[
\begin{array}{ccccc}
1 & 0 & 1 & 0 & 0 \\
1 & 1 & 0 & 0 & 1 \\
0 & 1 & 1 & 1 & 0 \\
0 & 0 & 0 & 1 & 1
\end{array}
\right]
\end{array}
$$

For a directed graph, the *node-arc incidence matrix* must account for the direction of the arc. Define

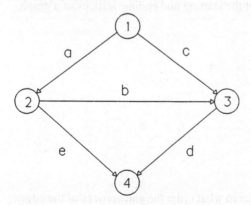

Fig. 2.2 A directed graph.

n_{ij} =   + 1 if the node associated with row i is the tail of the arc associated with column j

   =   − 1 if the node associated with row i is the head of the arc associated with column j

   =   0 otherwise

For a directed graph, this matrix is called the *node-arc incidence matrix of G* and each column contains exactly one + 1 and one − 1. The node-arc incidence matrix associated with the graph in Fig. 2.2 is

$$
\begin{array}{c@{\quad}c@{\quad}c@{\quad}c@{\quad}c@{\quad}c}
 & a & b & c & d & e \\
1 & \left[\begin{array}{ccccc} 1 \\ -1 \\ 0 \\ 0 \end{array}\right. & \begin{array}{c} 0 \\ 1 \\ -1 \\ 0 \end{array} & \begin{array}{c} 1 \\ 0 \\ -1 \\ 0 \end{array} & \begin{array}{c} 0 \\ 0 \\ 1 \\ -1 \end{array} & \left.\begin{array}{c} 0 \\ 1 \\ 0 \\ -1 \end{array}\right]
\end{array}
$$

Return to Chapter 1 and examine the conservation of flow equations associated with Fig. 1.13. Notice that the coefficients of the variables in these equations form the node-arc incidence matrix of the network.

The size of a vertex-edge or node-arc incidence matrix is nm. If the graph is very dense, that is, most pairs of nodes are connected by an arc, then the incidence matrix is much larger than the adjacency matrix. For sparse graphs, however, $mn << m^2$, and it may be beneficial to use this representation. Even so, only $2n$ elements will be nonzero.

Edge (Arc) List

An edge list of a network is simply a listing of the starting and ending vertices of a graph. Thus, for Fig. 2.1 we have the edge list

Edge	Start	End
a	1	2
b	3	2
c	1	3
d	4	3
e	2	4

Since the graph is undirected, it does not matter in what order the end vertices of the edges are listed. For a directed graph, of course, direction is important. Thus, the correct representation of the arc list for Fig. 2.2 is

Arc	Start	End
a	1	2
b	2	3
c	1	3
d	3	4
e	2	4

Only $2n$ storage spaces are needed to represent the graph. Thus edge lists and arc lists are quite useful for sparse graphs.

Adjacency Lists

An adjacency list maintains a pointer or index to a list of vertices that are adjacent to a given vertex. Such a structure is called a *linked list* in computer science terminology. The adjacency list corresponding to Fig. 2.1 is

Vertex	Adjacency List
1	2 3 0
2	1 3 4 0
3	1 2 4 0
4	2 3 0

The number 0 denotes the end of the list. Thus, to access all edges incident to vertex 3, we would search through the adjacency list corresponding to vertex 3, (1, 2, 4, 0), and find edges (3, 1), (3, 2), and (3, 4). When we reach 0 we know that we have found the end of the list. The size of the list is the number of incident edges plus one.

For a directed graph, we need only specify all nodes directed *from* a given node. Thus, for Fig. 2.2, the adjacency list would be

Node	Adjacency List
1	2 3 0
2	3 4 0
3	4 0
4	0

Adjacency lists exploit the sparsity of graphs, but it is difficult to locate a specific edge. We must maintain an array for each vertex or node. Such a structure is useful, however, when edges are added or deleted frequently.

Star

The star is a data structure that maintains a pointer for each vertex to an array containing the adjacency list. Thus, instead of keeping a separate array for each node, we maintain only two arrays, one for the pointers in the vertex list and a second for the combined adjacency list. The star data structure corresponding to Fig. 2.1 is given below.

Vertex	Pointer
1	1
2	3
3	6
4	9
5	11

Pointer	Adjacency List
1	2
2	3
3	1
4	3
5	4
6	1
7	2
8	4
9	2
10	3
11	0

Thus, to find all edges incident to vertex 2, one would first find the pointer corresponding to 2, which is 3, and the pointer corresponding to the *next* vertex, which is 6. Searching the adjacency list from position 3 to $5 = 6 - 1$ (since position 6 begins the list for the next vertex) will recover all the edges. Note that we have added a fictitious vertex with a number one larger than the highest vertex number in the graph in order to perform this search correctly.

The star data structure for a directed graph is generally called the *forward star* since only forward arcs (that is, arcs directed from a given node) are listed. The forward star for Fig. 2.2 is

Node	Pointer
1	1
2	3
3	5
4	6
5	6

Pointer	Adjacency List
1	2
2	3
3	3
4	4
5	4
6	0

Star data structures are not as easily updated as simple adjacency lists. However, they exploit the sparsity of graphs better and use less storage space than other data structures. They are extremely useful in algorithms that search for arcs locally from a given node.

While the data structures we have described are important for computer solution of optimization problems involving graphs and networks, we shall use the more intuitive pictorial representations whenever possible in explaining the procedures in this book.

2.3 ALGORITHMS

An *algorithm* is a sequence of steps designed to solve a problem. For example, the simplex algorithm for linear programming is a well-defined procedure that finds an optimal extreme point or determines that the problem is unbounded or infeasible. Much of the research in optimization centers around finding better and more efficient algorithms to solve problems.

To illustrate an algorithm for an graph problem, suppose we wish to determine if a directed path exists between a given pair of nodes in a directed graph, say from node s to node t. One algorithm for solving this problem is stated below.

0. Let NODE = "s".
1. Find all directed arcs from NODE. (We call this process "scanning" a node.) Mark the nodes at the head if these arcs as "reachable." Mark NODE as "scanned."

2. If node t is marked as reachable, then a directed path exists; stop. If not continue.
3. For all nodes marked as reachable and not yet scanned, select one and call it NODE. Return to step 1. If this set is empty, then no directed path exists; stop.

To apply this algorithm to the graph in Fig. 2.2, let $s = 1$ and $t = 4$. Since $(1, 2)$ and $(1, 3)$ are directed arcs, we mark nodes 2 and 3 as reachable and node 1 as scanned. We next select one of these, say node 2, and find all directed arcs from 2. These are $(2, 3)$ and $(2, 4)$. We would mark nodes 3 and 4 as reachable and node 2 as scanned. Since node 4 is now marked as reachable, we have found a directed path from node 1. How would the algorithm be modified if we wanted to construct this path after it is found to exist?

Clearly, there exists an uncountable number of possible graphs to which we might apply this algorithm. Each graph is an *instance* of the generic problem of determining the existence of a path between two vertices. Thus, a *problem* is the collection of all possible instances. An algorithm must work correctly on every instance of a problem. Proving this is often very difficult.

Formally, the *size* of an instance of a problem is the number of bits needed to represent it on a computer. For example, the number of bits required to encode the integer n is $\log_2 n$. We shall not get so detailed and will use the term size as a measure of the number of components of an instance. For example, we will say that a graph with n vertices is of size n, regardless of the numerical value of n.

For optimization problems, an *exact algorithm* is one which always produces an optimal solution if it exists. The simplex algorithm is an exact algorithm. A *heuristic algorithm*, or simply a heuristic, is one which produces a feasible and hopefully very good, but not necessarily optimal, solution. Heuristics generally are very fast and efficient. Often we use heuristics when an exact algorithm is not available or when it is computationally impractical. We shall have more to say about heuristics later in this chapter.

Figure 2.3 gives a taxonomy of algorithms. Direct algorithms do not work in iterations. For example, differentiation of a function is a direct algorithm. Iterative algorithms employ certain procedures that are repeated over and over again. The simplex method is an example of an iterative algorithm. Iterative algorithms that cannot be proved to converge to an optimum solution constitute the class of heuristics. Those that do converge to an optimum solution can be classified as approximation algorithms, which may converge only in a limiting sense, and finite algorithms, which converge in a finite number of steps. Finally, finite algorithms can be partitioned into path structure algorithms and tree structure algorithms. Path structure algorithms have the property that one iteration follows from the preceding one in a deterministic fashion. The simplex method is such an algorithm because it will always follow the same sequence of iterations to termination. In tree structure algorithms, on the other hand, the iteration sequences form a tree consisting of several parallel branches. Dynamic programming and branch and bound fall in this class. Nearly all of the algorithms that we shall discuss in this book are path structure algorithms and heuristics.

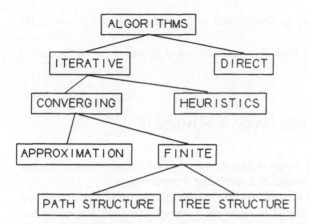

Fig. 2.3 A taxonomy of algorithms. (Adapted from H. Muller-Merbach, "Heuristics and their design: a survey," *European Journal of Operational Research*, Vol. 8, 1–23, 1981).

2.4 COMPUTATIONAL COMPLEXITY

Suppose that we have two or more exact algorithms or heuristics for a particular problem. How can we compare them? A measure often used to evaluate the performance of algorithms is computer execution time. Execution time, however, is a function of the type of computer, programming language, programmer skills, and so forth. Thus, it is very difficult to attempt to compare algorithmic performance solely on the basis of computer time. We need to be able to evaluate the algorithm on its own merits and not be influenced by how the algorithm is actually coded and whether one uses GO TO statements, WHILE . . . WEND loops, and so forth. The *principle of invariance* states that two different implementations of the same algorithm will not differ in efficiency by more than a multiplicative constant. That is, if two implementations take $t_1(n)$ and $t_2(n)$ seconds for an instance of size n, there exists a positive number c such that $t_1(n) \leq ct_2(n)$ for some n sufficiently large. Thus, while a 386-based machine will solve an instance faster than a 286-based machine, only a fundamental change in the algorithm itself will exhibit improvements that grow with the size of the instance.

A more appropriate way to measure algorithmic performance is to count the number of elementary operations that are required by the algorithm. These elementary operations include additions, subtractions, comparisons, multiplications, divisions, and transfers to storage locations. We generally take more of a macro view and count the number of steps that the algorithm must perform as a function of the size of the problem as we have defined it. We say that an algorithm takes time *of the order t(n)*, where t is a given function, if there exist a positive constant c and an implementation of the algorithm capable of solving every instance of the problem in a time bounded by $ct(n)$, where n is the size of the instance. We denote this as $O(t(n))$ and call this the *time complexity* of the algorithm.

For example, an algorithm to find the maximum of n numbers can be coded as follows:

```
MAXN = -999999
FOR I = 1 TO N
IF NUMBER(I) > MAXN THEN MAXN = NUMBER(I)
NEXT I
```

Since we must check each of the n numbers exactly once to find the maximum, the time complexity is $O(n)$. This is an example of *linear* time complexity.

As another example, consider the path-finding algorithm described above. Suppose that we use the node-arc incidence matrix to represent the graph. To find all directed arcs and their end nodes from a given node, we need to search each element of the row corresponding to that node to find all the 1s; this must be done m times. Then we must search each column to determine where the -1s are; this must be done $n - 1$ times. In the worst case, at each iteration, we need to search $m + m(n - 1) = mn$ entries in the matrix. Since we may have to do this for each node, the total number of operations would be $mn(n)$ or mn^2. Thus the time complexity of this procedure is $O(mn^2)$.

A change in data structure results in a fundamentally different algorithm and can result in substantial computational differences. Suppose that we use an adjacency list in the path-finding algorithm instead of a node-arc incidence matrix. Since the list maintains all nodes that are connected by an arc from a given node, we need only search one array for the given node. This will have at most $n - 1$ real entries plus the 0 entry. Since we may have to do this n times, the time complexity of this implementation would be $n(n - 1) = n^2 - n$. Since n^2 is the highest-order term, we say that the complexity is of the order n^2, or $O(n^2)$. An $O(n^2)$ algorithm clearly is superior to an $O(mn^2)$ algorithm. $O(mn^2)$ and $O(n^2)$ are examples of *polynomial* time complexity.

As a final example, suppose that we have a graph with n vertices and weights assigned to the edges and we seek a subset of vertices whose generated subgraph has the maximum weight. With no obvious algorithm to do this, we could exhaustively enumerate all subsets of two or more vertices. The number of such subsets is $2^n - n - 1$ (why?). This enumeration algorithm has complexity $O(2^n)$. This is *exponential* time complexity.

Of what benefit is such analysis of algorithm complexity? This can be seen through the following scenario. Suppose that we have three algorithms, A, B, and C, having time complexities respectively of n, n^2, and 2^n. Suppose further that one operation takes $1/1000$ second. Then in 1 second algorithm A can solve a problem of size 1000; algorithm B, a problem of size 31; and algorithm C, a problem of size 9. In 1 hour, however, algorithm A can solve a problem of size 3.6 million; algorithm B, a problem of size 1897; and algorithm C, a problem of size only 21. Figure 2.4 shows the relative rates of growth of these functions.

We must also give a word of caution. An algorithm with time complexity

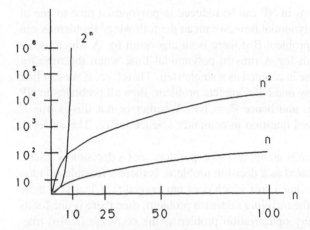

Fig. 2.4 Time complexity functions.

$34,367,291n^2$ would still be classified as $O(n^2)$ but clearly would not be as effective as one having complexity $0.002n^5$, except for extremely large problem instances. Thus, complexity analysis applies only to asymptotic behavior and need not be representative of relative performance on problems encountered in practice. (The simplex algorithm has been shown to have exponential growth for some contrived problems, yet it is very efficient for most realistic problems.)

Algorithms that have exponential time complexity cannot solve very large problems, no matter how powerful the computer may be. An algorithm is termed ''good'' or ''efficient'' if its time complexity is bounded by a polynomial function in the size of the problem. Such an algorithm is called a *polynomial time algorithm*.

One of the reasons that algorithm complexity has assumed increasing importance over the last several decades is that there exist problems for which polynomial time algorithms have not been discovered. Perhaps the most famous of these problems is the *traveling salesman problem*: Given a complete graph on n vertices with distances associated with the edges, find a cycle passing through all vertices exactly once having minimum total distance. Since very large problems cannot be solved by an exact algorithm, we must resort to heuristics to find good solutions. The traveling salesman problem will be studied further in Chapter 9.

The theory of computational complexity results from studying algorithms for decision problems. A decision problem is one having a yes or no answer. For example, does there exist a path between two vertices of a graph? Decision problems can be classified into one of three classes called P, NP, and NP-complete. Problems for which polynomial time algorithms are known are members of P. The class NP encompasses all problems in P as well as other problems which can be solved, but for which only exponential time algorithms are known. (There exist other problems not in NP that are provably unsolvable, but these will not be of concern to us.) The class NP-complete is a subset of NP

having the property that all problems in NP can be reduced in polynomial time to one of
them. By reducing a problem in polynomial time, we mean the following. Problem A can
be reduced in polynomial time to problem B if there is an algorithm for A which uses a
subroutine for B and the algorithm for A runs in polynomial time when the time for
executing each call of the subroutine is counted as a single step. Therefore, if we can find
a polynomial time algorithm for any one NP-complete problem, then all problems in NP
can be solved in polynomial time, and hence P = NP. Whether or not this is true is
perhaps the most important unsolved question in computer science today. The evidence
seems to indicate that P $\neq$ NP.

The traveling salesman problem is an optimization problem, not a decision problem.
However, it can be easily reformulated as a decision problem: Is there a feasible solution
to a traveling salesman problem having a cost which does not exceed C? Clearly, if there
is a polynomial time algorithm for the traveling salesman problem, then there is one for its
related decision problem. For many optimization problems, the converse is also true.
Because optimization problems are not in NP by definition, we use the term *NP-hard* to
describe optimization problems whose decision versions are NP-complete.

Since it is unlikely that we will be able to find a polynomial time algorithm for any
NP-hard problem, it is worthwhile to try to develop good heuristics for them. If a problem
is not known to be NP-hard, but no polynomial algorithm has been found, then there still
is hope that one may be discovered. Linear programming, for example, could not be
shown to be NP-hard, and only recently was a polynomial algorithm discovered.

2.5   HEURISTICS

Heuristics clearly are necessary for NP-complete problems if we expect to solve them in
reasonable amounts of computer time. Heuristics also are useful in speeding up the
convergence of exact optimization algorithms, typically by providing good starting
solutions. Heuristics can be classified into several categories.

Construction

Construction heuristics build a feasible solution one component at a time. In graph
applications, for example, one might add edges to a solution one at a time until a certain
structure such as a tree is realized. Such heuristics are often used to generate initial
feasible solutions to a problem.

Improvement

Improvement heuristics start with a complete feasible solution to a problem and successively improve it through a sequence of exchanges or mergers. For instance, one might
with begin with a cycle that represents a feasible solution to the traveling salesman

problem and try to remove some edges that can be replaced by others at a lower cost while still maintaining feasibility.

Partitioning and Decomposition

Partitioning and decomposition refer to methods that break a problem down into smaller components that can be solved in sequence or independently. We shall see one example of a heuristic for the traveling salesman problem that first finds a spanning tree in the graph and then solves an auxiliary problem to construct a feasible solution. Another example would be the partitioning of a planar graph into geographic regions, solving a subproblem within each region, and merging the solutions into one overall solution.

Specialized Methods

Many other special methods exist for certain problems that depend, for example, on solving a relaxation of an integer program, or applying probabilistic analysis. These usually are problem dependent and take advantage of special structures inherent in the problem.

Desirable characteristics of heuristics include

1. They run in reasonable computational time.
2. Solutions generally are close to optimal.
3. The probability of any one solution being very far from optimal is small.
4. Storage requirements are small.

The computational complexity of heuristics often can be determined in the same manner as for exact algorithms. In addition, we can sometimes compute worst-case bounds on the deviation of the solution from optimality. That is, we might be able to show that the heuristic solution will be no more than some fixed percentage from the optimal solution for any problem instance. Since the focus of this text is on understanding the graph and network algorithms themselves, we shall not focus on the derivation of such results but will state them whenever appropriate. Many good books and papers on heuristics are provided in the references.

APPENDIX: NETSOLVE – AN INTERACTIVE SOFTWARE PACKAGE FOR NETWORK ANALYSIS

NETSOLVE is a software package accompanying this book that was developed by Drs. James P. Jarvis and Douglas R. Shier for solving network problems. NETSOLVE has been designed to be

Interactive
Integrated

Flexible

Easy to use

In effect, NETSOLVE provides a "workbench" upon which a network can be assembled, validated, modified, and analyzed using various editing and optimization tools. Because the system is interactive, the user is given immediate feedback and can thus work more effectively. The integrated nature of the system means that the user who learns the commands once has immediate access to a number of editing and optimization tools. Several other features enhance the flexibility and user-friendliness of the system. For example, networks can be entered in a "free format" manner, descriptive names can be used (CHICAGO versus node 1072), an on-line library of "help" documents is available, and output can be selectively routed to a number of external data files for subsequent printing.

NETSOLVE will be used throughout this book to solve practical examples of optimization problems that use the algorithms we develop. The algorithms available in NETSOLVE and their chapter references in this text are listed below.

Algorithm	Optimization Problem	Chapter Reference
ASSIGN	Assignment problem	7
LPATH	Longest path	11
MATCH	Weighted matching	7
MAXFLOW	Maximum flow	6
MINFLOW	Minimum-cost flow	5
MST	Minimum spanning tree	3
SPATH	Shortest path	4
TRANS	Transportation	5
TSP	Traveling salesman	9

In an appendix at the end of each chapter we shall illustrate the use of NETSOLVE to solve problems in that chapter.

We will create a data file for the oil pipeline example in Chapter 1 (Fig. 1.15) to illustrate the ease with which NETSOLVE can be used. NETSOLVE commands are entered after the prompt ">". We begin by entering the command CREATE. NET-SOLVE then prompts you for additional information:

Enter name of network: PIPELINE

Directed or undirected? (D/U): U

Any node data? (Y/N): N

Any edge data? (Y/N): Y

Enter edge data in order: FROM, TO, COST, LOWER, UPPER. (Null line to end)

Required data: FROM   TO

Optional data: COST   LOWER   UPPER
Default value: 1.00   0.00   999999.00
(Unspecified values receive defaults)

At this point, you simply enter the beginning and ending vertices of each edge and their cost from Fig. 1.15. Since no lower or upper bounds are specified in this example, these data can be omitted. Therefore, we input the following data:

```
1,2,5
1,3,6
1,4,7
2,4,4
2,6,8
3,4,6
3,5,7
3,7,9
3,8,11
4,6,3
4,7,3
5,7,12
5,8,2
6,7,3
6,8,10
```

Be sure to signify the end of input with a null line (or return). To list the data, we type the command LIST EDGES at the prompt.

>LIST EDGES

FROM	TO	COST	LOWER	UPPER
1	2	5.00	0.00	999999.00
1	3	6.00	0.00	999999.00
1	4	7.00	0.00	999999.00
2	4	4.00	0.00	999999.00
2	6	8.00	0.00	999999.00
3	4	6.00	0.00	999999.00
3	5	7.00	0.00	999999.00
3	7	9.00	0.00	999999.00
3	8	11.00	0.00	999999.00
4	6	3.00	0.00	999999.00
4	7	3.00	0.00	999999.00
5	7	12.00	0.00	999999.00
5	8	2.00	0.00	999999.00
6	7	3.00	0.00	999999.00
6	8	10.00	0.00	999999.00

We see that we made an error in entering the last edge. The edge (6, 8) should have been (7, 8). Editing in NETSOLVE is very simple. We first delete the edge (6, 8) by entering the command

>DELETE EDGE 6 8

Then we add the correct edge as follows:

>ADD EDGE 7 8 COST = 10
>LIST EDGES

FROM	TO	COST	LOWER	UPPER
1	2	5.00	0.00	999999.00
1	3	6.00	0.00	999999.00
1	4	7.00	0.00	999999.00
2	4	4.00	0.00	999999.00
2	6	8.00	0.00	999999.00
3	4	6.00	0.00	999999.00
3	5	7.00	0.00	999999.00
3	7	9.00	0.00	999999.00
3	8	11.00	0.00	999999.00
4	6	3.00	0.00	999999.00
4	7	3.00	0.00	999999.00
5	7	12.00	0.00	999999.00
5	8	2.00	0.00	999999.00
6	7	3.00	0.00	999999.00
7	8	10.00	0.00	999999.00

We could use the EF (edge format) command to eliminate the listing of excess edge fields. For example, by issuing EF FROM TO COST, we do not print LOWER and UPPER fields since they are at default values.

To save this network, we enter the command

>SAVE
NETSOLVE responds with:
NETWORK SAVED IN DATASET: PIPELINE.NET

Appendix A at the end of this book is a comprehensive user's guide to the NET-SOLVE system. We suggest that you review this information before you use the software. The user's guide deals with the NETSOLVE command language, procedures for creating and editing networks, file manipulation, and advanced features of the system.

EXERCISES

1. Consider the graph shown below.

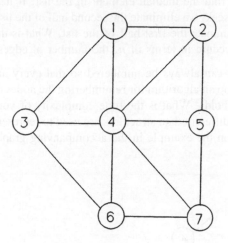

Show the following:
(a) Adjacency matrix
(b) Incidence matrix
(c) Edge list
(d) Adjacency list
(e) Star data structure

2. Repeat Exercise 1 for the directed graph shown below.

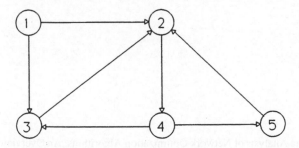

3. Show how to implement the path-finding algorithm using an adjacency matrix.

4. Show how to implement the path-finding algorithm using an incidence matrix.

5. In network algorithms, we often have to search an ordered list to determine if, for

instance, a particular edge exists. Suppose that the nodes are numbered $1, 2, \ldots, m$ and that the edges are ordered lexicographically. That is, edge (a, b) comes before (c, d) if $a < c$; and if $a = c$, then $b < d$. One search technique for locating a specific element is called *bisection search*. We find the median element in the list. If it is lexicographically larger than the one we seek, we eliminate the second half of the list. If it is lexicographically smaller, we eliminate the first half of the list. What is the worst-case time complexity of this procedure in terms of n, the number of edges?

6. The nodes of an acyclic directed graph can always be numbered so that every arc (x, y) has the property that $x < y$. Develop an algorithm for renumbering the nodes of any acyclic graph so that this property holds. What is the time complexity of your algorithm? Does the complexity vary if different data structures are used to represent the graph? Illustrate your algorithm(s) on the example in the accompanying graph.

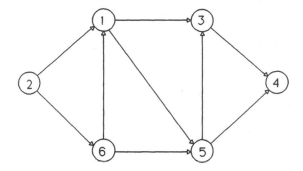

7. Develop algorithms for converting each of the following data structures to the others. Determine the time complexity of your algorithms.
    (a) Adjacency matrix
    (b) Incidence matrix
    (c) Edge list
    (d) Adjacency list
    (e) Star data structure

REFERENCES

Baker, E., and B. Golden, 1982. The Analysis of Network Optimization Algorithms: An Overview, Working Paper MS/S No. 82-022, College of Business and Management, University of Maryland.

Ball, M., and M. Magazine, 1981. The Design and Analysis of Heuristics, *Networks*, *11*, pp. 215–219.

Brassard, G., and P. Bratley, 1988. *Algorithmics, Theory and Practice*, Prentice-Hall, Englewood Cliffs, New Jersey.

Fisher, M., 1980. Worst-Case Analysis of Heuristic Algorithms, *Man. Sci.*, *26*, pp. 1–17.

Foulds, L. R., 1983. The Heuristic Problem-Solving Approach, *J. Oper. Res. Soc.*, *34*, pp. 927–934.

Fox, B. L., 1978. Data Structures and Computer Science Techniques in Operations Research, *Oper. Res.*, *26*, pp. 686–717.

Hu, T. C., 1982. *Combinatorial Algorithms*, Addison-Wesley, Reading, Massachusetts.

Johnson, D. S., and C. H. Papadimitriou, 1985. Computational Complexity. In *The Traveling Salesman Problem* (eds. E. L. Lawler, J. K. Lenstra, A. H. G. Rinnooy Kan, and D. B. Shmoys), Wiley, London.

Muller-Merbach, H., 1981. Heuristics and Their Design: A Survey, *Eur. J. Oper. Res.*, *8*, pp. 1–23.

Papadimitriou, C. H., and K. Steiglitz, 1982. *Combinatorial Optimization, Algorithms and Complexity*, Prentice-Hall, Englewood Cliffs, New Jersey.

Syslo, M. M., N. Deo, and J. S. Kowalik, 1983. *Discrete Optimization Algorithms with PASCAL Programs*, Prentice-Hall, Englewood Cliffs, New Jersey.

Zanakis, S. H., J. R. Evans, and A. A. Vazacopoulos, 1989. Heuristic Methods and Applications: A Categorized Survey, *Eur. J. Oper. Res.*, *43*, pp. 88–110.

3
TREE ALGORITHMS

3.1 INTRODUCTION AND EXAMPLES

A graph may contain many different trees. This chapter studies several algorithms used to construct trees (spanning trees in particular) with certain optimal properties. Trees have applications in many areas, such as facilities design and electrical networks and within algorithms for other problems, such as the traveling salesman problem.

Rumor Monger

Consider a small village in which some of the villagers have a daily chat with one another. Is it possible for a rumor to pass throughout the entire village? To answer this question, represent each villager by a vertex. Join two vertices by an edge if the corresponding two villagers have a daily chat with one another. If the resulting graph is connected, then it is possible for a rumor to pass through the entire village. Since a tree is necessarily connected, a graph that possesses no spanning tree cannot be connected. Therefore, we need only check if the graph is connected.

Let $G = (X,E)$ be an undirected graph in which each edge (x,y) is assigned a weight $a(x,y)$. We define the *weight of the tree* as the sum of the weights of the edges in the tree.

Highway Construction

The Department of Highways wishes to build enough new roads so that the five towns in a certain county will all be connected to one another either directly or via another town. The cost of constructing a highway between each pair of towns is known (see Fig. 3.1). We

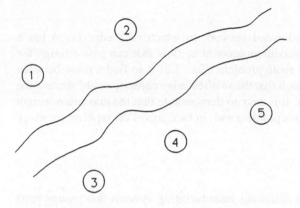

From/to	1	2	3	4	5
1	—	5	50	80	90
2		—	70	60	50
3			—	8	20
4				—	10
5					—

Fig. 3.1 Highway construction.

may represent this problem as a graph by letting each town correspond to a vertex, and each possible highway to be constructed correspond to an edge joining two vertices (towns). Associate a weight to each edge that is equal to the cost of constructing the corresponding highway.

Deciding which highways to build can be viewed as the problem of constructing a minimum-cost spanning tree for the corresponding graph. This follows since the edges of any spanning tree will connect each vertex (town) with every other vertex (town). Moreover, the minimum-weight spanning tree will represent the set of new highways of minimum total cost.

Note that if the highways are allowed to have junctions at places other than the five towns, the resulting problem is more complicated than a minimum spanning tree problem. This problem is called a *Steiner tree* problem and will be discussed later in this chapter.

You can probably think of many other similar problems related to physical systems. Some examples would be the design of cable TV networks, oil pipelines (recall the example described in Chapter 1), and electrical circuitry.

Maximum-Capacity Route

Suppose that $G = (X, E)$ is an undirected network in which each edge (x, y) has a capacity $c(x, y)$ that represents the maximum amount of flow that can pass through the edge (x, y). The maximum-capacity route problem (Fig. 3.2) is to find a route between two specified vertices, say 1 and 5, such that the smallest edge capacity on the route is the maximum among all possible routes. It is easy to demonstrate that the maximum-weight spanning tree provides a solution to this problem and, in fact, solves the problem for every pair of vertices in the graph.

Group Technology

Group technology is an approach to designing manufacturing systems that groups parts into *part families* and machines into *manufacturing cells* for the production of the part families. Spanning trees can be used to solve the subproblem of machine grouping. Typically, we are given machine-part incidence matrix A for which $a_{ij} = 1$ if part j must be processed by machine i and 0 otherwise. Denote by M_i the set of parts that require processing by machine i. For any two machines, i and k, define a distance measure $d(i, k)$ as follows:

$$d(i, k) = \frac{|M_i \oplus M_k|}{|M_i \cup M_k|}$$

The numerator is the symmetric difference, namely the number of parts requiring processingly *only* on machine i or machine k but not both. The denominator is the number of parts requiring processing on either machine or both. This distance function is a measure of the relative dissimilarity of two different machines with respect to their parts processing requirements. A value of zero indicates that the two machines have the same set of parts to process, while a value of one means that they process no parts in common.

We construct a complete graph in which each vertex represents a machine. Each edge

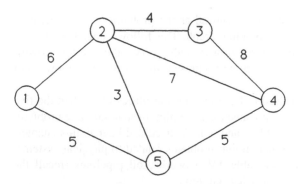

Fig. 3.2 Maximum-capacity route problem.

(i, k) is assigned the weight $d(i, k)$. The minimum-weight spanning tree in this graph minimizes the total dissimilarity that includes all machines. If we wish to group the machines into K cells, we delete the $K - 1$ largest edges of the spanning tree. The result will be a forest of K trees. The machines corresponding to the vertices in each tree provide the machine groupings. We shall illustrate this with an example using NETSOLVE in the appendix to this chapter.

3.2 SPANNING TREE ALGORITHMS

The *spanning tree algorithm*, given in the following discussion, is one of the most elegant algorithms that we shall encounter. The algorithm examines the edges in any arbitrary sequence and decides whether each edge will be included in the spanning tree. Thus, each edge will be labeled as either *assigned* to the spanning tree or *excluded* from the spanning tree.

When an edge is examined, the algorithm simply checks if the edge under consideration forms a cycle with the other edges already assigned to the tree. If so, then the edge under consideration is excluded from the tree; otherwise, it is assigned to the tree.

Clearly, if one were performing this algorithm by hand, it would be easy to determine visually when a cycle is formed. However, this may not be as straightforward for a computer implementation. There is a very simple way to accomplish this task. As the edges are assigned to the tree they form one or more connected components. The vertices belonging to a single connected component are collected together into what the algorithm terms a "bucket." An edge forms a cycle with the edges already assigned to the tree if both its endpoints are in the same connected component (bucket). A unique number can be assigned to each bucket and hence to all the vertices in the bucket. To check whether both endpoints are in the same bucket, we need only compare the bucket number assigned to those vertices.

The algorithm terminates with a spanning tree when the number of edges assigned to the tree equals the number of vertices less one, or, equivalently, when all vertices are in one bucket, since all spanning trees have the same number of edges. If the graph contains no spanning tree, which is equivalent to not being connected, then the algorithm terminates after examining all edges without assigning enough edges to the tree, and the vertices will be contained in more than one bucket.

Spanning Tree Algorithm

Initially all edges are unexamined and all buckets are empty.

> *Step 1*. Select any edge that is not a loop. Assign this edge to the tree and place both its endpoints in an empty bucket.
>
> *Step 2*. Select any unexamined edge that is not a loop. (If no such edge exists, then stop the algorithm as no spanning tree exists.)

One of four different situations must occur:

(a) Both endpoints of this edge are in the same bucket.
(b) One endpoint of this edge is in a bucket, and the other endpoint is not in any bucket.
(c) Neither endpoint is in any bucket.
(d) Each endpoint is in a different bucket.

If condition (a) occurs, then exclude the edge from the tree and return to step 2. If (b) occurs, assign the edge to the tree and place the unbucketed endpoint in the same bucket as the other endpoint. If (c) occurs, then assign the edge to the tree and place both endpoints in an empty bucket. If (d) occurs, then assign the edge to the tree and combine the contents of both buckets in one bucket, leaving the other bucket empty. Go to step 3.

Step 3. If all the vertices of the graph are in one bucket or if the number of assigned edges equals the number of vertices less one, stop the algorithm since the assigned edges form a spanning tree. Otherwise, return to step 2.

Note that each time a step of the algorithm is performed, one edge is examined. If there is only a finite number of edges in the graph, the algorithm must stop after a finite number of steps. Thus, the time complexity of this algorithm is $O(n)$, where n is the number of edges in the graph.

If the algorithm does not terminate with a spanning tree, then no spanning tree exists in the graph for the following reason. The algorithm will terminate with two nonempty buckets of vertices that have no edge joining a member of one set to a member of the other set. Otherwise, such an edge would have been assigned to the tree [condition (d) would have occurred] and the two buckets would have been combined by the algorithm. Hence, the algorithm does what it is supposed to do, that is, construct a spanning tree.

This algorithm has the property that each edge is examined at most once. Algorithms, like the spanning tree algorithm, which examine each entity at most once and decide its fate once and for all during that examination are called *greedy algorithms*. The advantage of performing a greedy algorithm is that you do not have to spend time reexamining entities. Clearly, such algorithms are extremely computationally efficient.

EXAMPLE 1

Let us construct a spanning tree for the graph in Fig. 3.3. Examine the edges in the following arbitrary order: (1, 2), (4, 5), (1, 4), (2, 5), (5, 3), (2, 3), (1, 3), and (3, 4). The results of each step of the algorithm are shown below.

Edge	Assigned?	Bucket No. 1	Bucket No. 2
(1, 2)	yes	1, 2	empty
(4, 5)	yes	1, 2	4, 5
(1, 4)	yes	1, 2, 4, 5	empty
(2, 5)	no	1, 2, 4, 5	empty
(5, 3)	yes	1, 2, 4, 5, 3	empty

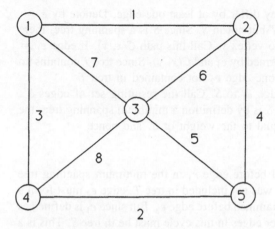

Fig. 3.3 Spanning tree example.

Since all vertices are in one bucket (and four edges have been assigned from the five-vertex graph), a spanning tree has been found.

Obviously, the spanning tree constructed by the algorithm depends on the order in which the edges are examined. If the edges in the example had been examined in the reverse order, the algorithm would have generated the spanning tree consisting of edges (3, 4), (1, 3), (3, 2), and (5, 3).

Now consider the problem of finding a spanning tree with the smallest possible weight or the largest possible weight, respectively called a *minimum spanning tree* and a *maximum spanning tree*. Obviously, if a graph possesses a spanning tree, it must have a minimum spanning tree and also a maximum spanning tree. These spanning trees can be readily constructed by performing the spanning tree algorithm with an appropriate ordering of the edges.

Minimum Spanning Tree Algorithm

Perform the spanning tree algorithm examining the edges in order of nondecreasing weight (smallest first, largest last). If two or more edges have the same weight, order them arbitrarily.

Maximum Spanning Tree Algorithm

Perform the spanning tree algorithm examining the edges in order of weight (largest first, smallest last). If two or more edges have the same weight, order them arbitrarily.

Proof of the Minimum Spanning Tree Algorithm. We shall prove by contradiction that the minimum spanning tree algorithm constructs a minimum spanning tree. Suppose that the algorithm constructs a tree T and some other tree S is in fact a minimum spanning tree.

Since S and T are not identical trees, they differ by at least one edge. Denote by $e_1 = (x, y)$ the first examined edge that is in T but not in S. Since S is a spanning tree, there exists in S a unique path from vertex x to vertex y. Call this path $C(x, y)$. If edge $e_1 = (x, y)$ is added to tree S, then a cycle is formed by e_1 and $C(x, y)$. Since tree T contains no cycles, this cycle must contain at least one edge e_2 not contained in tree T.

Remove edge e_2 from S and add edge e_1 to S. Call the resulting set of edges S'. Clearly, S' is also a spanning tree. Since S is by definition a minimum spanning tree, the weight of S' must be greater than or equal to the weight of S, and hence

$$a(e_1) \geq a(e_2)$$

Suppose that edge e_2 was examined before edge e_1 in the minimum spanning tree algorithm that generated tree T. Since e_2 was not included in tree T, edge e_2 must form a cycle with the edges of tree T that were examined before edge e_2. But since e_1 is defined as the first edge in T that is not in S, all other edges in this cycle must be in tree S. This is a contradiction since edge e_2 forms a cycle with these edges. Therefore, we must conclude that edge e_1 was examined before edge e_2, and $a(e_2) \geq a(e_1)$. Therefore, trees S and S' have the same total cost, and tree S' has one more edge in common with tree T than does spanning tree S.

The proof can now be repeated using S' as the minimum-cost spanning tree instead of S. This generates another minimum spanning tree S'' that has one more edge in common with tree T than did tree S'. Ultimately, a minimum spanning tree will be generated that is identical to tree T. Thus, tree T is a minimum spanning tree.   Q.E.D.

The proof for the maximum spanning tree algorithm is identical to the preceding proof except that minimum should everywhere be replaced by maximum.

EXAMPLE 2

We will construct a minimum-cost spanning tree for the highway problem given in Fig. 3.1. The edges in nondecreasing order of cost are $(1, 2)$, $(3, 4)$, $(4, 5)$, $(3, 5)$, $(1, 3)$, $(2, 5)$, $(2, 4)$, $(2, 3)$, $(1, 4)$, and $(1, 5)$. The results of the minimum spanning tree algorithm are:

Edge	Assigned?	Bucket No. 1	Bucket No. 2
$(1, 2)$	yes	1, 2	empty
$(3, 4)$	yes	1, 2	3, 4
$(4, 5)$	yes	1, 2	3, 4, 5
$(3, 5)$	no	1, 2	3, 4, 5
$(1, 3)$	yes	1, 2, 3, 4, 5	empty

Stop, since all vertices are in the same bucket and four edges have been assigned to the tree. A minimum-cost spanning tree consists of the edges $(1, 2)$, $(3, 4)$, $(4, 5)$, and $(1, 3)$.

The tree consisting of the edges (1, 2), (3, 4), (4, 5), and (2, 5) is also a minimum-cost spanning tree. What is the optimum cost?

The algorithm that we have presented for minimum spanning trees is due to Kruskal (1956). For a graph with m vertices and n edges, Kruskal's method may take up to n iterations. Each iteration requires the determination of the smallest edge available; this can be done in $O(\log n)$ using appropriate data structures. Hence the complexity is of order $O(n \log n)$. Notice that if the graph is dense, then the number of edges approaches $m(m - 1)/2$; hence the time complexity will be of order $O(m^2 \log n)$. On the other hand, if the graph is sparse, the number of edges is of order m and the time complexity will be $O(m \log n)$.

An alternative algorithm was developed by Prim (1957). It can be described as follows:

Step 1. Begin with any vertex. Select the edge of least weight that connects this vertex with another.

Step 2. At any intermediate iteration we have a subtree (a tree that is not spanning). Select the edge of least weight that connects some vertex in the subtree to a vertex not in the subtree.

Step 3. If the subtree spans all vertices, stop. Otherwise return to step 2.

EXAMPLE 3

Let us illustrate Prim's algorithm using the highway construction example in Fig. 3.1. We will arbitrarily select vertex 1 to begin. The edge of least weight incident to vertex 1 is (1, 2) having a weight of 5. Thus, the first subtree consists of the edge (1, 2). The remaining steps of the algorithm are summarized below.

Iteration	Vertices in Subtree	Edge Selected	Weight
1	1	(1, 2)	5
2	1, 2	(1, 3)	50
3	1, 2, 3	(3, 4)	8
4	1, 2, 3, 4	(4, 5)	10

Prim's algorithm takes $m - 1$ iterations to complete because one edge is added at each iteration. Finding the next edge to add can be done in $O(m)$ time. Thus, the overall complexity of this algorithm is $O(m^2)$.

EXAMPLE 4

To solve the maximum-capacity route problem, we find a maximum spanning tree. This can be done easily by selecting edges largest first. Why does the maximum spanning tree solve this problem? Any edge (x, y) not in the tree must satisfy

$$c(x, y) \leq \min\{c(x, 1), c(1, 2), \ldots , c(n, y)\}$$

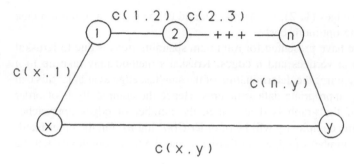

Fig. 3.4 Justification for the maximum capacity route solution.

(see Fig. 3.4). Otherwise we could replace some edge by (x, y) and have a larger minimum capacity from vertex x to vertex y. Thus, in the graph in Fig. 3.2, the maximum spanning tree is given by the edges (3, 4), (2, 4), (1, 2), and (4, 5). The maximum capacity on the route from vertex 1 to vertex 5 is 5; from vertex 2 to vertex 3, it is 7, and so on.

3.3   VARIATIONS OF THE MINIMUM SPANNING TREE PROBLEM

Many variations of the minimum spanning tree problem have been proposed and investigated. These typically involve adding restrictions on the structure of feasible solutions. In this section we briefly review some of these variations. The reader is referred to the references for more detail about these problems and algorithms for solving them.

The Capacitated Minimum Spanning Tree (Chandy and Lo, 1973)

The capacitated minimum spanning tree problem involves finding a spanning tree on a flow network that has applications in the design of teleprocessing networks. We are given an undirected graph $G = (X, E)$ with cost $d(x, y)$ and capacity $c(x, y)$ associated with edge (x, y). One of the vertices is a sink; the rest are sources. Each source i has a supply $a(i)$, and the sink has a demand equal to the sum of all supplies. The problem is to find a spanning tree in which all flow goes from the sources to the sink only along the edges of the tree at minimum cost and subject to the capacity constraints and, of course, conservation of flow. Fig. 3.5 shows an example for which the minimum spanning tree results in an infeasible solution. Figure 3.6 gives the optimal solution. No efficient algorithm is known for the general problem; thus heuristics must be used.

Degree-Constrained Spanning Trees (Gabow, 1978; Glover and Klingman, 1974)

Many applications of spanning trees occur in computer and communication networks. Many of these problems have constraints restricting the number of edges that are allowed

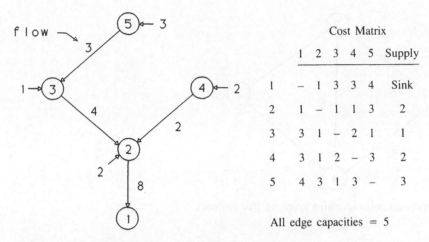

	1	2	3	4	5	Supply
1	–	1	3	3	4	Sink
2	1	–	1	1	3	2
3	3	1	–	2	1	1
4	3	1	2	–	3	2
5	4	3	1	3	–	3

All edge capacities = 5

Fig. 3.5 Capacitated minimum spanning tree example.

to be incident to a vertex. For example, one vertex might represent a central computing site. The other vertices might represent terminals which must be linked to the central site by cable paths. If we restrict the number of edges incident to the central site vertex to b, this guarantees that the computer's load is spread over b ports. Finding the minimum spanning tree guarantees that as little cable as possible is used. If only one vertex is constrained, as in this example, polynomial algorithms have been developed.

Optimum Communication Spanning Trees (Hu, 1974)

We are given a network with n vertices and a set of requirements $r(x, y)$ between the vertices. These might represent telephone calls, for example. We wish to design a

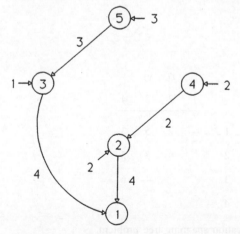

Fig. 3.6 Optimum solution to capacitated minimum spanning tree problem.

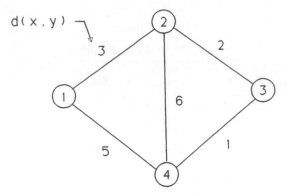

Fig. 3.7 Optimum communication spanning tree problem.

spanning tree so that the cost of communication is minimum among all spanning trees. We compute the cost of communication as the sum over all pairs of vertices of $r(x, y)$ times the distance $d(x, y)$ between vertices x and y. Recall that in a tree there is a unique path between any two vertices; thus these distances are unique. Figure 3.7 shows an example problem. A feasible solution is shown in Fig. 3.8. The cost of this solution is given by

$$r(1, 2)d(1, 2) + r(1, 3)d(1, 3) + r(1, 4)d(1, 4) + r(2, 3)d(2, 3)$$
$$+ \ r(2, 4)d(2, 4) + r(3, 4)d(3, 4)$$
$$= \ 2(3) + 5(5) + 3(9) + 3(2) + 3(6) + 6(8)$$
$$= \ 130$$

Except for the special cases in which all requirements are 1 or all distances are 1, the problem is not efficiently solved.

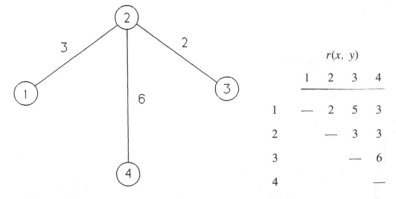

Fig. 3.8 Feasible solution to optimum communication spanning tree problem.

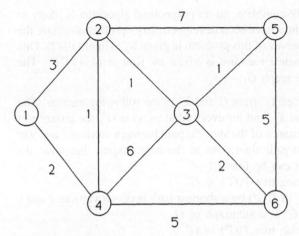

Fig. 3.9 Steiner tree example.

Steiner Trees

The Steiner problem on a graph is described as follows. Given a graph $G = (X, E)$ and a subset of vertices X', find a minimal-weight tree that includes all vertices of X' (and possibly others). The minimum spanning tree may not provide the optimal solution. For example, consider the graph in Fig. 3.9 and the subset of vertices {3, 4, 6}. If we find the minimum spanning tree only on the subgraph generated by vertices 3, 4, and 6, we get

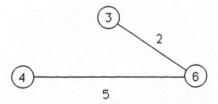

having a total weight of 7. However, by including vertex 2 in the set, we find

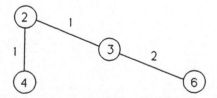

which has a weight of 4.

The Steiner graph problem is NP-complete, so no polynomial algorithm is likely to exist. Although various exact algorithms have been developed, large problems require the use of heuristics. A comprehensive survey of this problem is given by Winter (1987). One heuristic with a guaranteed performance measure is given by Kou et al. (1978). The algorithm proceeds as follows for a graph G.

Step 1. Construct a complete graph G' from G and X' in the following manner. The set of vertices in G' is the set X', and for every edge (x, y) in G', the distance of edge (x, y) is equal to the distance of the shortest path between vertices x and y in G. (We shall study shortest-path algorithms in the next chapter. For now, we assume that these distances can be found.)

Step 2. Find a minimum spanning tree $T(G')$ in G'.

Step 3. Replace each edge (x, y) in $T(G')$ by a shortest path between vertices x and y in G. The resulting graph G'' is a subgraph of G.

Step 4. Find a minimum spanning tree $T(G'')$ in G''.

Step 5. Delete edges in $T(G'')$, if necessary, so that all vertices having degree 1 in $T(G'')$ are members of X'. This represents a solution to the Steiner tree problem.

EXAMPLE 5

We will apply this algorithm to the graph in Fig. 3.9.

Step 1. The complete graph G' is shown below. Note that the shortest distance between vertices 3 and 4 is two.

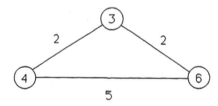

Step 2. The minimum spanning tree in G' consists of edges (3, 4) and (3, 6).

Step 3. We replace edge (3, 4) by its shortest path in G. The resulting graph, G'', is shown below.

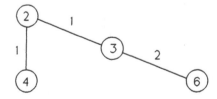

Step 4. The minimum spanning tree $T(G'')$ in G'' consists of the edges (2, 3), (2, 4), and (3, 6).

Step 5. Since no vertices having degree 1 are *not* members of X', we do not delete any edges.

The solution to the Steiner tree problem determined by this algorithm is (2, 3), (2, 4), and (3, 6).

It can be shown that the time complexity of this algorithm is $O(pm^2)$, where p is the cardinality of the set X' and m is the number of vertices in G. Moreover, it can be shown that the solution will be no more than $2(1 - 1/k)$ times the cost of the optimal solution, where k is the number of vertices of degree 1 in the *optimal* Steiner tree. Since k generally will not be known, in the worst case $k = m - 1$. Thus, the solution will never be more than $2[1 - 1/(m - 1)]$ times the cost of the optimal tree. The proof is simple but relies on a concept that we will not cover until Chapter 9, so we will not prove this result here. This is an example of a heuristic with *guaranteed worst-case performance*.

3.4 BRANCHINGS AND ARBORESCENCES

In Section 3.2, we considered tree-generating algorithms that ignore the direction of the arcs. In this section, we shall study an algorithm, due to Edmonds (1968), called the *maximum branching algorithm* that considers the direction of each arc.

Suppose the sales manager of a multinational company has a message that he or she wants conveyed to each of the district managers. What is the best way to accomplish this? One solution might be for the sales manager to phone each district manager personally. However, this might be very costly for the following reason: Suppose the sales manager is in Chicago and there are district managers in London and Paris. It would be far more expensive to phone each of them personally than it would be to phone London and have the London manager phone Paris.

Perhaps a better solution would be to have each district manager know in advance the selected district managers to whom he is to pass the message after receiving it, i.e., to have a "grapevine."

What properties should such a grapevine have? Obviously, each person should receive the message exactly once, and the cost of operating the grapevine should be as small as possible.

Construct a graph in which each vertex corresponds to a person in the grapevine and each arc represents the way a message may be transmitted between two people.

An *arborescence* is defined as a tree in which no two arcs are directed into the same vertex (see Fig. 3.10). Note that several arcs in an arborescence can share a common tail vertex. An arborescence can be thought of as a directed tree that can be used as a grapevine. The *root* of an arborescence is the unique vertex included in the arborescence that has no arcs directed into it. A *branching* is defined as a forest in which each tree is an arborescence.

A *spanning arborescence* is an arborescence that is also a spanning tree. A *spanning branching* is any branching that includes every vertex in the graph.

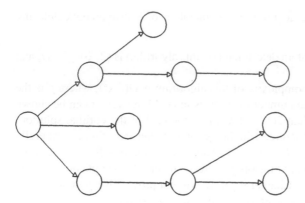

Fig. 3.10   An arborescence.

Associate a *weight* $a(x, y)$ to each arc (x, y). The *weight of an arborescence* (or *of a branching*) is defined as the sum of the weights of the arcs in the arborescence (or in the branching).

A *maximum branching of graph G* is any branching of graph G with the largest possible weight. A *maximum arborescence of graph G* is any arborescence of graph G with the largest possible weight. A *minimum branching* and a *minimum arborescence* are defined similarly.

The *maximum branching algorithm*, as its name suggests, constructs a maximum branching for any graph G. As shown in the following, the maximum branching algorithm can also be used to find (1) a minimum branching, (2) a maximum spanning arborescence (if one exists), (3) a minimum spanning arborescence (if one exists), (4) a maximum spanning arborescence rooted at a specified vertex (if one exists), and (5) a minimum spanning arborescence rooted at a specified vertex (if one exists).

1. A *minimum branching* is found by replacing each arc's weight by its negative. A maximum branching for the new arc weights corresponds to a minimum branching for the original arc weights.
2. The algorithm finds a *maximum spanning arborescence* as follows: Suppose a positive constant M is added to each arc's weight. Obviously, as M increases, a maximum branching for the graph will contain more and more arcs. Since no branching can contain more arcs than a spanning arborescence, the maximum branching produced by the algorithm will be a spanning arborescence (if one exists) for M large enough. Moreover, this spanning arborescence will be a maximum spanning arborescence.

Note that even if all arc weights are positive and the graph possesses a spanning arborescence, a maximum branching need not be a spanning arborescence (see, e.g., Fig. 3.11).

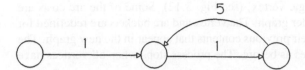

Fig. 3.11 Maximum branching has weight 5; spanning arborescence has weight 2.

3. A *minimum spanning arborescence* (if one exits) can also be generated by the algorithm. This is accomplished by replacing each arc's weight by its negative. Then, a large positive constant M should be added to each arc's weight. As mentioned in item 2, the effect of adding M to each arc's weight is to force the maximum branching algorithm to generate only branchings with the maximum possible number of arcs. Hence, the algorithm will generate a spanning arborescence (if one exists). Moreover, this spanning arborescence will be a minimum spanning arborescence.

4. and 5. The maximum branching algorithm can also be used to find a *maximum* (or *minimum*) *spanning arborescence rooted at a specified vertex*, say vertex a. (The sales manager mentioned at the beginning of this section is seeking a minimum-weight spanning arborescence rooted at Chicago.) This is accomplished by appending to the graph an additional vertex a' and an arc (a', a) with arbitrary weight. If the appended graph possesses a spanning arborescence, then this arborescence must be rooted at vertex a' since no arcs enter vertex a'. Any spanning arborescence rooted at a' must correspond to a spanning arborescence rooted at a in the original graph. As described in items 2 and 3, the arc weights can be altered so that the maximum branching algorithm will be forced to find a maximum (or minimum) spanning arborescence, if one exists.

As described in items 2 and 3, the arc weights can be altered so that the maximum branching algorithm will be forced to find a maximum (or minimum) spanning arborescence, if one exists.

We shall now proceed to describe the *maximum branching algorithm*. The maximum branching algorithm uses two buckets, the vertex bucket and the arc bucket. The vertex bucket contains only vertices that have been examined; the arc bucket contains arcs tentatively selected for the maximum branching. At all times, the arcs in the arc bucket form a branching. Initially both buckets are empty.

The algorithm successively examines the vertices in any arbitrary order. The examination of a vertex consists entirely of selecting the arc with the greatest positive weight that is directed into the vertex under examination (if any). If the addition of this arc to the arcs already selected for the arc bucket maintains a branching, then this arc is added to the arc bucket. Otherwise, this arc would form a cycle with some arcs already in the arc bucket. If this happens, then a new, smaller graph is generated by "shrinking" the arcs and vertices in this cycle into a single vertex, (see Fig. 3.12). Some of the arc costs are judiciously altered in the new, smaller graph. The vertex and arc buckets are redefined for the new graph as containing only their previous contents that appear in the new graph. The examination of each vertex continues as before. The process stops when all vertices have been examined.

Upon termination, the arc bucket contains a branching for the final graph. The final graph is expanded back to its predecessor by expanding out its "artificial" vertex into a cycle. All but one of the arcs in this cycle are added to the arc bucket. The arc that is not added to the arc bucket is carefully selected so that the contents of the arc bucket remain a branching. This process is repeated until the original graph is regenerated. The arcs in the arc bucket upon termination turn out to be a maximum branching.

Denote the original graph for which the maximum branching is sought by G_0, and denote each successive graph generated from G_0 by G_1, G_2, ... The vertex and arc buckets used for these graphs will be denoted by V_0, V_1, . . . and A_0, A_1, . . . , respectively. We are now ready to state the algorithm formally.

Maximum Branching Algorithm

Initially, all buckets V_0, V_1, . . . and A_0, A_1, . . . , are empty. Set $i = 0$.

Step 1. If all vertices of G_i are in bucket V_i, go to step 3. Otherwise, select any vertex v in G_i that is not in bucket V_i. Place vertex v into bucket V_i. Select an arc γ with the greatest positive weight that is directed into v. If no such arc exists, repeat step 1; otherwise, place arc γ into bucket A_i. If the arcs in A_i still form a branching repeat step 1; otherwise, go to step 2.

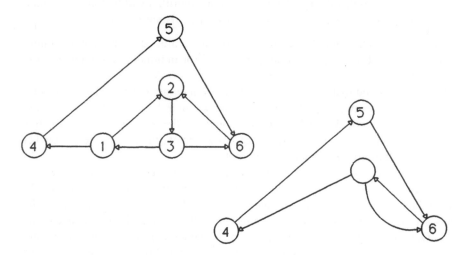

Fig. 3.12 Before and after shrinking cycle (1, 2), (2, 3), (3, 1).

Step 2. Since the addition of arc γ to A_i no longer causes A_i to form a branching, arc γ forms a cycle with some of the arcs in A_i. Call this cycle C_i. Shrink all the arcs and vertices in C_i into a single vertex called v_i. Call this new graph G_{i+1}. Thus, any arc in G_i that was incident to exactly one vertex in C_i will be incident to vertex v_i in graph G_{i+1}. The vertices of G_{i+1} are v_i and all the vertices of G_i not in C_i. Let the weight of each arc in G_{i+1} be the same as its weight in G_i except for the arcs in G_{i+1} that are directed into v_i. For each arc (x, y) in G_i that transforms into an arc (x, v_i) in G_{i+1}, let

$$a(x, v_i) = a(x, y) + a(r, s) - a(t, y) \tag{1}$$

where (r, s) is the minimum weight arc in cycle C_i, and where (t, y) is the unique arc in cycle C_i whose head is vertex y. [At this point, observe that

$$a(r, s) \geq 0$$

$$a(t, y) \geq a(r, s)$$

and

$$a(t, y) \geq a(x, y)$$

since arc (t, y) was selected as the arc directed into vertex y.]

Let V_{i+1} contain all the vertices in G_{i+1} that are in V_i, that is, $V_{i+1} = G_{i+1} \cap V_i$. (Thus, $v_i \in V_{i+1}$.) Let A_{i+1} contain all the arcs in G_{i+1} that are in A_i i.e., $A_{i+1} = G_{i+1} \cap A_i$. (Thus, A_{i+1} contains the arcs in A_i that are not in C_i.)

Increase i by one, and return to step 1.

Step 3. This step is reached only when all vertices of G_i are in V_i and the arcs in A_i form a branching for G_i. If $i = 0$, stop because the arcs in A_0 form a maximum branching for G_0. If $i \neq 0$, two cases are possible:

(a) Vertex v_{i-1} is the root of some arborescence in branching A_i.

(b) Vertex v_{i-1} is not the root of some arborescence in branching A_i.

If (a) occurs, then consider the arcs in A_i together with the arcs in cycle C_{i-1}. These arcs contain exactly one cycle in graph G_{i-1}, namely C_{i-1}. Delete from this set of arcs the arc in C_{i-1} that has the smallest weight. The resulting set of arcs forms a branching for graph G_{i-1}. Redefine A_{i-1} to be this set of arcs.

If (b) occurs, then there is a unique arc (x, v_{i-1}) in A_i that is directed into vertex v_{i-1}. This arc (x, v_{i-1}) corresponds in graph G_{i-1} to another arc, say arc (x, y), where vertex y is one of the vertices in cycle C_{i-1} that was shrunk to form vertex v_{i-1}.

Consider the set of arcs in A_i together with the arcs in cycle C_{i-1}. This set of arcs contains exactly one cycle in G_{i-1}, namely C_{i-1}, and exactly two arcs directed into vertex y, namely arc (x, y) and an arc in cycle C_{i-1}. Delete the latter arc from this set of arcs. The remaining arcs in this set form a branching in graph G_{i-1}. Redefine A_{i-1} to be this set of arcs. Having redefined A_{i-1}, decrease i by one unit and repeat step 3.

Proof of the Maximum Branching Algorithm. Consider any graph G_t produced by the algorithm and consider the branching A_t produced by step 3 for graph G_t. First, it will be shown that if A_t is a maximum branching for graph G_t, then branching A_{t-1} is a maximum branching for graph G_{t-1}.

To prove this, some definitions are needed. Let G' denote the subgraph consisting of all arcs in G_{t-1} not directed into a vertex in cycle C_{t-1}. Let G'' denote the subgraph consisting of all the arcs in G_{t-1} not in G'. Thus, every arc of G_{t-1} is present in exactly one of these subgraphs G' and G''. Let A'_{t-1} denote the arcs in A_{t-1} that are in G', and let A''_{t-1} denote the arcs of A_{t-1} that are in G''. Clearly, A'_{t-1} and A''_{t-1} are branchings in G' and G'', respectively.

If branching A_{t-1} is not a maximum branching for graph G_{t-1}, then there exists some branching B with greater total weight. Let B' denote the arcs in B that are in G', and let B'' denote the arcs of B that are in G''. Since B is a maximum branching, it follows that either

B' weighs more than A'_{t-1}

or

B'' weighs more than A''_{t-1}

Claim 1: A'_{t-1} is a maximum-weight branching for G'.
Claim 2: A''_{t-1} weights as much as B''.

If both Claims 1 and 2 are true, it follows that A_{t-1} must be a maximum branching for graph G_{t-1}.

Note that the branching A_i produced by the algorithm for the terminal graph G_i is a maximum branching since it contains a maximum positively weighted arc directed into each vertex in G_i if such an arc exists. Since the algorithm produces a maximum branching for the terminal graph G_i, then if both Claims 1 and 2 are true, the algorithm must produce a maximum branching A_{i-1} for graph G_{i-1}. By repeating this reasoning, we can conclude that if Claims 1 and 2 are true, then the branching A_0 produced by the algorithm is a maximum branching for the original graph G_0.

Hence, it remains only to show that Claims 1 and 2 are valid.

Proof of Claim 1: Suppose that cycle C_{t-1} contains n vertices. There is one arc with positive weight directed into each of these n vertices in graph G'. (Otherwise, the algorithm would not have formed cycle C_{t-1}.) Since there are only n vertices in G' that have arcs directed into themselves, a maximum branching for G' cannot contain more than n arcs. Moreover, no branching in G' can have weight exceeding the weight of cycle C_{t-1}, which consists of the maximum positive-weight arc directed into each of the n vertices in cycle C_{t-1}. However, at least one of the arcs in C_{t-1} must be absent from any maximum branching for G' since a branching cannot contain a cycle. Thus, at least one of these n vertices, say vertex $y \in C_{t-1}$, must either have no branching arc directed into it or else have an arc (x, y), $x \notin C_{t-1}$, directed into it.

For each vertex $z \in C_{t-1}$, construct a branching B_z in G' as follows:

(a) Include all arcs in cycle C_{t-1} except the arc in cycle C_{t-1} that is directed into vertex z.

(b) Include any maximum positive-weight arc (x, z), where $x \notin C_{t-1}$.

Select the branching B_z^* with the greatest weight. From equation (1), branching B_z^* is the branching A'_{t-1} generated by the algorithm.

Consider any branching B_1 in G' that is not of the form B_z. If only one of the arcs of C_{t-1} is not in B_1, it follows that B_1 cannot be a maximum branching for G' since it is not of the form B_z.

If two or more arcs of C_{t-1} are not in B_1, then either (1) each of these arcs is replaced by an arc of smaller weight directed into the same vertex or (2) no arc is directed into this vertex. In either case, this results in the decrease of the weight of arcs in the branching directed into the vertex. Hence, B_1 cannot be a maximum branching for G'.

Thus, A'_{t-1} is a maximum branching for G', and we can assume, without loss of generality, that A'_{t-1} is identical to B'. This concludes the proof of Claim 1.

Proof of Claim 2:   Two cases are possible:

(a) Branching A_t contains an arc (x, v_{t-1}) directed into vertex v_{t-1}.

(b) Branching A_t does not contain an arc directed into vertex v_{t-1}.

Case (a): By hypothesis, A_t is a maximum branching for G_t and contains an arc (x, v_{t-1}) directed into v_{t-1}. From Claim 1, B' is identical to A'_{t-1} and hence B' contains an arc (x, y), where $x \in C_{t-1}$ and $y \in C_{t-1}$. Since B is a branching in G_{t-1}, it follows that B'' cannot contain a path of arcs from a vertex in C_{t-1} to vertex x. Thus, B'' must be a maximum branching for G'' that does not contain a path of arcs from a vertex in C_{t-1} to vertex x.

Each arc in G'' corresponds to an arc with identical weight in G_t. Moreover, each branching in G'' corresponds to a branching in G_t with identical weight. Consequently, if A''_{t-1} is not a maximum branching in G'' that contains no path of arcs from a vertex in C_{t-1} to a vertex x, then A_t is not a maximum branching in G_t that contains arc (x, v_{t-1}), which is impossible. Hence, A''_{t-1} has the same weight as B'', which proves the claim for case a.

Case (b): Each arc in G'' corresponds to an arc with identical weight in G_t. By hypothesis A_t is a maximum branching for G_t. Since no arc in A_t is directed into v_{t-1}, it follows that every arc in A_t corresponds to an arc in A''_{t-1}. Moreover, any branching in G'' corresponds to a branching in G_t with the same weight. Hence if A''_{t-1} were not a maximum branching in G'', then A_t would not be a maximum branching in G_t, which is a contradiction.

Thus, A''_{t-1} must be a maximum branching in G'' and have the same weight as B'', which completes the proof of Claim 2.   Q.E.D.

EXAMPLE 6 (Maximum Branching Algorithm)

We shall now perform the maximum branching algorithm to find a maximum-weight branching for the graph in Fig. 3.13. The weight of each arc is shown next to the arc. The

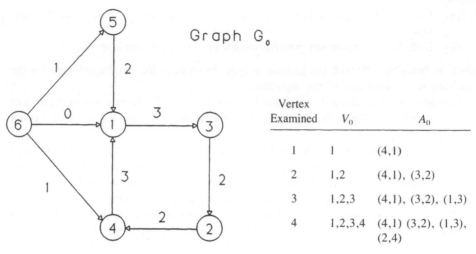

Fig. 3.13 Maximum branching algorithm.

algorithm will arbitrarily examine the vertices in numerical order. The result of the examination of the first four vertices 1, 2, 3, 4 is shown in Fig. 3.13.

After vertex 4 has been examined, the arcs in bucket A_0 no longer form a branching since they contain a cycle $(1, 3)$, $(3, 2)$, $(2, 4)$, $(4, 1)$. At this point, the algorithm shrinks this cycle into a vertex v_0. Figure 3.14 displays the new graph G_1 resulting from this shrinking. The calculations of the weights are shown next to each arc in graph G_1. Graph G_1 has only three vertices, 5, 6, and v_0. The result of the examination of each of these vertices is shown in Fig. 3.14.

After examining the three vertices in graph G_1, the algorithm has generated a maximum branching for graph G_1 consisting of arcs $(6, 5)$ and $(5, v_0)$. Using this

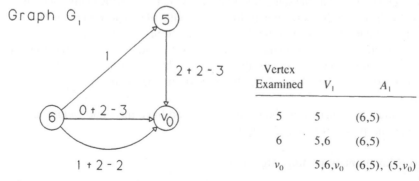

Fig. 3.14 Maximum branching algorithm.

branching, step 3 expands vertex v_0 back into its original cycle and adds arcs (1, 3), (3, 2), (2, 4), and (4, 1) to the arcs (6, 5) and (5, v_0) already in the branching. Next, arc (4, 1) is deleted from the branching so that only one branching arc, namely (5, 1), is directed into vertex 1. The resulting branching in graph G_0 consists of arcs (6, 5), (5, 1), (1, 3), (3, 2), and (2, 4). The total weight of this branching equals $1 + 2 + 3 + 2 + 2$ = 10, which is the maximum possible weight.

Note that this branching also happens to be a spanning arborescence of graph G_0 rooted at vertex 6.

APPENDIX: USING NETSOLVE TO FIND MINIMUM SPANNING TREES

The minimum spanning tree algorithm in NETSOLVE determines a minimum-cost (or maximum-value) spanning tree in an undirected or directed network. The "cost" associated with an edge is the value specified in the COST data field, and the cost of a spanning tree is simply the sum of all edge costs comprising the tree. If the network is connected, an optimal spanning tree is determined; otherwise, an optimal spanning forest is found. The command line to implement this algorithm is

>MST

or

>MST MIN

which will produce a minimum spanning tree or forest for the network. Using the command

>MST MAX

will produce a spanning tree or forest of maximum weight. If the terminal is currently ON, then in addition the (undirected) edges comprising an optimal spanning tree or forest will be displayed, together with the costs of the associated edges. Isolated nodes in the network will also be displayed.

In addition, the user will be asked whether a sensitivity analysis should be conducted on the problem. This analysis produces *tolerance intervals* for each relevant edge cost, such that varying the individual edge cost within that range will not change the optimality of the current tree.

The MST command can also be used to determine the connected components of an undirected graph. In this case any values defined for the COST data field (such as default settings) will suffice. The output from MST will group together nodes comprising the same connected components.

We will illustrate the use of the NETSOLVE minimum spanning tree algorithm first with the oil pipeline example from Chapter 1. Recall that in the appendix to Chapter 1 we demonstrated how to create, edit, and save the pipeline network using NETSOLVE. The network was saved using the name "PIPELINE." To retrieve the network data file, we use the GET command:

>GET PIPELINE

Once the data file has been retrieved, we simply type the command

>MST

The following output is obtained:

```
TOTAL COST OF MINIMUM SPANNING   TREE:        30.00
  EDGES IN THE MINIMUM SPANNING TREE
    FROM       TO              COST
    ----       --              --------
     1          2              5.00
     1          3              6.00
     2          4              4.00
     3          5              7.00
     4          6              3.00
     4          7              3.00
     5          8              2.00
```

Do you want sensitivity analysis? (Y/N): y

SENSITIVITY ANALYSIS FOR EDGE COSTS

FROM	TO	TREE EDGE	COST	TOLERANCE LOWER	UPPER
----	--	----	----	------	-----
1	2	YES	5.00	-999999.00	6.00
1	3	YES	6.00	-999999.00	6.00
1	4	NO	7.00	5.00	999999.00
2	4	YES	4.00	-999999.00	6.00
2	6	NO	8.00	4.00	999999.00
3	4	NO	6.00	6.00	999999.00
3	5	YES	7.00	-999999.00	10.00
3	7	NO	9.00	6.00	999999.00
3	8	NO	11.00	7.00	999999.00
4	6	YES	3.00	-999999.00	3.00
4	7	YES	3.00	-999999.00	3.00
5	7	NO	12.00	7.00	999999.00
5	8	YES	2.00	-999999.00	10.00
6	7	NO	3.00	3.00	999999.00
7	8	NO	10.00	7.00	999999.00

	1	2	3	4	5	6	7	8	9	10	11	12	13
1		1	1				1	1	1			1	1
2		1					1	1			1	1	
3	1				1								
4			1		1				1				
5		1					1	1	1			1	1
6					1								
7			1						1				
8			1						1				
9					1								

Fig. 3.15 Machine-part incidence matrix.

To interpret the sensitivity analysis information, we see that, for example, edge (1, 2) in the tree has a specified tolerance of minus infinity to 6. Since the edge cost currently is 5, this means that if the cost exceeds 6, then edge (1, 2) will no longer be contained in the minimum spanning tree. Similarly, edge (1, 4) not in the tree has a tolerance of 5 to plus infinity. The current cost is 7. If the cost of this edge is decreased below 5, then it would be beneficial to include this edge in the minimal spanning tree.

As a second example, we shall formulate and solve a group technology problem as described in Section 3.1. Figure 3.15 is a machine-part incidence matrix. The distance measure $d(i, k)$ was defined as the number of parts requiring processing only on one machine divided by the number of parts requiring processing on either machine. Thus, for $i = 1$ and $k = 2$, $d(1, 2) = 4/8 = 0.5$. A list of the edges in the graph created with NETSOLVE is shown in Fig. 3.16. Since lower and upper bounds are not necessary, NETSOLVE set these parameters at 0 and infinity, respectively. The results of the minimum spanning tree algorithm in NETSOLVE are shown in Fig. 3.17, and the spanning tree is shown in Fig. 3.18. Thus, if we wished to form three machine cells, we would remove the two largest edges from the tree. The resulting connected components would consist of the vertices {1, 2, 5}, {3, 9}, and {4, 6, 7, 8}. Can you find a natural assignment of parts to families that corresponds to these machine groups?

FROM	TO	COST	LOWER	UPPER
1	2	0.50	0.00	999999.00
1	3	1.00	0.00	999999.00
1	4	0.89	0.00	999999.00
1	5	0.14	0.00	999999.00
1	6	1.00	0.00	999999.00
1	7	1.00	0.00	999999.00
1	8	1.00	0.00	999999.00
1	9	1.00	0.00	999999.00
2	3	1.00	0.00	999999.00
2	4	1.00	0.00	999999.00
2	5	0.62	0.00	999999.00
2	6	1.00	0.00	999999.00
2	7	1.00	0.00	999999.00
2	8	1.00	0.00	999999.00
2	9	1.00	0.00	999999.00
3	4	1.00	0.00	999999.00
3	5	1.00	0.00	999999.00
3	6	1.00	0.00	999999.00
3	7	1.00	0.00	999999.00
3	8	1.00	0.00	999999.00
3	9	0.50	0.00	999999.00
4	5	0.87	0.00	999999.00
4	6	0.67	0.00	999999.00
4	7	0.75	0.00	999999.00
4	8	0.75	0.00	999999.00
4	9	1.00	0.00	999999.00
5	6	1.00	0.00	999999.00
5	7	1.00	0.00	999999.00
5	8	1.00	0.00	999999.00
5	9	1.00	0.00	999999.00
6	7	1.00	0.00	999999.00
6	8	1.00	0.00	999999.00
6	9	1.00	0.00	999999.00
7	8	0.00	0.00	999999.00
7	9	1.00	0.00	999999.00
8	9	1.00	0.00	999999.00

Fig. 3.16 Network parameters for group technology problem—NETSOLVE listing.

```
TOTAL COST OF MINIMUM SPANNING   TREE:        4.43
  EDGES IN THE MINIMUM SPANNING TREE
    FROM        TO              COST
    ----        --              --------
     1           2               0.50
     1           3               1.00
     4           5               0.87
     1           5               0.14
     4           6               0.67
     4           7               0.75
     7           8               0.00
     3           9               0.50
```

Fig. 3.17   NETSOLVE output for minimum spanning tree solution.

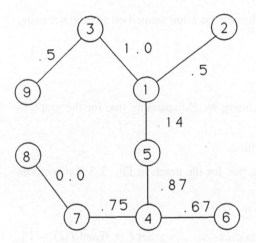

Fig. 3.18   Minimum spanning tree for group technology example.

EXERCISES

1. Apply the spanning tree algorithm to find a spanning tree for the following graph, taking the edges in order of their indexes.

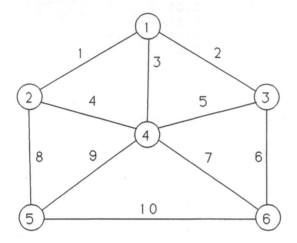

2. Describe how the spanning tree algorithm can be implemented on a computer using an
   (a) Incidence matrix
   (b) Adjacency matrix
   (c) Distance matrix

3. Construct a minimum-weight and maximum-weight spanning tree for the graph in Fig. 3.3 using Kruskal's algorithm.

4. Repeat Exercise 3 using Prim's algorithm.

5. Construct a minimum-weight spanning tree for the graph in Fig. 3.3 that includes edges (1, 2) and (3, 4).

6. Consider the following algorithm:

   Step 0. Arbitrarily label the edges of G as $e_1, e_2, \ldots, e_n$. Set $k = 0$ and $T(k) = \{\ \}$.
   Step 1. Let $k = k + 1$. If $k > n$ then terminate. Otherwise let $T(k) = T(k - 1) \cup e_k$.
   Step 2. If $T(k)$ is a forest, return to step 1. Otherwise $T(k)$ contains a cycle C with the edge e_k. Let e_s be the edge in C having the largest weight. Set $T(k) = T(k) - e_s$. Return to step 1.

   (a) Explain why this algorithm will find a minimum spanning tree. What is the time complexity of this procedure? Does it change for sparse or dense graphs?
   (b) Show that this algorithm is the same as Kruskal's if the edges are first ordered by nondecreasing weights.

(c) Apply this algorithm to the graph in Exercise 1 using the edge numbers as weights.

7. Find the maximum-capacity routes between all pairs of vertices in the following graph:

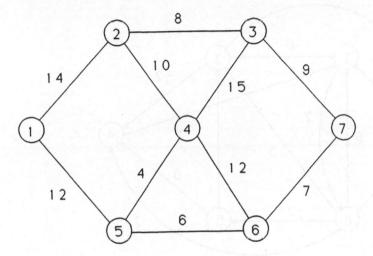

8. Find the minimum-weight spanning tree for the following machine-part incidence matrix:

	a	b	c	d	e	f	g	h	i	j	k	l	m	n	o	p	q	r	s	t
1				1	1		1						1							
2												1	1							
3				1	1															
4			1					1											1	1
5						1	1	1								1				
6	1	1												1						
7				1	1				1								1			
8						1	1	1							1					
9			1						1										1	1
10										1	1	1								

9. Develop and investigate the performance of heuristic procedures for solving the capacitated minimum spanning tree problem. Illustrate your algorithms using Fig. 3.6.

10. Develop and investigate the performance of heuristic procedures for solving the optimum communication spanning tree problem. Use Fig. 3.8 to illustrate your algorithms.

11. Given a graph with *m* vertices, *n* edges, and *p* vertices that are required to be in a Steiner tree, how many minimum spanning tree problems would have to be solved to find the optimum solution by exhaustive enumeration?

12. Construct a maximum-weight branching for the graph in Fig. 3.19.

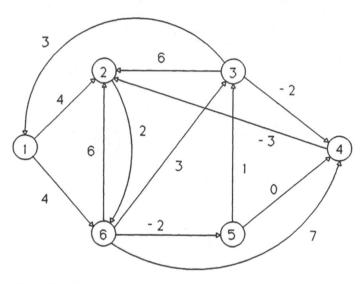

Fig. 3.19 Graph for Exercise 12.

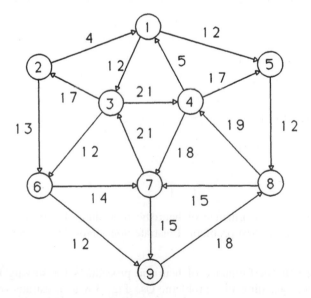

Fig. 3.20 Graph for Exercise 13.

13. Find a maximum-weight branching for the graph in Fig. 3.20.

14. Construct a maximum spanning arborescence for the graph in Fig. 3.13 that is rooted at vertex 1.

15. Suppose that you have just completed the maximum branching algorithm and discover that all your arc weights were understated by 5 units. Can you salvage your results, or is it necessary to repeat the algorithm? Explain.

16. Consider the following greedy maximum branching algorithm:

    Order the arcs according to their weights with the heaviest arc first. Select the first arc for the branching. Sequentially examine the rest of the arcs, selecting an arc for the branching if it forms a branching with the other arcs already selected for the branching.

    Show that this algorithm does not always terminate with a maximum branching.

17. Restate the maximum branching algorithm for the special case that all arcs have the same weight.

18. A pipeline system must be built to connect seven company-owned refineries with a port facility that receives imported crude oil. The cost of building the pipeline between any two points is $1000 per mile plus a $4000 setup cost for each segment. The distances between all pairs of points are given in the following table. Find the least-cost pipeline.

	P	R1	R2	R3	R4	R5	R6	R7
Port	0	5	6	8	2	6	9	10
Refinery 1		0	4	10	5	8	6	10
2			0	11	8	4	9	10
3				0	10	3	6	7
4					0	2	5	9
5						0	10	5
6							0	8
7								0

19. Suppose that the values in Exercise 18 represented the cost of shipping a year's supply of oil. Find the pipeline that would minimize the total yearly cost of shipments from the port to each refinery.

20. Each day, every researcher in a laboratory receives his or her assignment from one of his superiors. The researchers fall into four categories: senior, associate, assistant, and junior. There are, respectively, 1, 4, 6, and 5 persons currently in each category. Because of differences in sophistication and education, the times required to convey assignments are as follows:

From To:	Associate	Assistant	Junior
Senior	5	9	11
Associate		4	8
Assistant			4

What is the best way to disseminate the daily assignments?

REFERENCES

Chandy, K. M., and T. Lo, 1973. The Capacitated Minimum Spanning Tree, *Networks*, *3*, pp. 173–181.

Edmonds, J., 1968. Optimum Branchings. In *Mathematics of the Decision Sciences* (eds. G. Dantzig and A. Veinott). Lectures in Applied Mathematics, Vol. 2. American Mathematical Society, Providence, Rhode Island, pp. 346–361.

Gabow, H. N., 1978. A Good Algorithm for Smallest Spanning Trees with a Degree Constraint, *Networks*, *8*, pp. 201–208.

Glover, F., and D. Kingman, 1974. Finding Minimum Spanning Trees with a Fixed Number of Links at a Node, *Colloquia Mathematica Societatis Janos Bolyai 12. Progress in Operations Research*, Eger, Hungary, pp. 425–439.

Hu, T. C., 1961. The Maximum Capacity Route Problem, *Oper. Res.*, *9*, pp. 898–900.

Hu, T. C., 1974. Optimum Communication Spanning Trees, *SIAM J. Comput.*, *3*, pp. 188–195.

Kou, L., G. Markowsky, and L. Berman, 1978. A Fast Algorithm for Steiner Trees, Research Report RC 7390, IBM Thomas J. Watson Research Center, Yorktown Heights, New York.

Kruskal, J. B., 1956. On the Shortest Spanning Subtree of a Graph and the Traveling Salesman Problem, *Proc. AMS*, *7*, pp. 48–50.

Ng, S. M., 1989. A New Spanning Tree Algorithm for Group Technology Problems, Working Paper 1989-17, Industrial and Systems Engineering, University of Southern California, Los Angeles.

Pollock, M., 1960. The Maximum Capacity Route Through a Network, *Oper. Res.*, *8*, pp. 733–736.

Prim, R. C. 1957. Shortest Connection Networks and Some Generalizations, *Bell Syst. Tech. J.*, *36*, pp. 1389–1401.

Rosenstiehl, P., 1967. L'Arbre Minimum d'un Graphe, *Proceedings International Symposium on the Theory of Graphs in Rome*, 1966, Donod/Gordon-Breach, Rome.

Winter, P., 1987. Steiner Problem in Networks: A Survey, *Networks, 17*, pp. 129–167.

4
SHORTEST-PATH ALGORITHMS

4.1 INTRODUCTION AND EXAMPLES

Suppose that we are given a graph G in which each arc (x, y) has associated with it a number $a(x, y)$ that represents the *length* of the arc. In some applications, the length may actually represent cost or some other value. The *length of a path* is defined as the sum of the lengths of the individual arcs comprising the path. For any two vertices, s and t in G, it is possible that there exist several paths from s to t. The shortest-path problem involves finding a path from s to t that has the smallest possible length.

Among all classes of problem in network optimization, shortest-path problems have been some of the most extensively studied. They are commonly encountered in transportation, communication, and production applications and are often imbedded within other types of network optimization problems.

It is not possible here to discuss the many different variations of shortest-path algorithms that exist. Surveys by Deo and Pang (1984), and Gallo and Pallotino (1986) for example, list hundreds of references. Our main focus will be on some of the "classical" algorithms that form the basis for most of the different variations. We will, however, discuss some recent developments. First, let us present some applications.

Transportation Planning

Trucks that enter a state must file a route plan with the state government. They are not allowed to travel on roads with prohibitive weight restrictions. On the other hand, the trucker wants to reach his or her destination in the shortest possible time. This problem can be approached by constructing a graph corresponding to the highway network and deleting the arcs on which the truck is prohibited from traveling (or equivalently, giving them a very large length). By knowing the mileage corresponding to each arc as well as the legal speed limit, the travel time on each arc can be computed. Using these times as arc lengths, the trucker would want to solve a shortest-path problem from the point of entry to the final destination.

Salesperson Routing

Suppose that a salesperson in Boston plans to drive to Los Angeles to visit a client. She plans to use the interstate highway system and visit other clients along the route. She knows on the average how much commission she can earn from each proposed stop along the way from Boston to Los Angeles. What route should she take? In this situation, the length of each arc in the graph representing the highway system should be set equal to the net cost (driving less expected commission) of the corresponding highway segment. The salesperson should drive along the shortest path from the Boston vertex to the Los Angeles vertex.

Observe that in this case arc costs will be negative on any arc where the salesperson expects to make a profit and positive on any arc where the salesperson expects to incur a loss. In the previous example, all arc lengths were nonnegative. As we shall see later, these two situations require different solution algorithms.

Investment Planning

A small investor must decide how to invest his funds for the coming year in an optimal fashion. A variety of investments (money market funds, certificates of deposit, savings bonds, etc.) are available. For simplicity, suppose that investments can be made and withdrawn only on the first of each month. Create a vertex corresponding to the first day of each month. Place an arc from vertex x to vertex y for each investment that can be made at time x and mature at time y. Notice that this graph contains no directed cycles. Let the length of each arc equal the negative of the profit earned on the corresponding investment. The best investment plan corresponds to a shortest path (i.e., the path with the most negative total length) from the vertex for time zero to the vertex for one year from now.

Production Lot Sizing

A common problem encountered in production planning is to determine a schedule of production runs when the product's demand varies deterministically over time. We are given a set of demands $D_1, D_2, \ldots, D_n$ over an n-period planning horizon. At each period j, we incur a fixed setup cost A_j if we produce any positive amount as well as a unit production cost C_j. Any amount in excess of the demand for that period is held in inventory until the next period, incurring a per-unit holding cost of H_j.

This problem can be modeled as a mixed integer linear program. Let Q_j be the amount produced in period j and I_j be the amount held over in inventory to period $j + 1$. Y_j is a 0-1 variable that is equal to 1 if we produce in period j and equal to 0 otherwise. The model is

$$\min \; \sum_j (C_j Q_j + H_j I_j + A_j Y_j)$$

$$Q_j + I_{j-1} - I_j = D_j \quad \text{for } j = 1, \ldots, n$$

$$Q_j \leq M Y_j$$

$$Q_j \geq 0, \quad Y_j = 0, 1$$

It can be shown that an optimal solution will have the property that $Q_j = D_j + D_{j+1} + \cdots + D_k$ for some $k \geq j$. That is, production in any period must be a multiple of the demand for some set of future periods. Under these conditions, the problem can be reformulated as a shortest-path problem on a directed graph very similar to the equipment replacement example in Chapter 1. Let W_{jk} be the total cost associated with producing $D_j + D_{j+1} + \cdots + D_k$ units in period j.

$$W_{jk} = A_j + C_j Q_j + \Sigma H_j(I_j + \cdots + I_{k-1})$$

where

$$I_j = D_{j+1} + \cdots + D_k, I_{j+1} = D_{j+2} + \cdots + D_k, \ldots, I_{k-1} = D_k$$

The network for a four-period problem is shown in Fig. 4.1. Any directed path from node 1 to node n represents a sequence of production decisions. The minimum-cost path represents the optimal solution.

Knapsack Problem

The knapsack problem derives its name from the following situation. A camper has a set of n objects having weights w_i and values v_i to fit into a knapsack. The knapsack can hold a total weight of only W, which is less than the sum of all the weights. The camper must find a subset of objects that have the largest value and whose total weight does not exceed W. A more practical situation would be selecting experiments for space shuttle missions.

The knapsack problem can be formulated as a simple integer programming problem.

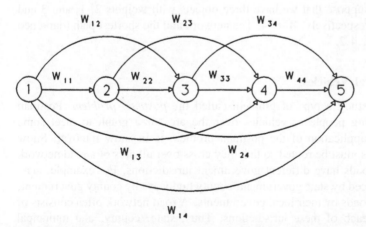

Fig. 4.1 Four-period production lot-sizing network.

Let $x_i = 1$ if object i is chosen and 0 otherwise. The problem is

$$\max \ \Sigma \ v_i x_i$$

subject to

$$\Sigma \ w_i x_i \leq W$$

$$x_i = 0, 1$$

We will show how to solve this problem as a shortest-path problem. First convert the problem to a minimization problem by multiplying the objective function by -1. Construction of the network is based on the following idea. Suppose we have some amount of weight $w \leq W$ available in the knapsack and decide whether or not to include object i. If we include object i in the knapsack, we will be left with a weight of $w - w_i$ remaining. If we do not include object i, a weight of w will still remain for other objects. Clearly if $w - w_i < 0$, we cannot consider including object i. We will begin with the available weight W and sequentially consider either including or not including each object in the knapsack. For each decision, we will know the amount of weight available in the knapsack for subsequent decisions.

The nodes of the network will consist of pairs (i, w), where i corresponds to object i and w corresponds to the current weight available given past decisions. We include nodes $(n + 1, w)$ for an artificial $(n + 1)$st object. For each node (i, w) we direct an arc to node $(i + 1, w)$ with cost 0 that corresponds to not selecting object i for the knapsack. We direct an arc from node (i, w) to $(i + 1, w - w_i)$ with cost $-v_i$ if $w - w_i \geq 0$ that corresponds to selecting object i. To allow the total weight of the chosen objects to be less than or equal to W, we add arcs directed from nodes $(n + 1, w)$ to $(n + 1, w - 1)$ with a cost of zero at the final stage. The shortest path from node $(1, W)$ to node $(n + 1, 0)$ represents the optimal solution.

To illustrate this, suppose that we have three objects with weights 2, 1, and 3 and values 30, 10, and 20, respectively. $W = 4$. The network and the shortest path (darkened arcs) are shown in Fig. 4.2.

Routing Snow Removal Vehicles

In Chapter 9 we will study a type of problem called the *postman problem*. Postman problems involve routing people or vehicles over the arcs of a graph to meet some objective. One useful application of the postman problem is in snow removal. Snow plows and salt spreaders must be routed so that they cross over all arcs of a road network that require service. Roads have different government jurisdictions. For example, state highways must be serviced by state government, county highways by county government, and city and township roads by their local governments. A road network often consists of a mix of roads from each of these jurisdictions. Thus, state, county, and municipal governments are each responsible for only a *subset* of arcs on a network. The vehicles, however, may travel along any of the road segments during their travels.

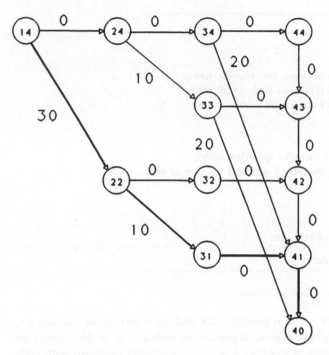

Fig. 4.2 Knapsack network.

One type of heuristic algorithm that is used to solve such problems involves a decision rule that selects the next arc on which to travel from a specific node. Suppose that a snowplow arrives at a node but no arcs incident to that node require service (either because they have already been serviced or because the vehicle is not responsible for that arc). To determine where to go next, one can solve a shortest-route problem from that node to any other node that has an incident arc requiring service. In a related situation, when a salt truck runs out of salt, returning to the salt depot along the shortest route would clearly be desirable. Using this approach, shortest-path problems must be solved repeatedly. This example illustrates how a shortest-path problem can be part of a larger algorithm for a more complex situation.

4.2 TYPES OF SHORTEST-PATH PROBLEMS AND ALGORITHMS

We can classify shortest-path problems into many different types. Figure 4.3 summarizes this classification. By ordinary path length, we mean that the path length is the sum of the lengths of the individual arcs. Generalized path length refers to problems in which the length is a more complicated function of arc lengths. For example, in transportation

---

I. **Ordinary path lengths**

    A. **Unconstrained**
        1. **Shortest path**
            a. **Shortest path between two specific nodes**
            b. **Shortest path from one node to all others**
            c. **Shortest path between all nodes**
        2. *k*-**shortest paths**

    B. **Constrained**
        1. **Shortest path that includes specified nodes**
        2. **Shortest path that includes a specified number of arcs**

II. **Generalized path lengths**

    A. **Turn penalties**

    B. **Length as a function of the path**

    C. **Algebraic related problems**

Fig. 4.3 Classification of shortest-path problems.

networks, one usually includes a "turn penalty" for making a turn at an intersection. Other real-valued functions might be used to determine the path length. In this chapter we shall be concerned with problems in category I.A.1: unconstrained problems with ordinary path lengths involving finding shortest paths between nodes or the *k*-shortest paths (i.e., the first-shortest, second-shortest, and so on) in a network. Constrained problems involve specifying intermediate nodes or a certain number of arcs that must be visited along the path. The reader is referred to the survey by Deo and Pang (1984) for a more detailed classification of shortest-path problems.

    Two general algorithmic approaches are used to solve shortest-path problems. *Label-setting* methods apply to networks with nonnegative arc lengths. *Label-correcting* methods apply to networks with negative arc lengths as well. Both approaches assign tentative *labels* to nodes at each step. These labels represent the shortest path distances that are known at that point of the algorithm. Label-setting methods set one or more permanent labels at each iteration. Permanent labels represent the actual shortest distances in the optimal solution. In label-correcting methods, all labels are temporary until the final iteration, at which time they all become permanent. As we shall see, algorithms for ordinary shortest-path problems have polynomial time complexity.

4.3 SHORTEST PATHS FROM A SINGLE SOURCE

In this section we shall consider algorithms that generate a path from node s to node t that has the smallest possible length. With simple modifications, they can also find the shortest paths from node s to all other nodes.

Dijkstra's Shortest-Path Algorithm

The first algorithm that we shall study is due to Dijkstra (1959) and provides the basis for the most efficient algorithms known for solving this problem. Most computational improvements for solving shortest-path problems have resulted from improving the data structures used to implement Dijkstra's algorithm. Dijkstra's algorithm is a label-setting algorithm in that a label is made permanent at each iteration.

The main idea underlying the Dijkstra shortest-path algorithm is quite simple. Suppose we know the k vertices that are closest in total length to vertex s in the graph and also a shortest path from s to each of these vertices. (Of course, a shortest path from vertex s to itself is the null path consisting of no arcs, which has length equal to zero.) Label vertex s and these k vertices with their shortest distance from s. Then the $(k + 1)$st closest vertex to x is found as follows. For each unlabeled vertex y, construct k distinct paths from s to y by joining the shortest path from s to x with arc (x, y) for all labeled vertices x (see Fig. 4.4). Select the shortest of these k paths and let it tentatively be the shortest path from s to y.

Which labeled vertex is the $(k + 1)$st closest vertex to s? It is the unlabeled vertex with the shortest tentative path from s as calculated above. This follows because the shortest path from s to the $(k + 1)$st closest vertex to s must use only labeled vertices as its intermediate vertices since all arc lengths are nonnegative.

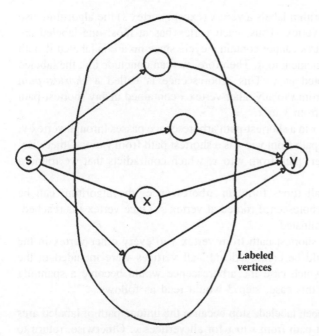

Labeled
vertices

Fig. 4.4 Illustration of Dijkstra's algorithm.

Therefore, if the k closest vertices to s are known, the $(k + 1)$st can be determined as above. Starting with $k = 0$, this process can be repeated until the shortest path from s to t has been found (that is, t is labeled). With this in mind as motivation, we can now formally state the Dijkstra shortest-path algorithm.

Step 1. Initially, all arcs and vertices are unlabeled. Assign a number $d(x)$ to each vertex x to denote the tentative length of the shortest path from s to x that uses only labeled vertices as intermediate vertices. Initially, set $d(s) = 0$ and $d(x) = \infty$ for all $x \neq s$. Let y denote the last vertex that was labeled. Label vertex s and let $y = s$.

Step 2. For each unlabeled vertex x, redefine $d(x)$ as follows:
$$d(x) = \min\{d(x), d(y) + a(y, x)\} \tag{1}$$
This can be performed efficiently by scanning the forward star of node y since only these nodes will be affected. If $d(x) = \infty$ for all unlabeled vertices x, then stop because no path exists from s to any unlabeled vertex. Otherwise, label the unlabeled vertex x with the smallest value of $d(x)$. Also label the arc directed into vertex x from a labeled vertex that determined the value of $d(x)$ in the above minimization. Let $y = x$.

Step 3. If vertex t has been labeled then stop, since a shortest path from s to t has been discovered. This path consists of the unique path of labeled arcs from s to t. If vertex t has not been labeled yet, repeat step 2.

Note that whenever the algorithm labels a vertex (except vertex s) the algorithm also labels an arc directed into this vertex. Thus, each vertex has at most one labeled arc directed into it, and the labeled arcs cannot contain a cycle since no arc is labeled if both its endpoints have a labeled arc incident to it. Therefore, we can conclude that the labeled arcs form an arborescence rooted at s. This arborescence is called a *shortest-path arborescence*. The unique path from s to any other vertex x contained in any shortest-path arborescence is a shortest path from s to x.

If the shortest path from s to x in a shortest-path arborescence passes through vertex y, it follows that the portion of this path from y to x is a shortest path from y to x. Otherwise, there exists another, even shorter, path from y to x, which contradicts that we found a shortest path from s to x.

Since the labeled arcs at all times form an arborescence, the algorithm can be regarded as the growing of an arborescence rooted at vertex s. Once vertex t is reached, the growing process can be terminated.

If you want to determine a shortest path from vertex s to every other vertex in the graph, the growing process could be continued until all vertices were included in the shortest-path arborescence, in which case the arborescence would become a spanning arborescence (if one exists). In this case, step 3 would read as follows:

Step 3. If all vertices have been labeled, stop because the unique path of labeled arcs from s to x is a shortest path from s to x for all vertices x. Otherwise, return to step 2.

EXAMPLE 1

Let us perform the Dijkstra shortest-path algorithm to find a shortest path from node s to node t in the graph in Fig. 4.5.

Step 1. Initially, only node s is permanently labeled, $d(s) = 0$. Assign tentative distances $d(x) = \infty$ for all $x \neq s$. Let $y = s$.

Step 2. Recompute tentative distances for the unlabeled nodes in the forward star of y as follows:

$$d(1) = \min\{d(1), d(s) + a(s, 1)\} = \min\{\infty, 0 + 4\} = 4$$
$$d(2) = \min\{d(2), d(s) + a(s, 2)\} = \min\{\infty, 0 + 7\} = 7$$
$$d(3) = \min\{d(3), d(s) + a(s, 3)\} = \min\{\infty, 0 + 3\} = 3$$

Since the minimum distance on any unlabeled node is $d(3) = 3$, we label node 3 and arc $(s, 3)$. The current shortest-path arborescence consists of arc $(s, 3)$ (see Fig. 4.6a). Let $y = 3$.

Step 3. Node t has not been labeled, so return to step 2.

Step 2.

$$d(4) = \min\{d(4), d(3) + a(3, 4)\} = \min\{\infty, 3 + 3\} = 6$$

The minimum tentative distance on the unlabeled nodes is $d(1) = 4$. Label node 1 and arc $(s, 1)$, which determined $d(1)$. The current shortest-path arborescence consists of arcs $(s, 3)$ and $(s, 1)$ (see Fig. 4.6b). Let $y = 1$.

Step 3. Vertex t has not been labeled, so return to step 2.

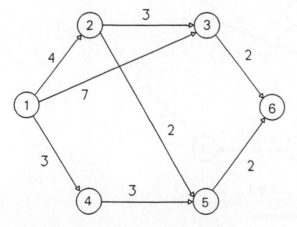

Fig. 4.5 Shortest-path example network.

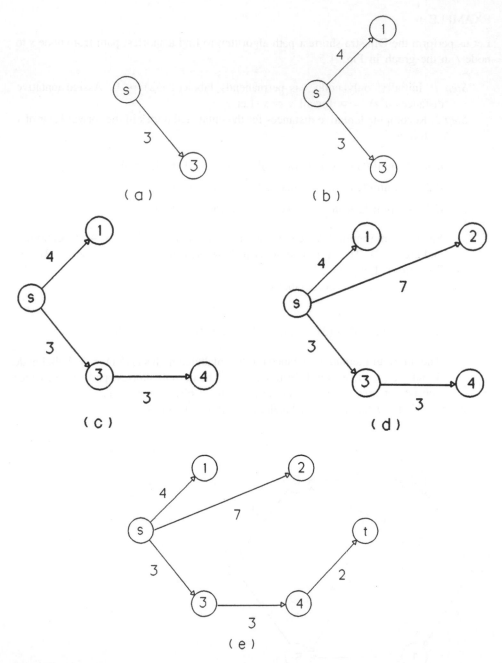

Fig. 4.6   Growing a shortest-path arborescence.

Step 2.

$$d(2) = \min\{d(2), d(1) + a(1, 2)\} = \min\{7, 4 + 3\} = 7$$
$$d(4) = \min\{d(4), d(1) + a(1, 4)\} = \min\{6, 4 + 2\} = 6$$

The minimum tentative distance on the unlabeled nodes is $d(4) = 6$. Label node 4 and either arc $(1, 4)$ or $(3, 4)$, since both determined $d(4)$. Let us arbitrarily select arc $(3, 4)$. Hence the shortest-path arborescence becomes arcs $(s, 3)$, $(s, 1)$, and $(3, 4)$ (see Fig. 4.6c). Let $y = 4$.

Step 3. Vertex t has not been labeled, so return to step 2.

Step 2.

$$d(t) = \min\{d(t), d(4) + a(4, t)\} = \min\{\infty, 6 + 2\} = 8$$

The minimum tentative distance label is $d(2) = 7$, so we label node 2 and arc $(s, 2)$, which determined $d(2)$. The current shortest-path arborescence consists of arcs $(s, 3)$, $(s, 1)$, $(3, 4)$, and $(s, 2)$ (see Fig. 4.6d). Let $y = 2$.

Step 3. Vertex t has not been labeled, so return to step 2.

Step 2.

$$d(t) = \min\{d(t), d(2) + a(2, t)\} = \min\{8, 7 + 2\} = 8$$

Thus, node t has been labeled at last. Also, arc $(4, t)$, which determined $d(t)$, is labeled. The final shortest-path arborescence consists of the arcs $(s, 3)$, $(s, 1)$, $(3, 4)$, $(s, 2)$, and $(4, t)$ (see Fig. 4.6e).

A shortest path from s to t consists of arcs $(s, 3)$, $(3, 4)$, and $(4, t)$ with a length of $3 + 3 + 2 = 8$. This path is not the only shortest path from s to t since the path $(s, 1)$, $(1, 4)$, and $(4, t)$ also has length 8. A shortest path from s to t will be unique if there is never any choice as to which arc to label.

Finally, note that if there were a tie for the node to be labeled—that is, two different nodes had the same minimum value of $d(x)$—the selection could be made arbitrarily. On the next iteration of step 2, the other vertex would be labeled.

To be able to trace the shortest path from s to t easily in a computer implementation, we define a *predecessor function*, $p(x)$. Whenever we label node x from node y, set $p(x) = y$. In Example 1, the algorithm terminates with the following predecessors:

$$p(1) = s$$
$$p(2) = s$$
$$p(3) = s$$
$$p(4) = 3$$
$$p(t) = 4$$

To trace the final path from node t, we need only apply this function recursively from node t. Thus, $p(t) = 4$; $p(4) = 3$; $p(3) = s$ represents the shortest path.

Initially, we assumed that all arc lengths were nonnegative. What would happen in the Dijkstra shortest-path algorithm if some of the arc lengths were negative? For example, consider the graph in Fig. 4.7. The shortest path from vertex s to vertex t is $(s, 1)$, $(1, t)$, whose length is $+ 2 - 2 = 0$. The reader can easily verify that if the Dijkstra shortest-path algorithm were applied to this graph, then the path (s, t) would erroneously be selected as the shortest path from vertex s to vertex t. Hence, there is no guarantee that the Dijkstra shortest-path algorithm will produce a shortest path when arc lengths are permitted to be negative.

All shortest-path algorithms consist of essentially two arithmetic operations, addition and minimization. To analyze the computational complexity of one of these algorithms, we need some way of comparing addition operations with minimization operations. Of course, this comparison varies between computers (human and mechanical), but for expediency we shall assume that these two operations require equivalent amounts of computational time.

Let us consider the computational complexity of the Dijkstra algorithm. At the first iteration of the Dijkstra algorithm, the $m - 1$ unlabeled vertices must be examined. From equation (1), this requires $m - 1$ additions, $m - 1$ minimizations, and the selection of the smallest of $m - 1$ numbers, i.e., another $m - 1$ minimizations. Thus, $3(m - 1)$ operations are required by the first iteration. Similarly, $3(m - 2)$ operations are required by the second iteration, etc. In total, $\sum_{i=1}^{m} 3(m - i) = 3m(m - 1)/2$ operations are required. Of course, at each iteration one must also determine which vertices are labeled and which are unlabeled. This requires additional work, but a clever programming technique can be used to avoid this. The details of this are beyond the scope of this text but can be found in Yen (1973) and Williams and White (1973). Thus, the Dijkstra algorithm requires $O(m^2)$ running time.

Ford's Algorithm

Fortunately, the Dijkstra shortest-path algorithm can be generalized to accommodate arcs with negative lengths. This generalization is due to Ford (1956) and is accomplished by three simple changes in the Dijkstra algorithm:

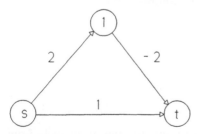

Fig. 4.7

1. In step 2, let equation (1) be applied to all vertices, not only the unlabeled vertices. Hence, a labeled as well as an unlabeled vertex can have its vertex number decreased.
2. If a labeled vertex has its vertex number decreased, then unlabel the labeled arc incident to it.
3. Terminate the algorithm only after all vertices are labeled and step 2 fails to lower any vertex numbers.

Proof. The proof of the Ford algorithm is achieved by contradiction. The Ford algorithm cannot terminate unless

$$d(x) + a(x, y) \geq d(y) \text{ for all } (x, y) \qquad (2)$$

Otherwise, vertex y would be unlabeled during the iteration of Step 2 occurring immediately after vertex x was labeled.

Suppose that upon the termination of the algorithm $d(y)$ does not equal the length of a shortest path from vertex s to vertex y for some vertex y. (If there is more than one such vertex y, then take y to be the vertex with the shortest path from s that contains the fewest arcs.) Whenever a vertex number $d(z)$ is finite, it equals the length of some path from vertex s to vertex z. Hence, it follows that upon termination $d(y)$ must equal the length of some path from s to y, and consequently $d(y)$ must exceed the length of a shortest path from s to y. Let vertex x be the next-to-last vertex on the shortest path from s to y. (If there is more than one such path, select the path with the fewest number of arcs.) Thus, $d(x)$ must equal the length of a shortest path from s to x, and $d(y) > d(x) + a(x, y)$. This contradicts expression (2). Q.E.D.

EXAMPLE 2

Let us apply the Ford algorithm to the graph in Fig. 4.7.

Step 1. Initially, only vertex s is labeled, $d(s) = 0$, $d(1) = \infty$, and $d(t) = \infty$.

Step 2. $y = s$

$$d(1) = \min\{d(1), d(s) + a(s, 1)\} = \min\{\infty, 0 + 2\} = 2$$

$$d(t) = \min\{d(t), d(s) + a(s, t)\} = \min\{\infty, 0 + 1\} = 1$$

Since $d(t) = \min\{d(1), d(t)\}$, vertex t is labeled and arc (s, t) is also labeled. The current shortest-path arborescence is (s, t).

Step 3. Since not all vertices are labeled, return to step 2.

Step 2. $y = t$. Since there are no arcs leaving t, all vertex numbers remain unchanged. Hence, vertex 1 is labeled and arc $(s, 1)$ is also labeled. The shortest-path arborescence now consists of arcs (s, t) and $(s, 1)$.

Step 3. Return to step 2 to try to lower vertex numbers.

Step 2. $y = 1$.

$$d(t) = \min\{d(t), d(1) + a(1, t)\} = \min\{1, 2 - 2\} = 0$$
$$d(s) = \min\{d(s), d(1) + a(1, s)\} = \min\{0, 2 + \infty\} = 0.$$

Since $d(t)$ is reduced from 1 to 0, vertex t and arc (s, t) are unlabeled. The shortest-path arborescence now consists of only arc $(s, 1)$. Vertex t is the only unlabeled vertex, and hence by default vertex t must be labeled and arc $(1, t)$ must also be labeled. The shortest-path arborescence is now $(s, 1)$ and $(1, t)$.

Step 3. Return to step 2 for $y = t$.

Step 2. $y = t$. Since there are no arcs leaving vertex t, no vertex numbers can be lowered. No vertices are unlabeled.

Step 3. Since all vertices are labeled and no vertex numbers can be lowered, the algorithm stops. The shortest path from s to t is $(s, 1)$, $(1, t)$, whose length is $2 - 2 = 0$.

Can the Ford algorithm fail? Yes, if the graph contains a directed cycle whose total length is negative. In this case, the cycle might be repeated infinitely many times, yielding a nonsimple path whose length is infinitely negative. For the salesperson routing example, a negative directed cycle corresponds to a profitable circular route that can be repeated infinitely many times yielding an infinite profit. In Example 2 the graph contained no directed cycles, so the Ford algorithm could be applied without hesitation.

What should we do if we are not certain whether the graph under consideration contains any negative directed cycles? Apply the Ford algorithm anyway, but keep count on the number of times each vertex is labeled. As soon as any vertex is labeled for the nth time, where n is the number of vertices in the graph, stop because the graph contains a negative directed cycle. Otherwise, the Ford algorithm terminates in a finite number of steps with correct results.

To show this, suppose that there are no negative directed cycles in the graph. When any vertex x receives its final vertex number value (i.e., the length of a shortest path from s to x), then at worst every other vertex can be labeled once more before another vertex receives its final vertex number value. Hence, no vertex can be labeled more than $m - 1$ times. From this, it follows that the Ford algorithm requires *at worst* $O(m^3)$ running time.

Ford's algorithm was the first label-correcting algorithm for the shortest-path problem. Label-correcting algorithms have the feature that no label is permanent until the algorithm terminates. Some nodes may have their labels corrected and be scanned more than once. The rationale behind this approach is to try to save computational time in searching for the unlabeled node with the minimum label at each iteration.

Partitioning Algorithms

A class of label-correcting algorithms called *partitioning algorithms* was proposed by Glover et al. (1985a). The steps of this algorithm are as follows:

Step 1. Set $d(s) = 0$ and $d(x) = \infty$ for all $x \neq s$. Set the predecessor function $p(x) = 0$ for all x. Set the iteration counter $k = 0$. Create two mutually exclusive and collectively exhaustive lists of unlabeled nodes called NOW and NEXT. Initially, NOW $= \{s\}$ and NEXT $= \{\ \}$.

Step 2. Select any node in NOW, say node u. If NOW is empty, go to step 4.

Step 3. Delete node u from NOW. For each node v in the forward star of u, if $d(u) + a(u, v) < d(v)$, then set $d(v) = d(u) + a(u, v)$, and $p(v) = u$. Add node v to NEXT if it is not already in NEXT or NOW. Return to step 2.

Step 4. If NEXT is empty, stop. Otherwise, set $k = k + 1$ and transfer all nodes from NEXT to NOW and return to step 2.

EXAMPLE 3

Let us illustrate this algorithm using the network in Fig. 4.5.

Step 1. $k = 0$. NOW $= \{1\}$, NEXT $= \{\ \}$, $d(1) = 0$, $d(x) = \infty$ for all $x \neq s$.

Step 2. Select node 1 from the set NOW; NOW $= \{\ \}$.

Step 3. Scan the forward star of node 1. Set $d(2) = 4$, $d(3) = 7$, and $d(4) = 3$. Set $p(2) = p(3) = p(4) = 1$. NEXT $= \{2, 3, 4\}$.

Step 2. NOW is empty. Go to step 4.

Step 4. $k = 1$. Transfer all nodes from NEXT to NOW. NOW $= \{2, 3, 4\}$, NEXT $= \{\ \}$.

Step 2. Select node 2 from NOW; NOW $= \{3, 4\}$.

Step 3. Scan the forward star of node 2. Set $d(5) = 6$, $p(5) = 2$. NEXT $= \{5\}$.

Step 2. Select node 3 from NOW; NOW $= \{4\}$.

Step 3. Scan the forward star of node 3. Set $d(6) = 9$, $p(6) = 3$. NEXT $= \{5, 6\}$.

Step 2. Select node 4 from NOW; NOW $= \{\ \}$.

Step 3. Scan the forward star of node 4. No labels are updated.

Step 2. NOW is empty. Go to step 4.

Step 4. $k = 2$. NOW $= \{5, 6\}$, NEXT $= \{\ \}$.

Step 2. Select node 5 from NOW; NOW $= \{6\}$.

Step 3. Scan the forward star of node 5. $d(6) = 8$, $p(6) = 5$. NEXT $= \{\ \}$. (Node 6 is not added to NEXT since it is already in NOW.)

Step 2. Select node 6 from NOW; NOW $= \{\ \}$.

Step 3. Scan the forward star of node 6. No labels are updated.

Step 2. NOW is empty. Go to step 4.

Step 4. NEXT is empty. Stop.

The partitioning shortest-path algorithm has time complexity $O(nm)$. For nonnegative arc lengths, step 4 can be replaced by step 4A:

Step 4A. If NEXT is empty, stop. Otherwise set $k = k + 1$ and transfer any subset of nodes from NEXT to NOW that includes some node i for which $d(i) =$ the current minimum distance label. Return to step 2.

Using this step, the algorithm still has time complexity $O(mn)$. A tighter computational bound of $O(m^2)$ can be shown if step 4 is replaced by the following:

Step 4B. If NEXT is empty, stop. Otherwise set $k = k + 1$ and select the subset transferred by step 4A to contain up to c nodes, where c is a predetermined constant (independent of n).

This can be accomplished by transferring up to $c - 1$ nodes from NEXT to NOW using some "threshold" value t, that is, nodes whose distance label satisfies $d(x) \leq t$, and then transferring a node with the smallest distance label of those nodes remaining on NEXT. If at some iteration, t is less than the minimum distance label of the nodes on NEXT, then t is recomputed as a weighted average of the minimum and average distance labels of the nodes currently in NEXT. When $c = 1$, this variation of the algorithm reduces to Dijkstra's algorithm. Glover et al. (1985b) subsequently developed several other variations of this basic algorithm. Using a variety of test problems, these algorithms have been shown to dominate most other algorithms in terms of computational performance, both by the original authors and in subsequent studies (Hung and Divoky, 1988).

Dijkstra's Two-Tree Algorithm

Dijkstra's two-tree algorithm is based on the idea of growing a shortest-path tree from both s and t simultaneously. When both trees have a node in common and meet certain conditions, the shortest path between s and t is found. The idea was first suggested by Dantzig (1960) and further developed by Nicholson (1966). Surprisingly, no one had seriously considered this approach from a computational viewpoint until Helgason et al. (1988). In their study, the original Dijkstra algorithm generated a shortest-path tree containing approximately 50% of the original nodes until a path from s to t was found. The two-tree method was found to contain only about 6% of the nodes. This can be of substantial benefit in algorithms that depend on repeated solution of shortest-path problems.

The two-tree shortest-path algorithm for a graph $G = (X, A)$ is described as follows. The subscripts s and t correspond to the trees grown from nodes s and t, respectively; all other notation is similar to that in the previous algorithms.

Step 1. Set $d_s(s) = 0$, $d_t(t) = 0$, $d_s(x) = \infty$ for $x \neq s$, $d_t(x) = \infty$ for $x \neq t$, $p_s(s) = p_t(t) = 0$. Define $R(s) = X$, $R(t) = X$.

Step 2. If $R(s) \cup R(t) \neq X$ then go to step 4. The shortest path has been found. Otherwise, let u_s be the minimum distance label for all nodes in the set $R(s)$ and let $Q(s)$ be the set of nodes having this minimum label. Set $R(s) = R(s) - Q(s)$. Similarly, let u_t be the minimum distance label for all nodes in the set $R(t)$ and let $Q(t)$ be the set of nodes having this minimum label. Set $R(t) = R(t) - Q(t)$.

Step 3. Scan the forward star for all nodes x in the set $Q(s)$. If $y \in R(s)$ and $u_s + a(x, y) < d_s(y)$, then set $d_s(y) = u_s + a(x, y)$ and $p_s(y) = x$. Scan the backward star for all nodes x in the set $Q(t)$. If $y \in R(t)$ and $u_t + a(y, x) < d_t(y)$, then set $d_t(y) = u_t + a(y, x)$, and $p_t(y) = x$. Return to step 2.

Step 4. Let $u = \min\{d_s(x) + d_t(x)\}$ for $x \in [X - R(s)] \cup [X - R(t)]$, and let J be the set of nodes x for which this minimum occurs. The shortest path from s to t has distance u and is given by the union of any paths from s to $x \in J$ and $x \in J$ to t as determined by the predecessor functions $p_s(\ )$ and $p_t(\ )$.

EXAMPLE 4

To illustrate this algorithm we will use the example in Fig. 4.5.

	s-tree	*t*-tree
Step 1.	$R(1) = \{1, 2, 3, 4, 5, 6\}$	$R(6) = \{1, 2, 3, 4, 5, 6\}$
	$d_s(1) = 0$	$d_t(6) = 0$
	$d_s(x) = \infty, x \neq s$	$d_t(x) = \infty, x \neq t$
Step 2.	$u_s = 0$	$u_t = 0$
	$Q(s) = \{1\}$	$Q(t) = \{6\}$
	$R(s) = \{2, 3, 4, 5, 6\}$	$R(t) = \{1, 2, 3, 4, 5\}$
Step 3.	$d_s(2) = 4, p_s(2) = 1$	$d_t(3) = 2, p_t(3) = 6$
	$d_s(4) = 3, p_s(4) = 1$	$d_t(5) = 2, p_t(5) = 6$
	$d_s(3) = 7, p_s(3) = 1$	
Step 2.	$u_s = 3$	$u_t = 2$
	$Q(s) = \{4\}$	$Q(t) = \{3, 5\}$
	$R(s) = \{2, 3, 5, 6\}$	$R(t) = \{1, 2, 4\}$
Step 3.	$d_s(5) = 6, \quad p_s(5) = 4$	$d_t(2) = 4, p_t(2) = 5$
		$d_t(4) = 5, p_t(4) = 5$
Step 2.	$u_s = 4$	$u_t = 4$
	$Q(s) = \{2\}$	$Q(t) = \{2\}$
	$R(s) = \{3, 5, 6\}$	$R(t) = \{1, 4\}$
Step 3.	No changes	$d_t(1) = 8, p_t(1) = 4$
Step 2.	At this point, $R(s) \cup R(t) \neq X$	
Step 4.	$[X - R(s)] \cup [X - R(t)] = \{1, 2, 3, 4, 5, 6\}$	
	$u = \min\{0 + 8, 4 + 4, 7 + 2, 3 + 5, 6 + 2, \infty\} = 8$	
	$J = \{1, 2, 4, 5\}$	

The trees that have been grown are shown in Fig. 4.8. Observe that there must be a node in J that is permanently labeled in both trees (in this case, it is node 2). Other nodes in J provide alternate optimal shortest paths.

4.4 ALL SHORTEST-PATH ALGORITHMS

The preceding section considered the problem of finding a shortest path from a specified vertex to every other vertex in the graph. In this section, we shall consider the problem of finding a shortest path between every pair of vertices in the graph. Of course, this second

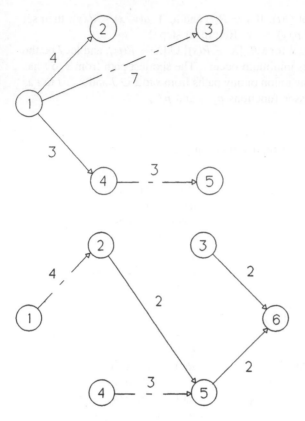

Fig. 4.8 Termination of Dijkstra's two-tree algorithm.

and more general problem could be solved by repeating the Dijkstra shortest-path algorithm once for each vertex in the graph taken as the initial vertex s. However, this would require a great many computations, and fortunately algorithms exist that are more efficient than repeating the Dijkstra shortest-path algorithm once for each vertex in the graph. This section presents two similar algorithms for finding all shortest paths. These algorithms are due to Floyd (1962) and Dantzig (1967). In both algorithms, arc lengths are permitted to be negative as long as no circuits have negative length.

Before presenting the algorithms, some notation is needed. Number the vertices 1, 2, . . . , m. Let d_{ij}^k denote the length of a shortest path from vertex i to vertex j, where only the first k vertices are allowed to be intermediate vertices. (Recall that an intermediate vertex is any vertex in the path, except the initial or terminal vertex of the path.) If no such path exists, then let $d_{ij}^k = \infty$. From this definition of d_{ij}^k it follows that d_{ij}^0 denotes the length of a shortest path from i to j that uses no intermediate vertices, i.e., the length of the shortest arc from i to j (if such an arc exists). Let $d_{ii}^0 = 0$ for all vertices i. Furthermore, d_{ij}^m represents the length of a shortest path from i to j.

Let D^k denote the $m \times m$ matrix whose i, jth element is d_{ij}^k. If we know the length of each arc in the graph, then we can determine matrix D^0. Ultimately, we wish to determine D^m, the matrix of shortest path lengths.

The Floyd shortest-path algorithm starts with D^0 and calculates D^1 from D^0. Next, the Floyd shortest-path algorithm calculates D^2 from D^1. This process is repeated until D^m is calculated from D^{m-1}.

The basic idea underlying each of these calculations is the following. Suppose we know

(a)   A shortest path from vertex i to vertex k that allows only the first $k - 1$ vertices as intermediate vertices.

(b)   A shortest path from vertex k to vertex j that allows only the first $k - 1$ vertices as intermediate vertices.

(c)   A shortest path from vertex i to vertex j that allows only the first $k - 1$ vertices as intermediate vertices. Since no cycles with negative length exist, the shorter of the paths given in items (d) and (e) must be a shortest path from i to j that allows only the first k vertices as intermediate vertices:

(d)   The union of the paths in items (a) and (b).

(e)   The path in item (c).

Thus,

$$d_{ij}^k = \min\{d_{ik}^{k-1} + d_{kj}^{k-1}, d_{ij}^{k-1}\} \tag{3}$$

From equation (3), we can see that only the elements of matrix D^{k-1} are needed to calculate the elements of matrix D^k. Moreover, these calculations can be done without reference to the underlying graph. We are now ready to state formally the *Floyd shortest-path algorithm* for finding a shortest path between each pair of vertices in a graph.

Floyd Shortest-Path Algorithm

Step 1. Number the vertices of the graph $1, 2, \ldots, m$. Determine the matrix D^0 whose i, jth element equals the length of the shortest arc from vertex i to vertex j, if any. If no such arc exists, let $d_{ij}^0 = \infty$. Let $d_{ii}^0 = 0$ for each i.

Step 2. For $k = 1, 2, \ldots, m$, successively determine the elements of D^k from the elements of D^{k-1} using the following recursive formula:

$$d_{ij}^k = \min\{d_{ik}^{k-1} + d_{kj}^{k-1}, d_{ij}^{k-1}\} \tag{3}$$

As each element is determined, record the path that it represents. Upon termination, the i, jth element of matrix D^m represents the length of a shortest path from vertex i to vertex j.

The optimality of this algorithm follows inductively from the fact that the length of a shortest path from i to j that allows only the first k vertices as intermediate vertices must be the smaller of (a) the length of a shortest path from i to j that allows only the first $k - 1$ vertices as intermediate vertices and (b) the length of a shortest path from i to j that allows only the first k vertices as intermediate vertices and uses the kth vertex once as an intermediate vertex.

Note that $d_{ii}^k = 0$ for all i and for all k. Hence, the diagonal elements of the matrices $D^1, D^2, \ldots, D^m$ need not be calculated. Moreover, $d_{ik}^{k-1} = d_{ik}^k$ and $d_{ki}^{k-1} = d_{ki}^k$ for all $i = 1, 2, \ldots, m$. This follows because vertex k will not be an intermediate vertex in any shortest path starting or originating at vertex k, since no cycles with negative length exist. Hence, in the computation of matrix D^k, the kth row and the kth column need not be calculated. Thus, in each matrix D^k only the $(m - 1)(m - 2)$ elements that are neither on the diagonal nor in the kth row or kth column need be calculated.

EXAMPLE 5 (Floyd Shortest-Path Algorithm)

For the graph in Fig. 4.9 the matrix D^0 of arc lengths is

$$D^0 = \begin{bmatrix} 0 & 1 & 2 & 1 \\ 2 & 0 & 7 & \infty \\ 6 & 5 & 0 & 2 \\ 1 & \infty & 4 & 0 \end{bmatrix}$$

The elements of D^1 and the corresponding shortest paths are calculated as follows:

$d_{ij}^1 = \min\{d_{i1}^0 + d_{1j}^0, d_{ij}^0\}$	Corresponding Path
$d_{11}^1 = d_{11}^0 = 0$	
$d_{12}^1 = d_{12}^0 = 1$	(1, 2)
$d_{13}^1 = d_{13}^0 = 2$	(1, 3)
$d_{14}^1 = d_{14}^0 = 1$	(1, 4)
$d_{21}^1 = d_{21}^0 = 2$	(2, 1)
$d_{22}^1 = 0$	
$d_{23}^1 = \min\{d_{21}^0 + d_{13}^0, d_{23}^0\} = \min\{2 + 2, 7\} = 4$	(2, 1), (1, 3)
$d_{24}^1 = \min\{d_{21}^0 + d_{14}^0, d_{24}^0\} = \min\{2 + 1, \infty\} = 3$	(2, 1), (1, 4)
$d_{31}^1 = d_{31}^0 = 6$	(3, 1)
$d_{32}^1 = \min\{d_{31}^0 + d_{12}^0, d_{32}^0\} = \min\{6 + 1, 5\} = 5$	(3, 2)
$d_{33}^1 = 0$	
$d_{34}^1 = \min\{d_{31}^0 + d_{14}^0, d_{34}^0\} = \min\{6 + 1, 2\} = 2$	(3, 4)
$d_{41}^1 = d_{41}^0 = 1$	(4, 1)
$d_{42}^1 = \min\{d_{41}^0 + d_{12}^0, d_{42}^0\} = \min\{1 + 1, \infty\} = 2$	(4, 1), (1, 2)
$d_{43}^1 = \min\{d_{41}^0 + d_{13}^0, d_{43}^0\} = \min\{1 + 2, 4\} = 3$	(4, 1), (1, 3)
$d_{44}^1 = 0$	

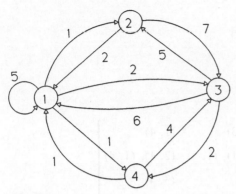

Fig. 4.9

In a similar way, matrices D^2, D^3, and D^4 and the corresponding shortest paths can be calculated. The results or these calculations are

$$D^2 = \begin{bmatrix} 0 & 1 & 2 & 1 \\ 2 & 0 & 4 & 3 \\ 6 & 5 & 0 & 2 \\ 1 & 2 & 3 & 0 \end{bmatrix}$$

Shortest paths for D^2:

$$\begin{bmatrix} & (1, 2) & (1, 3) & (1, 4) \\ (2, 1) & & (2, 1), (1, 3) & (2, 1), (1, 4) \\ (3, 1) & (3, 2) & & (3, 4) \\ (4, 1) & (4, 1), (1, 2) & (4, 1), (1, 3) & \end{bmatrix}$$

$$D^3 = \begin{bmatrix} 0 & 1 & 2 & 1 \\ 2 & 0 & 4 & 3 \\ 6 & 5 & 0 & 2 \\ 1 & 2 & 3 & 0 \end{bmatrix}$$

Shortest paths for D^3:

$$\begin{bmatrix} & (1, 2) & (1, 3) & (1, 4) \\ (2, 1) & & (2, 1), (1, 3) & (2, 1), (1, 4) \\ (3, 1) & (3, 2), & & (3, 4) \\ (4, 1) & (4, 1), (1, 2) & (4, 1), (1, 3) & \end{bmatrix}$$

$$D^4 = \begin{bmatrix} 0 & 1 & 2 & 1 \\ 2 & 0 & 4 & 3 \\ 3 & 4 & 0 & 2 \\ 1 & 2 & 3 & 0 \end{bmatrix}$$

Shortest path for D^4:

$$\begin{bmatrix} & (1, 2) & (1, 3) & (1, 4) \\ (2, 1) & & (2, 1), (1, 3) & (2, 1), (1, 4) \\ (3, 4), (4, 1) & (3, 4), (4, 1), (1, 2) & & (3, 4) \\ (4, 1) & (4, 1), (1, 2) & (4, 1), (1, 3) & \end{bmatrix}$$

Note that since the numbering of the vertices is arbitrary, the algorithm will find the shortest paths at earlier iterations if vertices with numbers close to one another are in fact vertices that are "close" to one another.

In the preceding numerical example, the actual arcs that comprise each shortest path were recorded as the Floyd algorithm was performed. For problems of any realistic size, it is obvious that this procedure would be impractical. Consequently, we need to develop a more efficient technique for determining the actual arcs that constitute the shortest path.

The next-to-last vertex in a path is called the *penultimate* vertex of that path. Let p_{ij} denote the penultimate vertex of the shortest path from vertex i to vertex j. (If there is more than one shortest path from vertex i to vertex j, then there may be more than one distinct penultimate vertex. In this case, let p_{ij} denote the set of all penultimate vertices. If, however, we are only interested in determining one shortest path from i to j, then we need only record one vertex for p_{ij}.) If p_{ij} is known for all vertices i and j, then all the vertices along a shortest path from i to j can be found as follows: Suppose that the penultimate vertex on a shortest path from i to j is vertex k, that is, $p_{ij} = k$. Then the second-to-last vertex on this path is the penultimate vertex on a shortest path from i to k, that is, p_{ik}. This process can be repeated until all the vertices on this path from i to j have been traced back. Hence, we need only know all p_{ij} to determine the actual shortest paths.

There are two methods for determining the p_{ij}:

1. *Tentative method:* Tentatively set p_{ij} equal to i for all j. Do this for all vertices i. Then, as the Floyd algorithm is performed, note whenever the minimum on the right side of equation (3) is $d_{ik}^{k-1} + d_{kj}^{k-1}$ rather than d_{ij}^{k-1}. When this occurs, set p_{ij} equal to p_{kj}. Otherwise, leave p_{ij} unchanged. [If there is a tie in equation (3), then both p_{kj} and the current value for p_{ij} may be recorded.] Upon termination of the Floyd algorithm, p_{ij} is the true penultimate vertex on the shortest path(s) from i to j.

2. *Terminal method:* After the Floyd algorithm has terminated, vertex p_{ij} is found as follows: Vertex p_{ij} is any vertex k such that $d_{ik}^m + d_{kj}^0 = d_{ij}^m$. Only the D^0 and D^m matrices are required to determine the value(s) of k.

If the Floyd algorithm has already been performed, one is forced to use the terminal method. However, if the Floyd algorithm is yet to be performed, then it is, of course, better to use the tentative method since this method requires little extra work when appended to the Floyd algorithm.

Another algorithm for finding a shortest path between each pair of vertices in a graph was proposed by Dantzig (1967). This algorithm is similar to the Floyd shortest-path algorithm in that the same calculations are performed. However, the order in which the calculations are performed is different.

Again, number the vertices $1, 2, \ldots, m$, and let d_{ij}^k denote the length of a shortest path from vertex i to vertex j that allows only the first k vertices to be intermediate vertices. Now, in contrast to the Floyd shortest-path algorithm, let D^k be an $k \times k$ matrix whose i,jth element is d_{ij}^k for $k = 1, 2, \ldots, m$. As before, we wish to calculate D^m, the matrix whose i,jth element denotes the length of a shortest path from vertex i to vertex j. As with the Floyd algorithm, the Dantzig algorithm calculates D^1 from D^0, etc., until D^m has been calculated from D^{m-1}.

What rationale underlies these calculations? First, note that each new matrix D^k contains one more row and one more column than its predecessor D^{k-1}. The $(k-1)^2$ elements of D^k that are also present in D^{k-1} are calculated from D^{k-1} as done by the Floyd algorithm. The new elements d_{ij}^k, $i = k$ or $j = k$, are calculated as follows: A shortest path from i to k (or from k to i) that allows only the first k vertices as intermediate vertices need never use vertex k as an intermediate vertex since all cycles have nonnegative length. Hence, a shortest path from i to k that allows only the first k vertices as intermediate vertices is any shortest path formed by taking a shortest path from i to some vertex $j < k$ that uses only the first $k - 1$ vertices as intermediate vertices joined with the shortest arc from j to k (if such an arc exists). Similarly, a shortest path from k to i that allows only the first k vertices to be intermediate vertices is the shortest path formed by joining a shortest arc from k to some vertex j, $j < k$ (if such an arc exists) to a shortest path from j to i that uses only the first $k - 1$ vertices as intermediate vertices. Lastly, let $d_{kk}^k = 0$.

With these ideas in mind, we can now formally state the Dantzig shortest path algorithm.

Dantzig Shortest-Path Algorithm

Step 1. Number the vertices of the graph $1, 2, \ldots, m$. Determine the matrix D^0 whose i,jth element d_{ij}^0 equals the length of the shortest arc from vertex i to vertex j. If no such arc exists, let $d_{ij}^0 = \infty$. Let D^k be an $k \times k$ matrix whose i,jth element is denoted by d_{ij}^k, for $k = 1, 2, \ldots, m$.

Step 2. For $k = 1, 2, \ldots, m$, successively determine each element of D^k from the elements of D^{k-1} as follows:

$$d_{kj}^k = \min_{i=1, 2, \ldots, k-1} \{d_{ki}^0 + d_{ij}^{k-1}\} \qquad (j = 1, 2, \ldots, k-1) \qquad (4)$$

$$d_{ik}^k = \min_{j=1, 2, \ldots, k-1} \{d_{ij}^{k-1} + d_{jk}^0\} \qquad (i = 1, 2, \ldots, k-1) \tag{5}$$

$$d_{ij}^k = \min \{d_{ik}^k + d_{kj}^k, d_{ij}^{k-1}\} \qquad (i, j = 1, 2, \ldots, k-1) \tag{6}$$

and lastly,

$$d_{ii}^k = 0 \text{ (for all } i \text{ and } k) \tag{7}$$

The paths corresponding to each element d_{ij}^m in D^m can be determined as before.

How many operations does the Dantzig algorithm require? To answer this question, note that the Dantzig algorithm performs essentially the same operations as the Floyd algorithm except in a different sequence. Equation (3) of the Floyd algorithm is identical to equation (6) of the Dantzig algorithm. Equations (4) and (5) of the Dantzig algorithm are simply $k - 1$ repetitions of equation (3) of the Floyd algorithm. Hence, both algorithms require at most the same number of calculations.

EXAMPLE 6 (Dantzig Shortest Path Algorithm)

We shall perform the Dantzig algorithm to find a shortest path between each pair of vertices in the graph in Fig. 4.9. The matrix of shortest arc lengths for this graph is

$$D^0 = \begin{bmatrix} 0 & 1 & 2 & 1 \\ 2 & 0 & 7 & \infty \\ 6 & 5 & 0 & 2 \\ 1 & \infty & 4 & 0 \end{bmatrix}$$

Clearly, $D^1 = [d_{11}^1] = [0]$. The elements of D^2 are calculated as follows:

	Corresponding Path
$d_{11}^2 = 0$	
$d_{12}^2 = \min\{d_{11}^1 + d_{12}^0\} = 0 + 1 = 1$	(1, 2)
$d_{22}^2 = 0$	
$d_{21}^2 = \min\{d_{21}^0 + d_{11}^1\} = 2 + 0 = 2$	(2, 1)

Note that these results are identical to those in the 2×2 upper left corner of D^2 calculated by the Floyd algorithm. The elements of D^3 are calculated as follows:

	Corresponding Path
$d_{13}^3 = \min\{d_{11}^2 + d_{13}^0, d_{12}^2 + d_{23}^0\} = \min\{0 + 2, 1 + 7\} = 2$	(1, 3)
$d_{23}^3 = \min\{d_{21}^2 + d_{13}^0, d_{22}^2 + d_{23}^0\} = \min\{2 + 2, 0 + 7\} = 4$	(2, 1), (1, 3)
$d_{31}^3 = \min\{d_{31}^0 + d_{11}^2, d_{32}^0 + d_{21}^2\} = \min\{6 + 0, 5 + 2\} = 6$	(3, 1)
$d_{32}^3 = \min\{d_{32}^0 + d_{22}^2, d_{31}^0 + d_{12}^2\} = \min\{5 + 0, 6 + 1\} = 5$	(3, 2)
$d_{12}^3 = \min\{d_{12}^2, d_{13}^3 + d_{32}^3\} = \min\{1, 2 + 5\} = 1$	(1, 2)
$d_{21}^3 = \min\{d_{21}^2, d_{23}^3 + d_{31}^3\} = \min\{2, 4 + 6\} = 2$	(2, 1)
$d_{11}^3 = d_{22}^3 = d_{33}^3 = 0$	

Thus,

$$D^3 = \begin{bmatrix} 0 & 1 & 2 \\ 2 & 0 & 4 \\ 6 & 5 & 0 \end{bmatrix}$$

and the shortest paths corresponding to D^3 are

$$\begin{bmatrix} & (1, 2) & (1, 3) \\ (2, 1) & & (2, 1), (1, 3) \\ (3, 1) & (3, 2) & \end{bmatrix}$$

Note that the matrix D^3 computed here is identical to the 3×3 upper left corner of the matrix D^3 computed by the Floyd shortest-path algorithm. It is left to the interested reader to repeat these calculations for D^4 and to verify that the matrix D^4 computed by the Dantzig algorithm is identical to the matrix D^4 computed by the Floyd algorithm.

Since the terminal method for constructing the arcs in a shortest path uses only the D^0 and D^n matrices, the terminal method is the same for both the Floyd and Dantzig algorithms.

However, the tentative method for constructing the arcs in a shortest path for the Dantzig algorithm is slightly different from the tentative method for the Floyd algorithm. Specifically, whenever equation (4) is performed in the Dantzig algorithm, let p_{kj} be the vertex that determines the right-side minimization. Whenever equation (5) is performed in the Dantzig algorithm, let p_{ik} be the vertex that determines the right-side minimization. Equation (6) of the Dantzig algorithm is identical to equation (3) of the Floyd algorithm, and the tentative method remains the same, namely let $p_{ij} = k$ if the minimum of the right side if $d_{ik}^k + d_{kj}^k$. Otherwise, the minimum of the right side is d_{ij}^{k-1} and p_{ij} must remain unchanged.

The Floyd algorithm must compute m matrices $D^1, D^2, \ldots, D^m$. Each of these matrices consists of m^2 elements. Hence, a total of m^3 elements must be computed by the Floyd algorithm. Each of these computations requires by equation (3) one addition and one minimization. Hence, the Floyd algorithm requires roughly m^3 additions and m^3 minimizations. [Strictly speaking, this is an overstatement, since some elements of D^i can be taken directly from D^{i-1} without using equation (3) as the elements ith row and ith column of D^{i-1} and D^i are identical.] The total amount of computation required by the Floyd algorithm is proportional to $2m^3$. Or, in more technical terminology, the Floyd algorithm requires $O(m^3)$ running time.

As suggested, the Floyd algorithm could be replaced by repeating the Dijkstra algorithm m times, once for each vertex as the initial vertex. This requires time proportional to $\frac{3}{2}m^3$, which is superior to the running time of the Floyd algorithm. If, however, some arc lengths are negative (but, of course, there are no negative cycles), then the Ford algorithm must replace the Dijkstra algorithm and at worst the running time is $O(m^4)$, which is inferior for large m to the Floyd algorithm running time. However, it must not be forgotten that most likely the Ford algorithm will terminate with far less than the worst possible number of operations. The interested reader is directed to the survey articles of Karp (1975) and Lawler (1971), and to Knuth (1973), a valuable reference for those interested in efficient programming techniques applicable to the algorithm described here.

4.5   THE k-SHORTEST-PATH ALGORITHM

The preceding sections considered the problem of finding various shortest paths. Often, however, knowledge of the second, third, fourth, etc., shortest paths between two vertices is useful. For example, an airline might want to know the runner-up shortest flight routes between Springfield and Ankara in case one of its clients could not take the shortest flight route due to visa difficulties, flight cancellations, or airline strikes along the shortest flight route.

This section first presents an algorithm, called the *double-sweep algorithm*, that finds the k shortest path lengths between a specified vertex and all other vertices in the graph. Next, this section presents two algorithms, called the *generalized Floyd algorithm* and the *generalized Dantzig algorithm*, that find the k shortest path lengths between every pair of vertices in the graph.

The Dijkstra, Floyd, and Dantzig algorithms were able to construct various shortest paths. These algorithms essentially consisted of performing a sequence of two arithmetic operations, addition and minimization. These two operations were performed on single numbers that represented either arc lengths or path lengths. For example, equation (1), which defines the Dijkstra algorithm, consists exclusively of addition and minimization. The same is true for equation (3), which defines the Floyd algorithm, and equations (4)–(7), which define the Dantzig algorithm.

The algorithms to be presented in this section (double-sweep algorithm, generalized Floyd algorithm, and generalized Dantzig algorithm) also consist exclusively of addition and minimization operations. However, these operations are performed not on single numbers (as with the previous algorithms) but on sets of k distinct numbers that represent the lengths of paths or arcs. With this as motivation, let R^k denote the set of all vectors $(d_1, d_2, \ldots, d_k)$ with the property that $d_1 < d_2 < \cdots < d_k$.[†] Thus, the components of a member of R^k are distinct and arranged in ascending order. For example, $(-3, -1, 0, 4, 27) \in R^5$.

Let $\mathbf{a} = (a_1, a_2, \ldots, a_k)$ and $\mathbf{b} = (b_1, b_2, \ldots, b_k)$ be two members of R^k. *Generalized minimization*, denoted by $+$, is defined as

$$\mathbf{a} + \mathbf{b} = \min_k\{a_i, b_i: i = 1, 2, \ldots, k\} \tag{8}$$

where $\min_k (X)$ means the k smallest distinct members of the set X.

[†] Let $\infty \leq \infty$.

Generalized addition, denoted by $\times$, is defined as

$$\mathbf{a} \times \mathbf{b} = \min_k\{a_i + b_j: i,j = 1, 2, \ldots, k\} \tag{9}$$

For example, if $\mathbf{a} = (1, 3, 4, 8)$ and $\mathbf{b} = (3, 5, 7, 16)$, then $\mathbf{a} + \mathbf{b} = \min_4(1, 3, 4, 8, 3, 5, 7, 16) = (1, 3, 4, 5)$ and $\mathbf{a} \times \mathbf{b} = \min_4(1 + 3, 1 + 5, 1 + 7, 1 + 16, 3 + 3, 3 + 5, 3 + 7, 3 + 16, 4 + 3, \ldots) = (4, 6, 7, 8)$. Note that the components of $\mathbf{a} + \mathbf{b}$ and $\mathbf{a} \times \mathbf{b}$ can be arranged so that $\mathbf{a} + \mathbf{b} \in R^k$ and $\mathbf{a} \times \mathbf{b} \in R^k$. Moreover, since members of R^k have their components arranged in ascending order, generalized minimization need not require more than k comparisons and generalized addition need not require more than $k(k - 1)$ additions and $k(k - 1)$ comparisons.

Extending our previous notation, let $d_{ij}^0 = (d_{ij1}^0, d_{ij2}^0, \ldots, d_{ijk}^0) \in R^k$ denote the lengths of the k shortest arcs from vertex i to vertex j. If two arcs from vertex i to vertex j have the same length, then this length appears only once in d_{ij}^0. If there are less than k arcs from i to j, then fill up the remaining components with ∞. For example, if vertex 5 is joined to vertex 3 by three arcs of lengths 9, 13, and 9, then for $k = 4$, $d_{53}^0 = (9, 13, \infty, \infty)$. If $i = j$, then suppose that there is an arc from vertex i to itself with length zero. Let D^0 denote the matrix whose i,jth element is d_{ij}^0.

Let $d_{ij}^l = (d_{ij1}^l, d_{ij2}^l, \ldots, d_{ijk}^l) \in R^k$ denote the k shortest distinct path lengths from vertex i to vertex j that use only vertices $1, 2, \ldots, l$ as intermediate vertices. (Recall that in Section 3.1 the vertices were numbered $1, 2, \ldots, m$.) Let D^l denote the matrix whose i,jth element is d_{ij}^l.

Lastly, let $d_{ij}^* = (d_{ij1}^*, d_{ij2}^*, \ldots, d_{ijk}^*) \in R^k$ denote the k shortest distinct path lengths from vertex i to vertex j. These path lengths are called the *optimal path lengths*. Let D^* denote the matrix whose i,jth element is d_{ij}^*.

Let matrix L be formed from matrix D^0 by replacing every component of every element d_{ij}^0 by ∞ whenever $i \leq j$. Let matrix U be formed from matrix D^0 by replacing every component of every element d_{ij}^0 by ∞ whenever $i > j$. Matrices L and U are called the upper and lower triangular portions of D^0. For example, for $k = 2$ and

$$D^0 = \begin{bmatrix} (0, 1) & (6, 10) & (2, 9) \\ (1, 8) & (0, \infty) & (3, \infty) \\ (\infty, \infty) & (2, 4) & (0, \infty) \end{bmatrix}$$

$$L = \begin{bmatrix} (\infty, \infty) & (\infty, \infty) & (\infty, \infty) \\ (1, 8) & (\infty, \infty) & (\infty, \infty) \\ (\infty, \infty) & (2, 4) & (\infty, \infty) \end{bmatrix}$$

and

$$U = \begin{bmatrix} (\infty, \infty) & (6, 10) & (2, 9) \\ (\infty, \infty) & (\infty, \infty) & (3, \infty) \\ (\infty, \infty) & (\infty, \infty) & (\infty, \infty) \end{bmatrix}$$

If we wish to know only the k shortest path lengths from a specific vertex (say vertex 1) to every other vertex in the graph, then we need only determine the first row $(d_{11}^*, d_{12}^*, \ldots, d_{1m}^*)$ of D^*. Note that this row consists of m members each belonging to R^k. Hence, altogether mk values and the corresponding paths must be determined.

The double-sweep algorithm (Shier, 1974, 1976) is an efficient way of computing these nk paths. The double-sweep algorithm will work as long as the graph contains no circuits with negative length. The double-sweep algorithm is initiated with any estimate $(d_{11}^{(0)}, d_{12}^{(0)}, \ldots, d_{1m}^{(0)})$ of $(d_{11}^*, d_{12}^*, \ldots, d_{1m}^*)$ in which each value is not less than the corresponding optimal value and for which $d_{111}^{(0)} = 0$. For example, all values could be set equal to ∞ except $d_{111}^{(0)}$, which equals zero. The algorithm then computes new improved (i.e., reduced) estimates of d_{1i}^* by seeing if any of the k values in $d_{1j}^{(0)} \times d_{ji}^0$ are less than any of the k values in the current estimate $d_{1i}^{(0)}$ of d_{1i}^*. If so, the smallest k values are chosen. This process is repeated for all j, yielding a new estimate $d_{1i}^{(1)}$ of d_{1i}^*. The algorithm terminates when two successive estimates $(d_{11}^{(i)}, d_{12}^{(i)}, \ldots, d_{1m}^{(i)})$ and $(d_{11}^{(i+1)}, d_{12}^{(i+1)}, \ldots, d_{1m}^{(i+1)})$ are identical in every component for $i \geq 1$. The terminal values of the estimates can be shown to equal the optimal values. See the proof of the double-sweep algorithm.

The double-sweep algorithm has the added efficiency that whenever possible in the revision of an estimate vector $d_1^{(i)} = (d_{11}^{(i)}, d_{12}^{(i)}, \ldots, d_{1m}^{(i)})$ to the next estimate $d_1^{(i+1)} = (d_{11}^{(i+1)}, d_{12}^{(i+1)}, \ldots, d_{1m}^{(i+1)})$, the values already computed in $d_1^{(i+1)}$ are used in the computation of the remaining values in $d_1^{(i+1)}$. This immediate use of updated estimate values can accelerate the discovery of the optimal values. With these ideas in mind, we can now formally state the double-sweep algorithm.

Double-Sweep Algorithm

Step 1. Initialization. Let the initial estimate $d_1^{(0)} = (d_{11}^{(0)}, d_{12}^{(0)}, \ldots, d_{1m}^{(0)})$ of d_i^* consist of values that equal or exceed the corresponding optimal values. However, let $d_{111}^{(0)} = 0$ since there is a path of zero length (the path without any arcs) from vertex 1 to itself.

Step 2. Given an estimate $d_1^{(2r)}$ of d_1^*, calculate new estimates $d_1^{(2r+1)}$ and $d_1^{(2r+2)}$ as follows:

Backward sweep:

$$d_1^{(2r+1)} = d_1^{(2r+1)}L + d_1^{(2r)} \tag{10}$$

Forward sweep:

$$d_1^{(2r+2)} = d_1^{(2r+2)}U + d_1^{(2r+1)} \quad (r = 0, 1, 2, \ldots) \tag{11}$$

Note that the addition and multiplication operations in the above matrix multiplications refer to generalized minimization and generalized addition, respectively, as defined by equations (8) and (9).

Terminate when two successive estimates $d_1^{(t-1)}$ and $d_1^{(t)}$ are identical for $t > 1$. The terminal estimate $d_1^{(t)}$ equals d_i^*. Stop.

At this point, the reader might wonder how to perform the computations required by equations (10) and (11) since the vector to be computed appears on both sides of each equation. For equation (10), first determine the last component $d_{1m}^{(2r+1)}$ of $d_1^{(2r+1)}$. (This is possible since the mth column of L consists entirely of infinite entries, and hence $d_1^{(2r+1)}$ is not needed to compute $d_{1m}^{(2r+1)}$.) Next, determine the second to last component $d_{1,m-1}^{(2r+1)}$ of $d_1^{(2r+1)}$. [This is possible since the $(m-1)$th column of L contains infinite entries except in its last row, and consequently only $d_{1m}^{(2r+1)}$, which has already been calculated, is needed to determine $d_{1,m-1}^{(2r+1)}$.] Next, determine $d_{1,m-2}^{(2r+1)}$, etc. Hence, equation (10) computes the components of $d_1^{(2r+1)}$ from last to first and is called the *backward sweep*. In a similar way, equation (11) computes the components of $d_1^{(2r+2)}$ from first to last and is called the *forward sweep*.

The double-sweep algorithm determines path lengths. How can the actual path (or paths) corresponding to the past length be found? Since the double-sweep algorithm is initialized with arbitrary larger-than-optimal estimates that do not correspond to any particular path, we cannot carry along at each iteration the path (or paths) corresponding to each new estimate as was possible with the Floyd algorithm and Dantzig algorithm. The path corresponding to each path length can be retrieved as follows: Suppose we wish to find the pth shortest path from vertex 1 to vertex j. This path must consist of the lth ($l \leq p$) shortest path from vertex 1 to some vertex i together with an arc from vertex i to vertex j. The length of this path plus the length of this arc must equal d_{ijp}^*. Vertex i is called the penultimate vertex on the pth shortest path from vertex 1 to vertex j. Vertex i is determined by searching through the arc length matrix D^0 and the set of shortest path lengths d_1^*. Once vertex i has been located, this process can be repeated to find the penultimate vertex on the lth shortest path from vertex 1 to vertex i. Ultimately, the entire pth shortest path from vertex 1 to vertex j can be traced back by repeating this process.

It is possible that the penultimate vertex is not unique. In this case, all penultimate vertices can be recorded and all the paths of equal length that tie for the pth shortest path from vertex 1 to vertex j can be traced back (if desired).

If all paths are to be determined, it is computationally best to determine first all shortest paths, then all second shortest paths, etc., since knowledge of the lth shortest paths is needed to determine the pth ($l \leq p$) shortest paths.

Proof of the Double-Sweep Algorithm. Equation (10) is a generalized minimization between corresponding components of $d_1^{(2r+1)}L$ and $d_1^{(2r)}$. Similarly, equation (11) is a generalized minimization between corresponding components of $d_1^{(2r+2)}U$ and $d_1^{(2r+1)}$. Hence, no value is ever replaced with a larger value in the succeeding estimate. Thus, $d_1^{(0)}$, $d_1^{(1)}$, . . . , $d_1^{(t)}$ form mk nonincreasing sequences. Moreover, no value can ever be less than its optimal value (which we will assume to be finite or $+\infty$) for the following reason: Suppose that the *first* less-than-optimal value to be calculated is for the pth shortest path from vertex 1 to vertex j. By equations (10) and (11), this less-than-optimal path length was computed as the sum of a number not less than the length of a shortest path from vertex 1 to some vertex i, plus the length of an arc from vertex i to vertex j. This implies that this path length is less than optimal, which contradicts our assumption that we are computing the first less-than-optimal path length to appear.

Thus, the sequence of estimates $d_1^{(0)}$, $d_1^{(1)}$, . . . , $d_1^{(t)}$ forms mk nonincreasing sequences of path lengths that are always greater than or equal to their corresponding optimal values. It remains to show that each of these mk sequences converges to its optimal value in a finite number of steps. This will be accomplished by showing that each succeeding estimation in this sequence contains at least one more optimal value than its predecessor.

First, some lemmas are needed.

Lemma 4.1. If vertex i is the penultimate vertex of some pth shortest path from vertex 1 to vertex j, then the portion P' of P from vertex 1 to vertex i is one of the p shortest paths from vertex 1 to vertex i.

Proof. If path P' were not among the p shortest paths from vertex 1 to vertex i, there would be p paths from vertex 1 to vertex i with distinct lengths that are shorter than P'. Each of these paths could be extended to vertex j by the addition of a single arc. This would result in p paths from vertex 1 to vertex j that are shorter than path P, which is a contradiction. Q.E.D.

Lemma 4.2. If the p shortest path lengths from vertex 1 to every other vertex are known, then the length of the $(p + 1)$st shortest path length from vertex 1 to itself is determined during the next double sweep for $p \geq 0$.

Proof. Consider any $(p + 1)$st shortest path P from vertex 1 to itself. Let i denote the penultimate vertex of path P. From Lemma 1, it follows that the portion P' of P from vertex 1 to vertex i is among the $(p + 1)$st shortest paths from vertex 1 to vertex i. Path P' cannot be an $(p + 1)$st shortest path; otherwise, there would be $p + 1$ shortest paths from vertex 1 to itself that have vertex i as their penultimate vertex, which contradicts the fact that $d_{111}^* = 0$ and corresponds to a path without arcs.

By assumption, the p shortest path lengths from vertex 1 to vertex i are known. Also, the lengths of the k shortest arcs from vertex i to vertex 1 are known. Hence, the length of P will be determined during the next double sweep. Q.E.D.

We shall now employ these two lemmas to show that at each double sweep at least one more optimal value is calculated. Suppose that prior to the rth double sweep, the length of the p shortest paths from vertex 1 to every other vertex has been determined for some $p \geq 0$. Let j be any vertex for which the optimal value $d_{1j,p+1}^*$ of the $(p + 1)$st shortest path from vertex 1 to vertex j has not yet been determined.

Let i denote the penultimate vertex of some $(p + 1)$st shortest path from vertex 1 to vertex j. If the length of the $(p + 1)$st shortest path from vertex 1 to vertex i has already been determined, the next double sweep will determine the optimum value of the length of the $(p + 1)$st shortest path from vertex 1 to vertex j, since this path is the merger of one of the $(p + 1)$st shortest paths from vertex 1 to vertex i and an arc from i to j.

If the length of the $(p + 1)$st shortest path from vertex 1 to vertex i has not yet been determined by the algorithm, we cannot be certain that the next double sweep will generate the optimum value of the length of the $(p + 1)$st shortest path from vertex 1 to vertex j. If the next double sweep does in fact generate this path length, then the proof is complete. If not, then the $(p + 1)$st shortest path from vertex 1 to vertex j must consist of the $(p + 1)$st shortest path from 1 to i together with an arc from i to j, by Lemma 4.1.

Repeat the preceding argument, replacing vertex i by vertex j. Either successive repetitions of this argument will lead to a situation in which an optimum value is calculated during the next double-sweep iteration of the algorithm or else the $(p + 1)$st shortest path from vertex 1 to vertex j will be traced back to vertex 1. If the latter occurs, then by Lemma 4.2, the next double sweep determines an optimal value for the $(p + 1)$st shortest path from vertex 1 to itself, which concludes the proof that the algorithm terminates finitely.   Q.E.D.

As a postscript to the proof, note that if the graph contained no cycles with negative length that are accessible from vertex 1, then after $r \geq m$ double sweeps, the first component of each estimate vector $d_{11}^{(2r)}$, $d_{12}^{(2r)}$, . . . , $d_{1m}^{(2r)}$ would no longer decrease. Similarly, after $r \geq 2m$ double sweeps the first two components of each estimate vector $d_{11}^{(2r)}$, $d_{12}^{(2r)}$, . . . , $d_{1m}^{(2r)}$ would no longer decrease. In general, after $r \geq cm$ double sweeps, the first c components of each estimate vector $d_{11}^{(2r)}$, $d_{12}^{(2r)}$, . . . , $d_{1m}^{(2r)}$ would no longer decrease. (Note that the superscript $2r$ is used since each double sweep determines two estimates.)

Hence, we can detect the existence of a cycle with negative length that is accessible from vertex 1 by noticing any decrease (or for that matter any change at all) in the first c components of any estimate vector $d_{11}^{(2r)}$, $d_{12}^{(2r)}$, . . . , $d_{1m}^{(2r)}$ after the rth double sweep ($r \geq c$) for any $c = 1, 2, . . . , m$.

Since the double-sweep algorithm calculates at least one additional optimal value at each double sweep, at most mk single sweeps are required. In practice, far fewer sweeps are required since more than one optimal value is frequently computed during one double sweep.

How many generalized additions and generalized minimizations does a double sweep require? There are $m^2 - m$ elements off the diagonal of matrix D^0. Hence, matrices L and U each have $\frac{1}{2}(m^2 - m)$ elements that are possibly finite. Each of these elements requires one generalized addition and one generalized minimization. Thus, each double sweep requires roughly m^2 generalized additions and m^2 generalized minimizations. Hence, at most $\frac{1}{2}km^3$ generalized additions and minimizations are required to complete the double-sweep algorithm.

For $k = 1$, the double-sweep algorithm requires at most $\frac{1}{2}m^3$ additions and $\frac{1}{2}m^3$ minimizations. Thus, for $k = 1$, the double-sweep algorithm requires $O(m^3)$ running time and is superior to the worst running time for the Ford algorithm but inferior to the running time of $O(m^2)$ for the Dijkstra algorithm. Of course, all comparisons must be tempered with the reminder that we are considering the worst possible running times for the double-sweep and Ford algorithms.

EXAMPLE 6 (Double-Sweep Algorithm)

Let us use the double-sweep algorithm to find the lengths of the three shortest paths from vertex 1 to every other vertex in the graph in Fig. 4.10. The matrix D^0 of path lengths is

$$
\begin{bmatrix}
(0, \infty, \infty) & (1, \infty, \infty) & (\infty, \infty, \infty) & (\infty, \infty, \infty) \\
(2, 4, \infty) & (0, \infty, \infty) & (-1, \infty, \infty) & (\infty, \infty, \infty) \\
(2, \infty, \infty) & (\infty, \infty, \infty) & (0, \infty, \infty) & (-1, \infty, \infty) \\
(\infty, \infty, \infty) & (3, \infty, \infty) & (2, \infty, \infty) & (0, \infty, \infty)
\end{bmatrix}
$$

Matrix L is

$$
\begin{bmatrix}
(\infty, \infty, \infty) & (\infty, \infty, \infty) & (\infty, \infty, \infty) & (\infty, \infty, \infty) \\
(2, 4, \infty) & (\infty, \infty, \infty) & (\infty, \infty, \infty) & (\infty, \infty, \infty) \\
(2, \infty, \infty) & (\infty, \infty, \infty) & (\infty, \infty, \infty) & (\infty, \infty, \infty) \\
(\infty, \infty, \infty) & (3, \infty, \infty) & (2, \infty, \infty) & (\infty, \infty, \infty)
\end{bmatrix}
$$

Matrix U is

$$
\begin{bmatrix}
(\infty, \infty, \infty) & (1, \infty, \infty) & (\infty, \infty, \infty) & (\infty, \infty, \infty) \\
(\infty, \infty, \infty) & (\infty, \infty, \infty) & (-1, \infty, \infty) & (\infty, \infty, \infty) \\
(\infty, \infty, \infty) & (\infty, \infty, \infty) & (\infty, \infty, \infty) & (-1, \infty, \infty) \\
(\infty, \infty, \infty) & (\infty, \infty, \infty) & (\infty, \infty, \infty) & (\infty, \infty, \infty)
\end{bmatrix}
$$

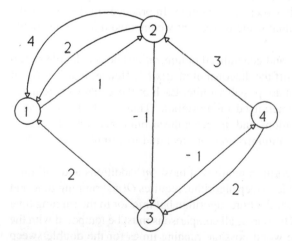

Fig. 4.10 Double-sweep algorithm example.

From equation (10), the backward sweep becomes

$$d_{14}^{(2r+1)} = d_{14}^{(2r)}$$
$$d_{13}^{(2r+1)} = (2, \infty, \infty) \times d_{14}^{(2r+1)} + d_{13}^{(2r)}$$
$$d_{12}^{(2r+1)} = (3, \infty, \infty) \times d_{14}^{(2r+1)} + d_{12}^{(2r)}$$
$$d_{11}^{(2r+1)} = (2, 4, \infty) \times d_{12}^{(2r+1)} + (2, \infty, \infty) \times d_{13}^{(2r+1)} + d_{11}^{(2r)}$$

From equation (11), the forward sweep becomes

$$d_{11}^{(2r+2)} = d_{11}^{(2r+1)}$$
$$d_{12}^{(2r+2)} = (1, \infty, \infty) \times d_{11}^{(2r+2)} + d_{12}^{(2r+1)}$$
$$d_{13}^{(2r+2)} = (-1, \infty, \infty) \times d_{12}^{(2r+2)} + d_{13}^{(2r+1)}$$
$$d_{14}^{(2r+2)} = (-1, \infty, \infty) \times d_{13}^{(2r+2)} + d_{14}^{(2r+1)}$$

The following table gives the results of the first five sweeps. The results of the fourth and fifth sweeps are identical in every component and hence represent the optimal path lengths.

	$d_{11}^{(r)}$	$d_{12}^{(r)}$	$d_{13}^{(r)}$	$d_{14}^{(r)}$	
$r = 0$	$(0, \infty, \infty)$	(∞, ∞, ∞)	(∞, ∞, ∞)	(∞, ∞, ∞)	$\leftarrow$ (backward)
$r = 1$	$(0, \infty, \infty)$	(∞, ∞, ∞)	(∞, ∞, ∞)	(∞, ∞, ∞)	$\leftarrow$ (forward)
$r = 2$	$(0, \infty, \infty)$	$(1, \infty, \infty)$	$(0, \infty, \infty)$	$(-1, \infty, \infty)$	$\leftarrow$ (backward)
$r = 3$	$(0, 2, 3)$	$(1, 2, \infty)$	$(0, 1, \infty)$	$(-1, \infty, \infty)$	$\leftarrow$ (forward)
$r = 4$	$(0, 2, 3)$	$(1, 2, 3)$	$(0, 1, 2)$	$(-1, 0, 1)$	$\leftarrow$ (backward)
$r = 5$	$(0, 2, 3)$	$(1, 2, 3)$	$(0, 1, 2)$	$(-1, 0, 1)$	

Note that the double-sweep algorithm terminated after 5 sweeps or 2½ double sweeps compared to the theoretical maximum of $\frac{1}{2}mk = 6$ double sweeps.

In the remainder of this section, we shall consider the more general problem of finding the k shortest paths from *each* vertex to every other vertex. The Floyd algorithm and Dantzig algorithm of Section 4.3 solved this problem for $k = 1$. These algorithms will now be extended to $k > 1$.

Of course, this problem could be solved by performing the double-sweep algorithm once for each vertex as the initial vertex. However, computationally this is not very efficient since much information regarding shortest paths from a vertex that is not the initial vertex would never be utilized.

Before proceeding to the generalized Floyd algorithm and the generalized Dantzig algorithm for finding the k shortest paths between every pair of vertices, an additional definition is needed.

Let $d_{ii} \in R^k$ denote the lengths of any k paths from vertex i to itself including the empty path of length zero. These paths are cycles. The *convolution* d_{ii}^C of d_{ii} is defined as the lengths of the k distinct shortest paths from vertex i to itself that can be formed by combining paths represented by a components of d_{ii}. Thus, each member of d_{ii}^C corresponds to a path formed by repeating one or more of the cycles in d_{ii}. Mathematically,

$$d_{ii}^C = d_{ii} \times d_{ii} \times d_{ii} \times \cdots \times d_{ii} \in R^k \qquad (k \text{ times}) \qquad (12)$$

(Recall that $\times$ denotes generalized addition.)

The ideas underlying the generalized Floyd algorithm and the generalized Dantzig algorithm are identical to those underlying the Floyd algorithm and the Dantzig algorithm, except that addition is replaced by generalized addition and minimization is replaced by generalized minimization. Also, one must consider the possibility of generating paths that begin with a cycle attached to the initial vertex or terminate with a cycle attached to the terminal vertex.

The generalized Floyd algorithm is the same as the original Floyd algorithm except that d_{ij}^p now represents a vector in R^k signifying the lengths of the k shortest paths from vertex i to vertex j that use only the first m vertices as intermediate vertices. Equation (3) becomes

$$d_{pp}^p = (d_{pp}^{p-1})^C \qquad (13)$$

$$d_{ip}^p = d_{ip}^{p-1} d_{pp}^p \text{ (for all } i \neq p) \qquad (14)$$

$$d_{pi}^p = d_{pp}^p d_{pi}^{p-1} \text{ (for all } i \neq p) \qquad (15)$$

$$d_{ij}^p = d_{ip}^p d_{pj}^p + d_{ij}^{p-1} \text{ (for all } i, j \neq p) \qquad (16)$$

The generalized Dantzig algorithm is the same as the original Dantzig algorithm except that d_{ij}^p now represents a vector in R^k signifying the lengths of the k shortest paths from vertex i to vertex j that use only the first p vertices as intermediate vertices. Equation (7) becomes

$$d_{pp}^p = (d_{pp}^0 + \sum_{i=1}^{p-1} \sum_{j=1}^{p-1} d_{pi}^0 d_{ij}^{p-1} d_{jp}^0)^C \qquad (17)$$

Equation (4) becomes

$$d_{pi}^p = \sum_{j=1}^{p-1} d_{pp}^p d_{pj}^0 d_{ji}^{p-1} \qquad (i = 1, 2, \ldots, p - 1) \qquad (18)$$

Equation (5) becomes

$$d_{ip}^p = \sum_{j=1}^{p-1} d_{ij}^{p-1} d_{jp}^0 d_{pp}^p \qquad (i = 1, 2, \ldots, p - 1) \qquad (19)$$

Equation (6) becomes

$$d_{ij}^p = d_{ip}^p d_{pj}^p + d_{ij}^{p-1} \quad (i, j = 1, 2, \ldots, p - 1) \qquad (20)$$

Note that in each generalized algorithm d_{pp}^p is the first element of D^p to be calculated. Next, d_{pp}^p is used to calculate d_{ip}^p and d_{pi}^p for all $i \neq p$. Lastly, the elements d_{ij}^p ($i \neq p$, $j \neq p$) are calculated.

In the generalized Floyd algorithm, d_{pp}^p is calculated in equation (13) by taking the smallest k combinations of all simple cycles from vertex p to itself that use only the first p vertices as intermediate vertices.

Then, d_{ip}^p is calculated in equation (14) by taking the smallest k combinations of paths d_{ip}^{p-1} from i to p coupled with cycles d_{pp}^p at p. In a similar way, equation (15) calculates d_{pi}^p. Lastly, the calculation of d_{ij}^p in equation (16) is merely the generalized-operator restatement of equation (3).

In the generalized Dantzig algorithm, d_{pp}^p is calculated in equation (17) by taking the smallest k combinations of (a) loops from vertex p to itself and (b) cycles that start at vertex p, proceed to some other vertex i ($i < p$), then proceed to some other vertex j ($j < p$, possibly $j = 1$), and then return to vertex p, that is, $d_{pi}^0 d_{ij}^{p-1} d_{jp}^0$ for all i, $j < p$. Equation (18) is equation (4) restated in terms of generalized operations and allowing for cycles d_{pp}^p attached to vertex p. Similarly, equation (19) is equation (5) restated in terms of generalized operations and allowing cycles d_{pp}^p attached to vertex p. Lastly, equation (20) is equation (6) restated in terms of generalized operations.

As with the original algorithms, the optimality of the generalized algorithms can be proved by an induction on m, the number of vertices in the graph.

Let us determine how many generalized operations are required by the generalized Floyd algorithm. The algorithm must calculate m matrices $D^0, D^1, \ldots, D^m$. Each matrix requires one performance of equation (13), $m - 1$ performances of equation (14), $m - 1$ performances of equation (15), and $(m - 1)^2$ performances of equation (16).

Equation (13) requires one convolution, which by equation (12) consists of k generalized additions. Hence, equation (13) requires a total of k generalized additions.

Equations (14) and (15) each require one generalized addition. Equation (16) requires one generalized addition and one generalized minimization.

Thus, each matrix $D^1, D^2, \ldots, D^m$ requires $m^2 + k - 2$ generalized additions and $(m - 1)^2$ generalized minimizations, or roughly m^2 generalized additions and m^2 generalized minimizations. In total, the generalized Floyd algorithm requires about m^3 generalized additions and m^3 generalized minimizations. Since the generalized Dantzig algorithm is essentially a resequencing of the operations performed by the generalized Floyd algorithm, it requires approximately the same number of operations. Hence, each has a running time of about $2m^3$ or $O(m^3)$.

How does this compare with the number of operations required by the double-sweep algorithm to do the same job? As shown before, the double-sweep algorithm requires at most a running time of $O(m^3)$, using mk generalized additions and m^3k generalized minimizations. If the double-sweep algorithm were performed m times (once for each vertex as the initial vertex), then at most a running time of km^4 or $O(m^4)$ is required. However, since the double-sweep algorithm usually terminates early, the actual running time will be much less.

Hence, we can conclude that the choice of algorithm depends on (1) the comparison of km^4 and $2m^3$, (2) our estimates of how prematurely the double-sweep algorithm will terminate with an optimal solution, and (3) the observation that fewer operations are required to perform a generalized operation in the double-sweep algorithm since the vectors in the double-sweep algorithm usually consist of many components equal to infinity. Computational experience with the double-sweep algorithm has been reported by Shier (1974). Some improved algorithms for the k-shortest path problem are given by Shier (1979).

4.6 OTHER SHORTEST PATHS

All the shortest-path algorithms studied thus far can be regarded as well-defined sequences of addition operations and minimization operations. Let us consider a different path problem, namely the bottleneck problem. The *bottleneck* of a path is defined as the length of the shortest arc in the path. The *bottleneck problem* is the problem of finding a path between two vertices with the largest possible bottleneck.

For example, suppose that a bridge collapses when its weight limit is exceeded. We might wish to know the maximum weight that a vehicle can carry between points a and b so that it does not exceed the weight limit of any bridge that it crosses. If we consider the bridges as arcs and let the arc lengths equal the corresponding weight limits, this problem becomes the problem of finding a path from a to b with the largest bottleneck.

In the shortest-path problem, we associate a number to each path called its length. This number is found by adding together all arc lengths in this path. In the bottleneck problem, we associate a number to each path called its bottleneck. This number is found by taking the minimum of all arc lengths in this path. In the shortest-path problem, one path is preferred to another path if its length is smaller than the length of the other path. This requires a minimization operation. In the bottleneck problem, one path is preferred to another path if its bottleneck is larger than the bottleneck of the other path. This requires a maximization operation. Thus, the bottleneck problem can be regarded as similar to the shortest-path problem except that minimization replaces addition and maximization replaces minimization.

If the Floyd algorithm or Dantzig algorithm or their generalizations or the double-sweep algorithm were performed with traditional minimization as the addition operation and with traditional maximization as the minimization operation, these algorithms would produce the best bottleneck paths. Moreover, if traditional minimization was retained as the minimization operation, these algorithms would produce the worst bottleneck paths.

However, if any of these algorithms are used to solve the bottleneck problem, we must remember that the absence of an arc must be interpreted as a bottleneck of $-\infty$, whereas in the shortest-path problem the absence of an arc was interpreted as a length of $+\infty$. Similarly, the null path (path without any arcs) has a bottleneck of $-\infty$, whereas the null path has a length of zero. Thus, for the bottleneck problem, the matrix D^0 that initialized the Floyd algorithm and the Dantzig algorithm consists of arc lengths and $-\infty$ wherever no arc appears. Similarly, for the bottleneck problem, the vector $d_1^{(0)}$ that initialized the double-sweep algorithm should consist of entries that are smaller than the corresponding optimal entries (since now minimization has been changed to maximization).

The Floyd, Dantzig, and double-sweep algorithms can be extended to even another path problem, the gain problem. Associate a real number, called the *gain factor*, with each arc in the graph. Let the *gain of a path* be defined as the product of the gain factors of the individual arcs in the path. (If an arc is repeated in a path, then repeat its gain factor in the calculation of the path's gain.) The *gain problem* is the problem of finding a path with the largest gain between two vertices.

To illustrate this situation consider the finance officer of a large company who has decided to invest the company's surplus funds in bonds that mature within the next five years. A variety of bonds that mature within the next five years are available. Funds released by bonds that mature early can be reinvested in other bonds as long as the funds are ultimately released within five years. How should the finance officer spread the investments over the next five years?

Let the beginning of each investment year be depicted as a vertex. Let each variety of bond be depicted as an arc from the vertex representing its date of issue to the vertex depicting its maturity date. Let the gain factor of each arc equal the value at maturity of one dollar invested in the corresponding bond. When presented in graph terms, the finance officer's problem can be viewed as the problem of finding a path from vertex 0 to vertex 5 with the greatest gain.

As another example, a certain amount of pilferage occurs during each segment of a cargo's journey from the factory to the sales outlet. How should the shipment be routed to minimize the amount of pilferage? This problem can be rephrased as a gain problem if we set the gain factor of each arc equal to the fraction of the cargo that is not pilfered during shipment across that arc.

If we replace arc length by arc gain factor, and if we replace addition by multiplication and minimization by maximization, then the Floyd, Dantzig, and double-sweep algorithms will calculate the paths with the largest gains instead of the shortest lengths. However, just as these shortest-path algorithms failed when cycles with negative length were present, these algorithms cannot find the path with the greatest gain when there is a cycle with gain greater than one. This follows since this cycle may be repeated infinitely often to form a (nonsimple) path with infinite gain.

Similarly, if we wanted to find a path between two vertices with the smallest gain, we would retain the minimization operation in the preceding algorithm instead of replacing it with maximization. Again, the algorithms will fail if the graph contains a cycle with gain less than -1. This follows since this cycle could be repeated infinitely often to produce a (nonsimple) path whose gain approaches $-\infty$.

In summary, the Floyd, Dantzig, and double-sweep algorithms can be modified to handle additional problems such as the bottleneck and gains problems. This is accomplished by substituting different operations for the addition and minimization operations in these algorithms and by changing appropriately the arc length matrix.

APPENDIX: USING NETSOLVE TO SOLVE SHORTEST-PATH PROBLEMS

The NETSOLVE shortest-path algorithm determines shortest-length paths between various node pairs in either a directed or undirected network. The "length" of an edge is the value specified by the COST data field, and the length of a path is the sum of all edge lengths comprising the path. The edge lengths can have any algebraic sign, but it is necessary for all directed cycles to have nonnegative length. (However, any negative-length cycle will be detected and reported by the algorithm.)

The NETSOLVE command to implement the shortest-path algorithm is SPATH. Several variants of the SPATH command are available for performing shortest-path analyses, depending on whether one needs to determine shortest paths

From a given source node to a given destination node
From a given source node to all network nodes
From all network nodes to a given destination node
From any network node to any other network node

For example, to determine a shortest path from the node named ATL to the node named PHX, the following command would be entered:

>SPATH ATL PHX

The length of the shortest path is displayed at the terminal. If the terminal is currently ON, the edges comprising a shortest path are also displayed, together with the length of each such edge. Also, the user will be asked whether a sensitivity analysis should be conducted on the problem. This analysis produces a *tolerance interval* for each edge on the shortest path, such that varying the individual edge length within that range will not change the shortest path.

If the command line

>SPATH ATL *

is entered, then shortest paths are calculated from ATL to all other nodes in the network. If the terminal is ON, then the "distance" (shortest path length) to each accessible node is displayed, together with its predecessor node, the node immediately preceding a given node on the shortest path from the origin node to the given node. In addition, the user will be asked whether a sensitivity analysis should be conducted on the problem. This analysis produces *dual variables* for each (accessible) node and *reduced costs* for each edge. Also *tolerance intervals* are produced for each relevant edge length, such that varying the individual edge length within that range will not change the optimality of the current shortest-path tree.

In a similar way, the command line

>SPATH * ATL

generates information on shortest paths from all network nodes to ATL. In this case, however, a successor node is displayed, indicating the node immediately following a given node on a shortest path from that given node to the destination node.

Finally, the command line

>SPATH * *

produces information on shortest paths between all pairs of nodes. Because the output generated for this problem can be rather substantial, the user may wish to turn the terminal OFF and turn the journal ON when solving this type of problem for large networks.

We will illustrate the use of NETSOLVE for an investment problem that we will model as a knapsack problem. An investor has $50,000 to invest in five possible instruments. The following table shows the investment amount, yield, and value in thousands of dollars.

Investment	Denomination	Yield	Value
A	$10,000	20%	2.0
B	20,000	14	2.8
C	30,000	18	5.4
D	10,000	9	0.9
E	20,000	13	2.6

We assume that all investments have the same maturity and that we can select at most one of each. The problem is a simple knapsack problem in which we seek to maximize the total value (or minimize the negative of the total value) subject to a budget constraint of $50,000.

The network representation of this problem is shown in Fig. 4.11. Figure 4.12 gives

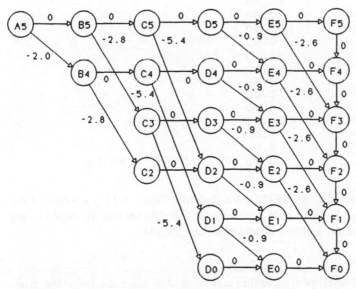

Fig. 4.11 Knapsack network for investment problem.

FROM	TO	COST
A5	B4	-2.00
A5	B5	0.00
B4	C2	-2.80
B4	C4	0.00
B5	C3	-2.80
B5	C5	0.00
C2	D2	0.00
C3	D0	-5.40
C3	D3	0.00
C4	D1	-5.40
C4	D4	0.00
C5	D2	-5.40
C5	D5	0.00
D0	E0	0.00
D1	E0	-0.90
D1	E1	0.00
D2	E1	-0.90
D2	E2	0.00
D3	E2	-0.90
D3	E3	0.00
D4	E3	-0.90
D4	E4	0.00
D5	E4	-0.90
D5	E5	0.00
E0	F0	0.00
E1	F1	0.00
E2	F0	-2.60
E2	F2	0.00
E3	F1	-2.60
E3	F3	0.00
E4	F2	-2.60
E4	F4	0.00
E5	F3	-2.60
E5	F5	0.00
F1	F0	0.00
F2	F1	0.00
F3	F2	0.00
F4	F3	0.00
F5	F4	0.00

Fig. 4.12   NETSOLVE Data.

the listing of arc parameters for the NETSOLVE data file. Figure 4.13 is a listing of the output for the shortest path between node A,5 and node F,0. We see that we should invest in instruments A, C, and D, yielding a total return of $8,300.

EXERCISES

1. Discuss how the shortest-path algorithm can be used to solve the following problems:

```
SHORTEST PATH LENGTH FROM A5         TO F0      :       -8.30

   EDGES IN THE SHORTEST PATH
      FROM       TO                   COST
      ----       --              --------
      A5         B4                   -2.00
      B4         C4                    0.00
      C4         D1                   -5.40
      D1         E0                   -0.90
      E0         F0                    0.00
```

Do you want sensitivity analysis? (Y/N): y

SENSITIVITY ANALYSIS FOR EDGE COSTS

```
                              EDGE            TOLERANCE
      FROM       TO           COST        LOWER         UPPER
      ----       --           ----        -----         -----
      A5         B4          -2.00   -999999.00         -1.90
      B4         C4           0.00   -999999.00          0.10
      C4         D1          -5.40   -999999.00         -5.30
      D1         E0          -0.90   -999999.00         -0.80
      E0         F0           0.00   -999999.00          0.30
```

Fig. 4.13   Optimal solution to knapsack problem.

(a)   You must have an automobile at your disposal for the next five years until your retirement. There are currently various automobiles that you may purchase with different expected lifetimes and costs, and various leasing arrangements are available. What should you do?

(b)   An airline is approached by a passenger who wishes to fly from Springfield, Illinois to Ankara, Turkey, spending as little time as possible in the air since he is afraid of flying. How should the airline route this passenger?

2.   Routing of traffic often includes delays at intersections or "turn penalties." For example, in the network below, every turn incurs a penalty of 3 in addition to the travel times on the arc.

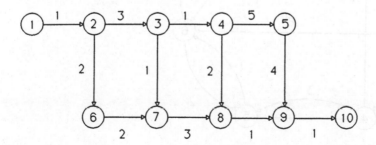

Show how to reformulate the network as a shortest-path problem without turn penalties. (Hint: Create a new network in which the nodes correspond to arcs in the original network.)

3. Construct the shortest-path network for the following production lot sizing problem.

Period	1	2	3	4
Demand	60	100	140	200
Unit cost	7	7	8	7
Holding cost	1	1	2	2

4. Suppose a message is to be sent between two nodes of a communication network. Assume that the individual communication links (edges) operate independently with a known set of probabilities. What path between the two nodes has the maximum reliability, that is, probability that all of its edges are operative? Show how to formulate this problem as a shortest-path problem and apply it to the following network.

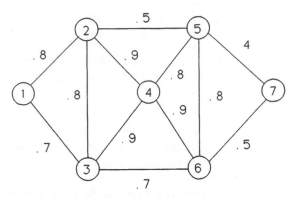

5. Consider the following network:

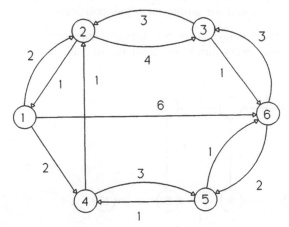

Find the shortest path from vertex 1 to every other vertex using the following algorithms:
(a)   Dijkstra
(b)   Ford

6.   Two sites are to be connected by a new highway. The highway can be thought of as a series of connected links, with alternative routings between potential intersections as shown in the following network. Determine the least expensive highway configuration connecting the origin and destination using
(a)   Partitioning algorithm
(b)   Dijkstra two-tree algorithm

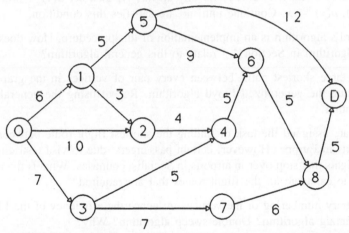

7.   Exercise 6 in Chapter 2 showed that the nodes of an acyclic directed graph can be numbered so that for each arc (x, y), $y > x$. Show how Dikjstra's algorithm can be simplified for such networks. Does this change the time complexity? Apply your modified algorithm to the network in Exercise 6.

8.   Develop a procedure for determining sensitivity analysis results for shortest-path problems analogous to those found in linear programming. For example, what are the cost ranges on arcs in the shortest-path arborescence which will maintain optimality of the arborescence? By how much can the distance of an arc not in the shortest-path arborescence change before it would enter a shortest path?

9.   Using the network in Exercise 5, calculate the shortest path between every pair of vertices using
(a)   Floyd's algorithm
(b)   Dantzig's algorithm

10.   Prove the following propositions about Dijkstra's algorithm:
(a)   During the execution of Dijkstra's algorithm, if $d(x) < \infty$, there is a path from s to x of length $d(x)$.

(b) $d(x)$ is always greater than or equal to the length of a shortest path from s to x at any stage of the algorithm.

11.   Count the number of additions and comparisons required for Dijkstra's and Floyd's algorithms. What is the time complexity relative to these elementary operations?

12.   Discuss the computational implications of implementing Dijkstra's algorithm using a distance matrix rather than a star data structure.

13.   A generic label-correcting algorithm can be stated as follows:

   Step 1. Set $d(s) = 0$, $p(s) = 0$. Set $d(x) = \infty$ for all $x \neq s$.
   Step 2. If $d(y) > d(x) + a(x, y)$ for any arc (x, y), then set $d(y) = d(x) + a(x, y)$, $p(y) = x$. Continue until no arc satisfies this condition.

   Show that Ford's algorithm is an implementation of this procedure. How does the partitioning algorithm in Section 4.3 relate to this general algorithm?

14.   Calculate the three shortest paths between every pair of vertices in the graph in Exercise 5 using the generalized Floyd algorithm. Repeat using the generalized Dantzig algorithm.

15.   Suppose you are assigned the task of finding the shortest flight route between all pairs of airports in Europe. However, certain passengers, due to visa restrictions, cannot take flights that stop over in airports in Socialist countries. What is the most efficient way to calculate all the flight routes that are required?

16.   Does the arbitrary numbering of the vertices influence the efficiency of the Floyd algorithm? Dantzig algorithm? Double-sweep algorithm? Why?

17.   Calculate the three shortest paths from vertex 1 to each other vertex in the graph in Exercise 5.

18.   Show that a nonoptimal solution may be repeated during two consecutive sweeps of the double-sweep algorithm but may not be repeated for three consecutive sweeps of the double-sweep algorithm.

19.   Suppose that the generalized Floyd algorithm was used to calculate the best three bottlenecks between every pair of vertices in a graph. How can this information be used to determine the path corresponding to each bottleneck?

20.   In the gains problem, what values should be assigned to the components of the vector d_{ii}^0?

21.   The spillage rates between the four tanks in a petroleum refinery are shown in the table.

Tank	A	B	C	D
A	0.00	0.13	0.14	0.15
B	0.08	0.00	0.13	0.08

Tank	A	B	C	D
C	0.17	0.12	0.00	0.18
D	0.10	0.06	0.13	0.00

Find the two best ways to ship petroleum from tank C to tank D.

22. Upon terminating the double-sweep algorithm, you discover that one of your arc lengths was incorrect. Under what conditions can you salvage some of your results from the double-sweep algorithm?

23. A pharmaceutical firm wishes to develop a new product within the next 12 months to compete with a similar product recently marketed by their chief competitors. The development of a new product requires four stages, each of which can be performed at a slow, normal, or fast pace. The times and costs of these are shown in the table.

	Theoretical Research[a]	Laboratory Experiments[a]	Government Approval[a]	Marketing[a]
Slow	5, 5	3, 6	6, 1	5, 8
Normal	4, 7	2, 8	4, 1	4, 10
Fast	2, 10	1, 12	2, 3	3, 15

[a]Time in months, cost in thousands.

What is the best way for the firm to have the new product ready within 12 months without spending more than $25,000? What is the second-best way to achieve the same goal?

24. A hotel manager must make reservations for the bridal suite for the coming month. The hotel has received a variety of reservation requests for various combinations of arrival and departure days. Each reservation would earn a different amount of revenue for the hotel due to a variety of rates for students, employees, airline personnel, etc. How can the Dijkstra algorithm be used to find the best way to schedule the bridal suite with maximum profits to the hotel?

(Hint: Represent each reservation request by an arc joining its arrival date to its departure date. The resulting graph will contain no cycles. A variation of the Dijkstra algorithm can be applied to this situation.)

REFERENCES

Dantzig, G. B., 1960. On the Shortest Route Through a Network, *Man. Sci. 6*, pp. 187–190.

Dantzig, G. B., 1967. All Shortest Routes in a Graph, *Theory of Graphs*, International Symposium, Rome, 1966, Gordon and Breach, New York, pp. 91–92.

Deo, N., and C. Pang, 1984. Shortest-Path Algorithms: Taxonomy and Annotation, *Networks, 14*, pp. 275–323.

Dijkstra, E. W., 1959. A Note on Two Problems in Connexion with Graphs, *Numer. Math.*, *1*, pp. 269–271.

Dreyfus, S. E., 1969. An Appraisal of Some Shortest Path Algorithms, *Oper. Res.*, *17*, pp. 395–412.

Floyd, R. W., 1962. Algorithm 97, Shortest Path, *Commun. ACM*, *5*, p. 345.

Ford, L. R., 1956. Network Flow Theory, Rand Corporation Report, p. 923.

Gallo, G., and S. Pallotino, 1986. Shortest Path Methods: A Unifying Approach, *Math. Program. Study*, *26*, pp. 38–64.

Glover, F., D. Klingman, and N. Phillips, 1985a. A New Polynomially Bounded Shortest Path Algorithm, *Oper. Res.*, *33*, pp. 65–73.

Glover, F., D. Klingman, N. Phillips, and R. Schneider, 1985b. New Polynomial Shortest Path Algorithms and the Computational Attributes, *Manage. Sci.*, *31*, pp. 1106–1128.

Helgason, R. V., J. L. Kennington, and B. D. Stewart, 1988. Dijkstra's Two-Tree Shortest Path Algorithm, Technical Report 88-OR-13, Department of Operations Research and Engineering Management, Southern Methodist University, Dallas; revised July 1988.

Hung, M., and J. J. Divoky, 1988. A Computational Study of Efficient Shortest Path Algorithms, *Comput. Oper. Res.*, *15*, pp. 567–576.

Karp, R. M., 1975. On the Computational Complexity of Combinatorial Problems, *Networks*, *5*, pp. 45–68.

Knuth, D. E., 1973. *The Art of Computer Programming*, Vol. 3, *Sorting and Searching*, Addison-Wesley, Reading, Massachusetts.

Lawler, E. L., 1971. The Complexity of Combinatorial Computations: A Survey, *Proc. 1971 Polytechnic Institute of Brooklyn Symposium on Computers and Automata*.

Minieka, E. T., 1974. On Computing Sets of Shortest Paths in a Graph, *Commun. ACM*, *17*, pp. 351–353.

Minieka, E. T., and D. R. Shier, 1973. A Note on an Algebra for the k Best Routes in a Network, *J. IMA*, *11*, pp. 145–149.

Nicholson, T. A., 1966, Finding the Shortest Route Between Two Points in a Network, *Comput. J.*, *9*, pp. 275–280.

Shier, D. R., 1974. Computational Experience with an Algorithm for Finding the k Shortest Paths in a Network, *J. Res. Natl. Bur. Stand.*, *78B*(July–September), pp. 139–165.

Shier, D. R., 1976. Iterative Methods for Determining the k Shortest Paths in a Network, *Networks*, *6*, pp. 205–2320.

Shier, D. R., 1979. On Algorithms for Finding the *k* Shortest Paths in a Network, *Networks*, *9*, pp. 195–214.

Williams, T. A., and G. P. White, 1973. A Note on Yen's Algorithms for Finding the Length of All Shortest Paths in *N*-Node Nonnegative Distance Networks, *J. ACM*, *20*(3), pp. 389–390.

Yen, J. Y., 1970. An Algorithm for Finding Shortest Routes from All Source Nodes to a Given Destination in General Networks, *Q. Appl. Math.*, *27*, pp. 526–530.

Yen, J. Y., 1973. Finding the Lengths of All Shortest Paths in N-Node Nonnegative Distance Complete Networks Using $1/2N^3$ Additions and N^3 Comparisons, *J. ACM*, *19*(3), pp. 423–424.

5
MINIMUM-COST FLOW ALGORITHMS

5.1 INTRODUCTION AND EXAMPLES

In Chapter 1 we defined a flow in a network and presented some basic concepts about flows. One of the most useful types of network optimization problems consists of finding minimum-cost flows in networks. In a minimum-cost flow problem, we are given a directed network $G = (X, A)$ in which each arc (x, y) has a cost per unit of flow $c(x, y)$ and possibly lower and/or upper bounds on the flow values. Each node x has a net supply/demand $b(x)$. If $b(x) > 0$, $b(x)$ represents the supply at node x; if $b(x) < 0$, $-b(x)$ represents the demand at node x. If $b(x) = 0$, the node is a transshipment node. A feasible flow in the network is one in which the flow on arc (x, y), $f(x, y)$, satisfies any lower- and upper-bound constraints and conservation of flow is preserved at all nodes. Recall that conservation of flow means that the flow into a node must equal the flow out of the node. (Supplies are considered as external flows into a node; demands are considered as flows out.) The objective is to find a feasible flow that minimizes the total cost, $\Sigma\ c(x, y)f(x, y)$, where the summation is taken over all arcs in the network.

Many practical problems can be modeled as minimum-cost flow problems. We present some examples next.

Distribution

Several warehouses (supply points) must ship a single commodity to a number of retail outlets (demand points). Shipping costs are proportional to the distance shipped and the amount shipped. Figure 5.1 shows an example with three warehouses and three retail outlets. Figure 5.2 is a network flow representation. The minimum-cost solution is to ship 3 units from A to 1; 7 units from A to 2; 5 units from B to 1; 1 unit from C to 2; and 6 units

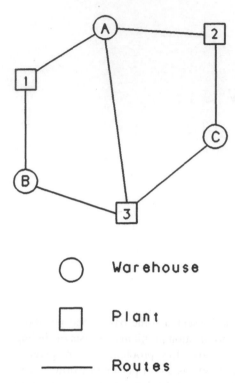

Fig. 5.1   A distribution network.

from C to 3. The total cost is 5(3) + 4(7) + 4(5) + 6(1) + 6(6) = 105. This type of problem is commonly known as a *transportation problem*. The transportation problem is a special case of a general minimum-cost network flow problem in which the underlying graph is bipartite.

Statistical Matching of Large Data Files (Barr and Turner, 1981)

The Office of Tax Analysis of the Department of the Treasury is charged with evaluating the effects of proposed tax code revisions. Such evaluations are based on models which predict the effects of the proposed revisions on the tax liability of families having specified characteristics. These models require a statistical profile of the population to be analyzed. For practical reasons, these statistical profiles generally are derived from existing databases. Typically the required data must be obtained by merging data from several sources.

Existing databases generally are statistically based rather than including the entire population. For example, the *Statistics of Income (SOI)* file represents a sample of roughly

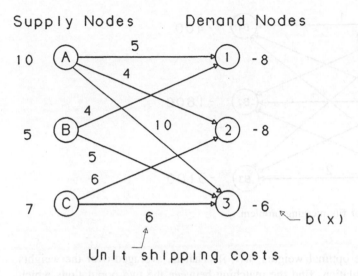

Fig. 5.2 Network flow representation.

50,000 income tax returns, each with 100 to 200 data items. The *Current Population Survey*, generated monthly by the Bureau of the Census, represents interviews with approximately 67,000 households and includes some items in common with SOI and some not in common. Few, if any, records based on the same individual are likely to be included in both files to be merged. Common data items which form the basis for a merge can be miscoded, inaccurate, or misrepresented. Therefore, exact matching of records is not practical.

Suppose file A has 3 records reflecting populations of sizes 2000, 100, and 3000, respectively. File B has 3 records reflecting populations of 400, 1600, and 3100. Both files represent the same population of 5100. Files A and B share one data item: annual income. These are 10, 15, and 22 for file A and 11, 18, and 20 for file B (in thousands). The proximity or closeness of a match between record i from file A and record j from file B is measured by the absolute value of the difference in the annual incomes:

$$c(i, j) = |\text{income(A)} - \text{income(B)}|$$

In order to preserve the weights associated with the populations underlying each file, associate a weight $w(i, j)$ with the matching or record i of file A and record j of file B which preserves the initial weights. For example,

$$w(1, 1) + w(1, 2) + w(1, 3) = 2000$$

and

$$w(1, 2) + w(2, 2) + w(3, 2) = 1600$$

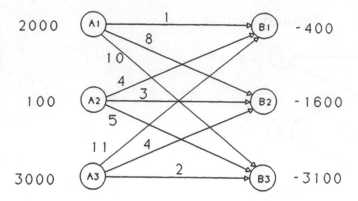

Fig. 5.3 Network model of file merging problem.

We seek to determine the optimal weights, $w(i, j)$. That is, averaged over the weights associated with each population, find the matching between the two populations which minimizes the difference between paired records.

This problem can be modeled as a transportation problem. The flows on the arcs represent the weights $w(x, y)$. A small example is shown in Fig. 5.3. The solution to this problem is $w(1, 1) = 400$, $w(1, 2) = 1500$, $w(1, 3) = 100$, $w(2, 2) = 100$, and $w(3, 3) = 3000$. The merged data file thus has five distinct records. The first is formed by merging record 1 from file A with record 1 from file B and assigning a weight of 400; the remaining are listed below:

A Record	B Record	Weight
1	2	1500
1	3	100
2	2	100
3	3	3000

A real example solved in 1978 as a result of the Carter Tax Reform Initiative consisted of over 100,000 constraints and 25 million variables. It was decomposed into six problems each having over 4 million variables and solved in about 7 hours using specialized network computer codes. Evidence suggests that ordinary linear programming would have taken 7 months to solve!

Production Scheduling (Bazaraa et al., 1990; Bowers and Jarvis, 1989)

Polyester fiber is used in manufacturing a textile fabric. An initial inventory of 10 units is on hand. The maximum inventory capacity is 20 units. An inventory of 8 units is required at the end of the next four-month planning period. The current unit price of the fiber stock is 6 but is expected to increase by 1 during each of the next several months. The unit cost of keeping the fiber in inventory is 0.25 per month. Contractual obligations require production which will consume 12, 19, 15, and 20 units of fiber over the next four months. Early shipment of the finished product is possible but is constrained by a maximum utilization of 18 units of fiber per month and a minimum utilization of 14 units per month. How much fiber should be purchased in each month to minimize total cost?

A network flow representation of this problem is shown in Fig. 5.4. Arcs from node S to Mi represent the purchasing of stock in month i. Node S has an unlimited supply. Node II has a supply of 10 and represents the initial inventory. Arcs from Mi to M$i+1$ correspond to inventory holding from month i to month $i+1$. Node FI has a demand equal to the final inventory required at the end of month 4. Arcs from node Mi to node Pi represent production in month i. Lower and upper bounds on the flow specify the minimum and maximum utilization restrictions. Arcs from nodes Pi to Rj represent the shipment of the finished product to meet the demands at the nodes Rj. This is an example of a general minimum-cost network flow problem.

Determining Service Districts (Cooper, 1972; Larson and Odoni, 1981)

A region has two hospitals which serve as bases for emergency medical vehicles. Two vehicles are located at hospital 1 and three vehicles at hospital 2 (see Fig. 5.5). The

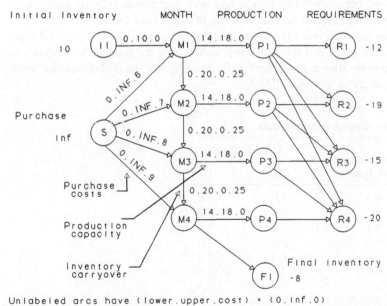

Unlabeled arcs have (lower.upper.cost) = (0.Inf.0)

Fig. 5.4 Network flow model of production scheduling.

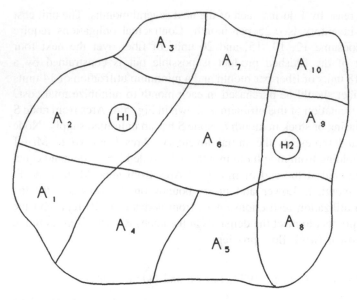

Fig. 5.5 Service district assignment example.

hospitals operate independently. The region has been partitioned into 10 areas, each independently generating calls for service at a rate of r_i per hour for service area i. The average service time for a call is 0.5 hour, and the total rate at which calls arrive is 3.0 per hour. Specific data are given in Table 5.1. How should areas be assigned to hospitals so that the average response time is minimized while ensuring that the probability that a call incurs a queueing delay is less than 10%?

Table 5.1 Service district example data

Area	Call Rate	Average Time to	
		H_1	H_2
A_1	0.2	3.0	6.0
A_2	0.3	2.0	5.5
A_3	0.1	2.0	5.0
A_4	0.2	3.0	4.0
A_5	0.4	4.0	2.0
A_6	0.3	3.0	2.5
A_7	0.1	4.0	3.5
A_8	0.5	5.0	1.5
A_9	0.6	5.5	2.0
A_{10}	0.3	6.5	4.0

The assignment of areas to hospitals can be modeled by considering each area within the region as a supply point with a supply r_i and adding arcs from each area to each hospital with costs equal to the response time from the area to the hospital. The total demand is the overall service rate of 3.0 calls per hour. The delay probability depends on the total call rate assigned to each hospital. Using a simple queueing model with a service time of 0.5 hour, the maximum call rate yielding a delay probability of 10% is approximately 1.0 to two servers and 2.1 for three servers. Hence we have a capacity of 1.0 on the flows to hospital 1 and 2.1 for hospital 2. The network is shown in Fig. 5.6. Finding the minimal-cost flows from the sources A_1 through A_{10} to the sink T will provide the optimal assignments.

Natural Gas Distribution

To study the effects of new sources, regulation and deregulation, pipeline capacity, supply, and demand on distribution and allocation of natural gas in the United States, a model called GASNET3 was developed. This model includes the physical transport of gas among producers, pipelines, distributors, and consumers; sales by producers to pipeline companies to distributors to consumers including exchanges of supply; pipeline capacity, pumping stations, and losses in transmission.

Networks provide a natural framework for modeling the natural gas distribution system. The key difference relative to other flow networks is the loss along a transmission line. Because of this, we no longer have conservation of flow. (Electrical transmission networks have similar characteristics.) To model this, we need to introduce an additional arc parameter, the *gain* on the arc. If an arc has a gain g, then a flow with value f entering

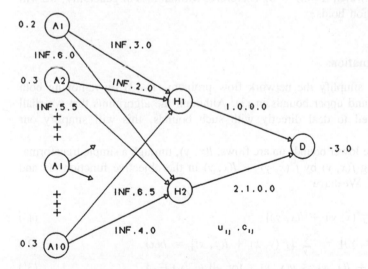

Fig. 5.6 Network flow model of service district problem.

the arc leaves the arc with a value *gf*. *Flows with gains* represent a special class of minimum-cost network flow problems having numerous applications in production, financial modeling, scheduling, and blending problems.

We have seen that a variety of different problems can be modeled as minimum-cost network flow problems. In this chapter we will study algorithms for solving these types of problems.

5.2   BASIC PROPERTIES OF MINIMUM-COST NETWORK FLOW PROBLEMS

A minimum-cost network flow problem defined on a directed graph $G = (X, A)$ can be formulated as the linear program

$$\min \sum_{(x, y) \in A} c(x, y) f(x, y) \tag{1}$$

$$\sum_y f(x, y) - \sum_y f(y, x) = b(x), \text{ for all } x \in X \tag{2}$$

$$l(x, y) \le f(x, y) \le u(x, y), \text{ for all } (x, y) \in A \tag{3}$$

We assume that $\Sigma\, b(x) = 0$; that is, the sum of the supplies equals the sum of the demands. If $\Sigma\, b(x) < 0$, then clearly there is no feasible solution. If $\Sigma\, b(x) > 0$, we create a fictitious demand node having a demand $\Sigma\, b(x)$ and extend an arc from each supply node to this fictitious node. The cost associated with each of these arcs is zero. This results in a network for which $\Sigma\, b(x) = 0$. Therefore, without loss of generality, we will assume that this condition holds.

Two Simple Transformations

We will show how to simplify the network flow problem (1)–(3) by removing both nonzero lower bounds and upper bounds in (3). Although the algorithms that we shall present can be modified to deal directly with such bounds, this will simplify our presentation.

We may remove the lower bounds on arc flows, $l(x, y)$, through a simple transformation. Consider replacing $f(x, y)$ by $f'(x, y) + l(x, y)$ in the objective function (1) and constraints (2) and (3). We have

$$\min \sum_{(x, y) \in A} c(x, y) \, [f'(x, y) + l(x, y)] \tag{1'}$$

$$\sum_y [f'(x, y) + l(x, y)] - \sum_y [f'(y, x) + l(y, x)] = b(x) \tag{2'}$$

$$l(x, y) \le f'(x, y) + l(x, y) \le u(x, y), \quad \text{for all } (x, y) \in A \tag{3'}$$

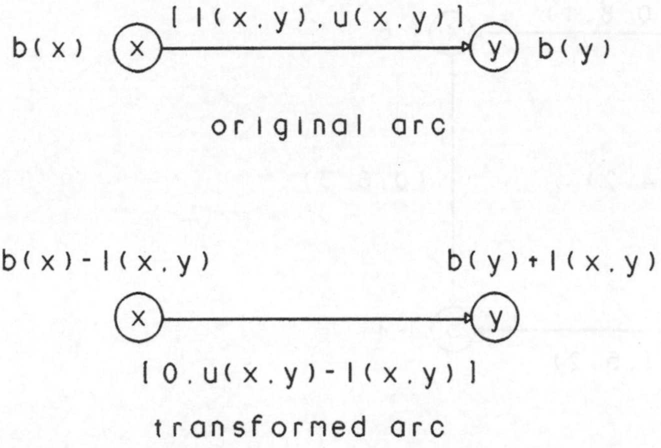

$b(x)$ (x) $\xrightarrow{\hspace{1cm} [\; l(x,y), u(x,y) \;]}$ (y) $b(y)$

original arc

$b(x) - l(x,y)$ $b(y) + l(x,y)$

(x) $\xrightarrow{\hspace{4cm}}$ (y)

$[\; 0, u(x,y) - l(x,y) \;]$

transformed arc

Fig. 5.7 Removing lower bounds on arc flows.

which simplifies to

$$\min \sum_{(x,\,y)\in A} c(x,y)f'(x,y) + \sum_{(x,\,y)\in A} c(x,y)l(x,y) \qquad (1')$$

$$\sum_{y} f'(x,y) - \sum_{y} f'(y,x) = b(x) + \sum_{y} l(x,y) - \sum_{y} l(y,x) \qquad (2')$$

$$0 \le f'(x,y) \le u(x,y) - l(x,y), \text{ for all } (x,y) \in A \qquad (3')$$

Notice that the objective function, right-hand side of (2), and upper-bound constraints are changed by constant terms. This transformation has the simple network interpretation shown in Fig. 5.7. For each arc (x, y), we subtract $l(x, y)$ from $b(x)$, add it to $b(y)$, and subtract it from $u(x, y)$.

EXAMPLE 1

Consider the network flow problem shown in Fig. 5.8. Applying the transformation given by (2') and (3') to each node gives the network shown in Fig. 5.9. After solving for the minimum-cost flow in this network, flows in the original network are recovered from

$f(1, 3) = f'(1, 3) + 2$

$f(3, 2) = f'(3, 2) + 2$

$f(3, 5) = f'(3, 5) + 3$

The true minimum cost would be found by adding the constant term in (1') to the minimum cost in the transformed problem. For this example, the cost would be increased by $2(5) + 2(2) + 3(2) = 20$.

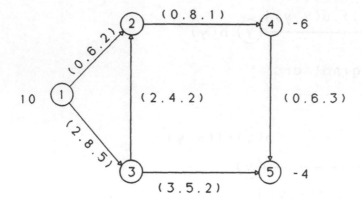

(lower . upper . cost)

Fig. 5.8 Network for Example 1.

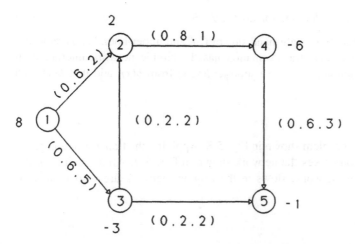

(lower . upper . cost)

Fig. 5.9 Network for Example 1 with lower bounds removed.

We may employ a different transformation to remove the upper bound from (3). Suppose that we add a slack variable $s(x, y)$ to the upper-bound constraint and multiply by -1:

$$-f(x, y) - s(x, y) = -u(x, y) \tag{4}$$

Now solve (4) for $f(x, y)$ and substitute in the conservation of flow equation at node y:

$$\sum_x f(y, x) - \sum_x f(x, y) = b(y)$$

$$\sum_x f(y, x) - \sum_x [u(x, y) - s(x, y)] = b(y)$$

$$\sum_x f(y, x) + \sum_x s(x, y) = b(y) + \sum_x u(x, y) \tag{5}$$

Adding equation (4) to (1)–(3) is equivalent to creating a new node z in the network having a demand of $u(x, y)$ with two arcs directed into it: $f(x, y)$ and $s(x, y)$. From equation (5) we see that $s(x, y)$ is directed out of node y, and the net supply of node y has been modified to $b(y) + \sum_x u(x, y)$. The cost associated with $s(x, y)$ is zero. Note that the net supply in the entire network remains zero since $u(x, y)$ is both added and subtracted at some node. This transformation has a simple interpretation in the network as shown in Fig. 5.10. Essentially, each upper-bounded arc is split into two by the transformation.

EXAMPLE 2

We will remove the upper bounds in the network in Fig. 5.9 using the transformation described above. The complete transformed network is shown in Fig. 5.11. Notice that since every original node has arcs directed only out of it and every new node has arcs directed only into it, the network is bipartite as shown in Fig. 5.12. Thus, this transformation converts any network flow problem with upper bounds on arc flows into an equivalent transportation problem.

Since we can remove lower and upper bounds through these simple transformations,

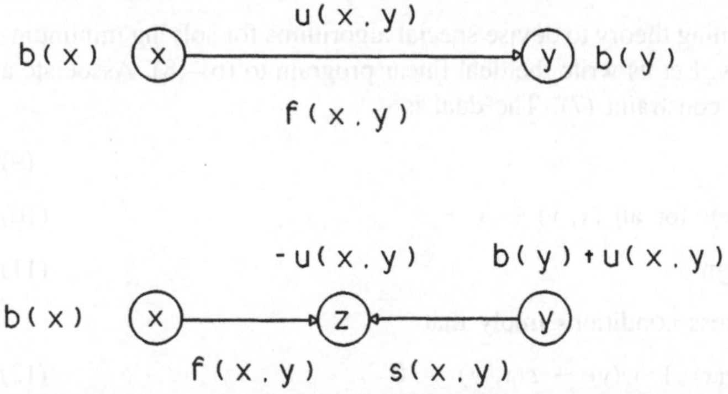

Fig. 5.10 Removing upper bounds on arc flows.

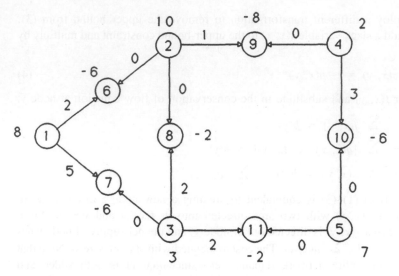

Fig. 5.11 Network in Fig. 5.9 with upper bounds removed.

we will henceforth assume that all lower bounds are zero and no upper bounds exist. The minimum-cost network flow problem thus can be simplified to

$$\min \sum_{(x,\,y)\in A} c(x,\,y)f(x,\,y) \tag{6}$$

$$\sum_y f(x,\,y) - \sum_y f(y,\,x) = b(x), \text{ for all } x \in X \tag{7}$$

$$f(x,\,y) \geq 0, \text{ for all } (x,\,y) \in A \tag{8}$$

Network Flow and Linear Programming Theory

We can use linear programming theory to devise special algorithms for solving minimum-cost network flow problems. Let us write the dual linear program to (6)–(8). Associate a dual variable $w(x)$ to each constraint (7). The dual is

$$\max \sum_{x\in X} b(x)w(x) \tag{9}$$

$$w(x) - w(y) \leq c(x,\,y), \text{ for all } (x,\,y) \in A \tag{10}$$

$$w(x) \text{ unrestricted in sign} \tag{11}$$

The complementary slackness conditions imply that

$$\text{If } f(x,\,y) > 0, \text{ then } w(x) + w(y) - c(x,\,y) = 0 \tag{12}$$

$$\text{If } w(x) - w(y) < c(x,\,y), \text{ then } f(x,\,y) = 0 \tag{13}$$

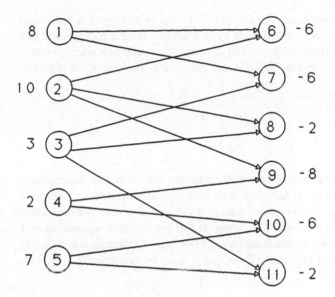

Fig. 5.12 Bipartite structure of Fig. 5.11.

Linear programming theory states that any solutions to the primal and dual problems are optimal if the following conditions hold:

(a) Primal feasibility
(b) Dual feasibility
(c) Complementary slackness

We define $\bar{c}(x, y) = c(x,y) - w(x) + w(y)$ to be the *reduced cost* associated with $f(x, y)$. Observe that $\bar{c}(x, y) > 0$ is equivalent to dual feasibility. These optimality conditions will be important as we develop algorithms for the minimum-cost flow problem.

The left-hand side of (7) defines the node-arc incidence matrix A of the underlying network. That is, each variable $f(x, y)$ in (7) has a coefficient of $+1$ in the row corresponding to node x and of -1 in the row corresponding to node y. This matrix of coefficients has dimension $m \times n$. If we add all the constraints of (7) we obtain $0 = 0$. Therefore, the rows of the node-arc incidence matrix are linearly dependent.

Theorem 5.1. The rank of a node-arc incidence matrix is $m - 1$; that is, the maximum number of linearly independent columns in the matrix is $m - 1$.

Proof. We need only show that there exists an $(m - 1) \times (m - 1)$ submatrix that is nonsingular. Let T be any spanning tree in the network. We know that T consists of m

nodes and $m - 1$ arcs that do not form a cycle. Let A_T be the submatrix of A associated with the arcs in T. Since T is a tree, there must be at least one node x having only one incident arc. Therefore, the corresponding row in A_T must have only one nonzero entry. Permute the rows and columns so that this nonzero entry is in the upper left position. Thus, A_T becomes

$$\begin{bmatrix} \pm 1 & 0 \\ q & A_{T'} \end{bmatrix}$$

The matrix $A_{T'}$ corresponds to the tree obtained by removing node x and its incident arc. Since T' is also a tree, it must have at least one node with only one incident arc. Permute the rows and columns of $A_{T'}$ in the same manner as A_T so that the nonzero entry corresponding to that node is in the upper left corner. If this procedure is repeated $m - 1$ times and then delete the last row, the resulting $(m - 1) \times (m - 1)$ matrix will be lower triangular with nonzero diagonal elements. Therefore it must be nonsingular.   Q.E.D.

EXAMPLE 3

Consider the network in Fig. 5.13. Let T consist of arcs $(1, 2)$, $(1, 3)$, and $(3, 4)$. The matrix A_T is

$$
\begin{array}{c}
 \\ 1 \\ 2 \\ 3 \\ 4
\end{array}
\begin{array}{ccc}
(1, 2) & (1, 3) & (3, 4) \\
\end{array}
$$

$$
\begin{array}{c}
1 \\ 2 \\ 3 \\ 4
\end{array}
\begin{bmatrix}
1 & 1 & 0 \\
-1 & 0 & 0 \\
0 & -1 & 1 \\
0 & 0 & -1
\end{bmatrix}
$$

Row 2 has only one nonzero entry. Permute the rows and columns to place this entry in the upper left corner.

$$
\begin{array}{c}
2 \\ 1 \\ 3 \\ 4
\end{array}
\begin{array}{ccc}
(1, 2) & (1, 3) & (3, 4) \\
\end{array}
$$

$$
\begin{array}{c}
2 \\ 1 \\ 3 \\ 4
\end{array}
\begin{bmatrix}
-1 & 0 & 0 \\
1 & 1 & 0 \\
0 & -1 & 1 \\
0 & 0 & -1
\end{bmatrix}
$$

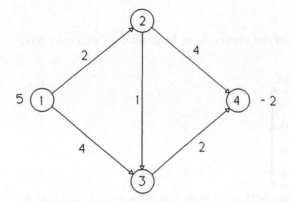

Fig. 5.13 Network for Example 3.

Deleting the last row leaves the 3×3 matrix

$$
\begin{array}{cccc}
 & (1, 2) & (1, 3) & (3, 4) \\
2 & \begin{bmatrix} -1 \\ 1 \\ 0 \end{bmatrix} & \begin{matrix} 0 \\ 1 \\ -1 \end{matrix} & \begin{matrix} 0 \\ 0 \\ 1 \end{matrix} \end{array}
$$

which is in lower triangular form.

We may transform the matrix A into a full row rank matrix by adding a column consisting of all zeros and a single 1. By convention, we will position the 1 at the last row (node m). We may think of this column as "half an arc"—one whose head is not incident to any node. This additional arc may be regarded as an artificial variable in the linear programming formulation. We call this artificial arc the *root arc* and its incident node the *root node*. Any spanning tree in the network plus this artificial arc (called a *rooted spanning tree*) therefore will correspond to a set of m linearly independent columns, that is, a *basis*. For any feasible solution, this artificial variable must be basic and the flow on this arc must be zero (why?). Thus, a basis for a minimum-cost network flow problem is equivalent to a rooted spanning tree in the network.

The proof of Theorem 5.1 characterizes another important property of minimum-cost network flow problems, namely that every basis matrix (defined by A_T plus the artificial column) is triangular. In linear programming, we need to solve the system of equations

$$B\mathbf{x}_B = \mathbf{b}$$

to find the basic solution $\mathbf{x}_B$. If the basis matrix B is triangular, this system of equations is solved easily by substitution.

EXAMPLE 4

Consider the triangular representation of the matrix A_T in Example 3 to which we have added a root arc:

$$
\begin{array}{cccc}
 & (1,2) & (1,3) & (3,4) & \text{Root} \\
2 & \begin{bmatrix} -1 & 0 & 0 & 0 \\ 1 & 1 & 0 & 0 \\ 0 & -1 & 1 & 0 \\ 0 & 0 & -1 & 1 \end{bmatrix}
\end{array}
\quad
\begin{array}{c} \\ 1 \\ 3 \\ 4 \end{array}
$$

Suppose the net supplies at each node are $b(1) = 5$, $b(2) = 0$, $b(3) = -3$, and $b(4) = -2$. Then the basic solution is found by solving the system of equations

$$
\begin{bmatrix} -1 & 0 & 0 & 0 \\ 1 & 1 & 0 & 0 \\ 0 & -1 & 1 & 0 \\ 0 & 0 & -1 & 1 \end{bmatrix}
\begin{bmatrix} f(1,2) \\ f(1,3) \\ f(3,4) \\ \text{root} \end{bmatrix}
=
\begin{bmatrix} 0 \\ 5 \\ -3 \\ -2 \end{bmatrix}
$$

From the first row, we immediately find that $f(1, 2) = 0$. Substituting this value in row 2, we solve for $f(1, 3)$ as follows:

$$
\begin{aligned}
f(1, 3) &= 5 - f(1, 2) \\
        &= 5
\end{aligned}
$$

Substituting this value in the third row yields

$$
\begin{aligned}
f(3, 4) &= -3 + f(1, 3) \\
        &= -3 + 5 \\
        &= 2
\end{aligned}
$$

Finally, from the last row, we see that the flow on the root arc is zero (as it should be).

We observe that all computations are additions and subtractions and thus very efficient. Further, if the supplies and demands are integers, the value of all basic variables will be integral. Thus every extreme point of the minimum-cost network flow linear program will be integral. Since an optimal solution to a linear program will always occur at an extreme point, an all-integer optimal solution will exist.

5.3   COMBINATORIAL ALGORITHMS FOR MINIMUM-COST NETWORK FLOWS

In this section we shall present two algorithms for finding minimum-cost network flows by seeking cycles and paths in a graph that satisfy certain properties. Both of these algorithms use the optimality conditions of the linear programming formulation that we presented in Section 5.2 but are quite different from the simplex algorithm of linear programming. In the next section we will see how the simplex algorithm can be specialized to solve minimum-cost flow problems very efficiently.

The first algorithm, called the *negative-cycle algorithm*, begins with a feasible solution to the primal problem and seeks to find a feasible solution to the dual problem by adjusting the flow in the network and the complementary dual solution while maintaining primal feasibility. Given any feasible flow in the network $G = (X, A)$, we create a *residual network* $R = (X, A')$. The residual network consists of the same node set as G. For every arc $(x, y) \in A$, there is an identical arc in A' of R with cost $c(x, y)$ and capacity of infinity. In addition, if $f(x, y) > 0$, then there is an arc $(y, x) \in A'$ having a cost $-c(x, y)$ and capacity of $f(x, y)$. Increasing the flow on arc (y, x) in R is equivalent to decreasing the flow on arc (x, y) of G. The negative-cycle algorithm is based on the following result.

Theorem 5.2. A feasible flow in G is optimal if and only if R has no directed cycle with negative cost.

Proof. We first show that if we have a minimum-cost flow in G, then R can contain no negative-cost directed cycle. If we have an optimal flow in G, then the reduced cost on each arc must be nonnegative. In particular, the sum of the reduced costs around any directed cycle C would be

$$0 \leq \sum_C \bar{c}(x, y) = \sum_C [c(x, y) - w(x) + w(y)]$$
$$= \sum_C c(x, y) + \sum_C [-w(x) + w(y)]$$

The last term sums to zero since dual variable cancels itself out around any directed cycle. Therefore $\Sigma\, c(x, y) \geq 0$ and no negative cycle exists. Conversely, suppose that we have a feasible solution with no negative-cost directed cycle in the residual network. From Chapter 4 we know that the shortest path between any nodes must be finite. Let $d(x)$ be the shortest distance from node 1 to node x. Then

$$d(y) \leq d(x) + c(x, y)$$

or $c(x, y) + d(x) - d(y) \geq 0$. Define $w(x) = -d(x)$. Then

$$c(x, y) - w(x) + w(y) = \bar{c}(x, y) \geq 0$$

which is equivalent to dual feasibility. Complementary slackness conditions are satisfied implicitly for the following reasons. If $\bar{c}(x, y) > 0$ and $f(x, y) > 0$ for some arc (x, y) in G, then the residual network would contain arc (y, x) with $c(y, x) = -c(x, y)$. But then $\bar{c}(y, x) < 0$, which would be a contradiction.

The *negative-cycle algorithm* proceeds as follows:

Step 1. Find a feasible flow in the network G.

Step 2. Construct the residual network R and determine if it contains a negative-cost cycle. If not, then stop; the current flow in G is optimal. If a negative cycle exists, go to step 3.

Step 3. Let δ be the minimum arc capacity on the negative-cost cycle. Push δ units of flow along the arcs of the cycle in G as follows. If (x, y) in G is in the same direction as in the cycle, then increase the flow on (x, y) by δ. If (x, y) in G is in the opposite direction to that in the cycle, then decrease the flow on (x, y) by δ units.

A feasible flow can be found in step 1 by using the path-finding algorithm described in Chapter 2. Select any pair of nodes s and t for which $b(s) > 0$ and $b(t) < 0$ [if all $b(x) = 0$ then stop]. Find a directed path from s to t (if this cannot be done for every such pair of nodes, then the problem is infeasible). Send $v = \min\{b(s), -b(t)\}$ units of flow along this path. Set $b(s) = b(s) - v$ and $b(t) = b(t) + v$ and repeat.

EXAMPLE 5

Consider the problem in Fig. 5.13. To find a feasible flow, let us first choose $s = 1$ and $t = 4$. Suppose the path-finding algorithm identifies the path $(1, 2), (2, 4)$. Send 2 units of flow along this path. Set $b(1) = 5 - 2 = 3$ and $b(4) = -2 + 2 = 0$. Next, select $s = 1$ and $t = 3$. A path from s to t is the arc $(1, 3)$. Send 3 units of flow along this path. We now have a feasible flow to initiate the negative-cycle algorithm. In the networks below, flows are circled beside the arcs. The initial network and its residual network are shown below.

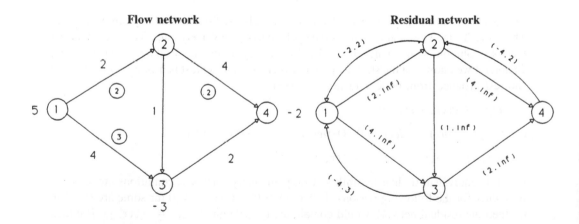

Using a shortest-path algorithm that will detect a negative cycle if its exists (for example, Ford's algorithm), we might identify the cycle 1-2-3-1 in the residual network:

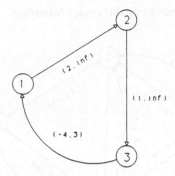

The minimum capacity along this cycle is 3; thus we increase the flow by 3 units along this cycle. In the original network, since the directions of arcs (1, 2) and (2, 3) are the same as in the cycle, we increase the flow. However, arc (1, 3) is in the opposite direction, so we decrease the flow in G. The new solution and the corresponding residual network are shown below.

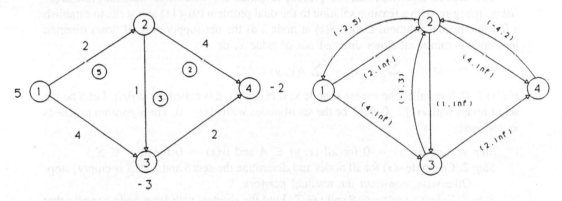

The shortest-path algorithm detects the negative cycle 2-3-4-2:

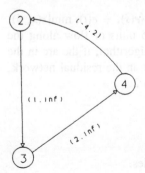

The minimum capacity along this cycle is 2, so we send 2 units along this cycle. Adjusting the flows in G, we have the following solution:

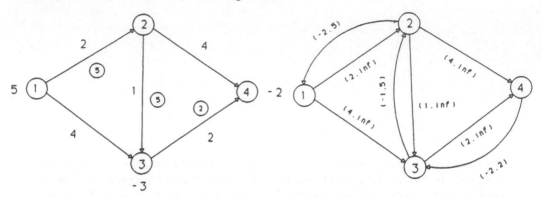

The residual network contains no negative cycle; therefore, the current flow is optimal.

The second algorithm that we present is called the *successive shortest-path algorithm*. It begins with a feasible solution to the dual problem (9)–(11) and seeks to establish a feasible primal solution. Define $e(x)$ at node x as the net supply plus all flows directed into node x minus all flows directed out of node x, or

$$e(x) = b(x) + \sum_y f(y, x) - \sum_y f(x, y)$$

If $e(x) > 0$, it is called the *excess* at node x; if $e(x) < 0$, it is called the *deficit*. Let S be the set of nodes with $e(x) > 0$ and T be the set of nodes with $e(x) < 0$. The algorithm proceeds as follows:

Step 1. Let $f(x, y) = 0$ for all $(x, y) \in A$ and $w(x) = 0$ for all $x \in X$.

Step 2. Compute $e(x)$ for all nodes and determine the sets S and T. If S is empty, stop. Otherwise, construct the residual network.

Step 3. Select a node $s \in S$ and $t \in T$. Find the shortest path from node s to all other nodes in the residual network using the reduced costs $\bar{c}(x, y)$. Let $d(x)$ be the shortest distance from s to x.

Step 4. Set $w(x) = w(x) - d(x)$ for all x. Compute $\delta = \min\{e(s), -e(t), \min[r(x, y),$ for (x, y), on the shortest path from s to $t]\}$. Send δ units of flow along the shortest path from s to t. As with the negative-cycle algorithm, if the arc in the original network is in the opposite direction to the arc in the residual network, decrease the flow on the original arc. Return to step 2.

EXAMPLE 5

Let us again consider the network in Fig. 5.13.

Step 1. We begin with a zero flow and zero dual variables.

Step 2.

$$e(1) = 5$$
$$e(3) = -3$$
$$e(4) = -2$$

Step 3. $S = \{1\}; T = \{3, 4\}$. The flow network and residual network are shown below.

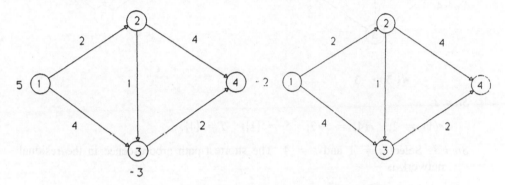

Select $s = 1$ and $t = 3$. The shortest path arborescence is

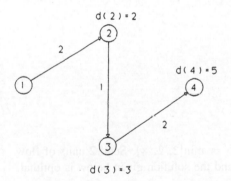

Step 4. We update the dual variables:

$$w(2) = 0 - 2 = -2$$
$$w(3) = 0 - 3 = -3$$
$$w(4) = 0 - 5 = -5$$

$\delta = \min\{5, 3, \infty\} = 3$. Therefore, we send 3 units of flow along the shortest path from node 1 to node 3.

Step 2. The new flows and the residual network are

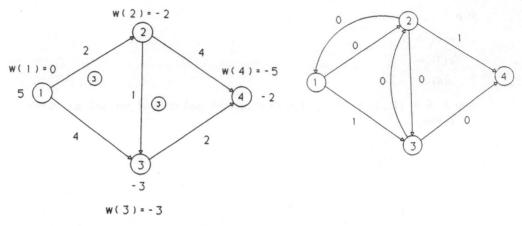

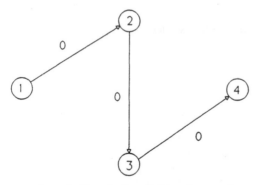

Step 2.

$$e(1) = 2; \quad e(4) = -2; \quad S = \{1\}; \quad T = \{4\}.$$

Step 3. Select $s = 1$ and $t = 4$. The shortest-path arborescence in the residual network is

Step 4. The dual variables do not change. $\delta = \min\{2, 2, \infty\}$. Send 2 units of flow along this path. All $e(x)$ are now zero and the solution given below is optimal.

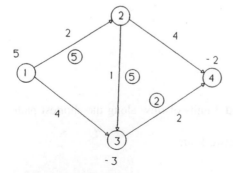

Neither the negative-cycle algorithm nor the successive shortest-path algorithm has polynomially bounded time complexities since it is not possible to bound the number of iterations as a function of the size of the network. However, the number of iterations can be shown to be bounded by a function in both the network size and the problem parameters. Such time bounds are called *pseudopolynomial*.

5.4   THE SIMPLEX METHOD FOR MINIMUM-COST NETWORK FLOWS

The simplex algorithm can be applied to problem (1)–(3) in a very specialized fashion that lends itself to fast and efficient computation. We call this specialization the *network simplex algorithm*. Computer implementations of the network simplex algorithm are hundreds of times faster than general-purpose linear programming codes.

The groundwork for the network simplex algorithm was laid in Section 5.2. We showed that a basis for the minimum-cost network flow problem (1)–(3) is characterized as a rooted spanning tree in the network. All we need to do is to show how to perform the steps of the simplex algorithm using this network representation of the basis rather than a matrix representation.

Let us quickly review the simplex algorithm:

Step 1. Begin with some basic feasible solution.

Step 2. Compute the complementary dual solution, that is, a dual solution that satisfies complementary slackness.

Step 3. Price out the nonbasic variables by computing the reduced costs $\bar{c}(x, y)$. If $\bar{c}(x, y) \geq 0$, stop; the current basis is optimal. Otherwise select a nonbasic variable with $\bar{c}(x, y) < 0$ to enter the basis.

Step 4. Determine the basic variable to leave the basis and pivot, bringing the new variable into the basis and updating the values of all other basic variables. Return to step 2.

Let us assume that we have an initial basic feasible solution in the form of a spanning tree on the network. The complementary slackness condition is

If $f(x, y)$ is basic,   then $c(x, y) - w(x) + w(y) = 0$

If the network has n nodes, there are n dual variables $w(x)$. However, only $n - 1$ flows $f(x, y)$ are basic. Thus, we have $n - 1$ complementary slackness conditions in n variables. We can arbitrarily set some $w(x)$ to any value. It is customary to set the dual variable corresponding to the root node to zero. Once this is done, the remaining dual variables can be uniquely determined as follows. Select any node x for which the dual variable $w(x)$ is known. Consider the set of arcs (x, y) such that $w(y)$ has not yet been determined. Each of these arcs corresponds to a complementary slackness condition. Since $w(x)$ is known, we can solve uniquely for $w(y)$ as

$$w(y) = w(x) - c(x, y)$$

We continue in this fashion until all dual values have been assigned. Since there is a unique path between any pair of nodes in a spanning tree, all remaining dual variables are uniquely determined from the root. Notice that all computations are simple additions and subtractions.

We previously saw that the reduced costs are easily computed as

$$\bar{c}(x, y) = c(x, y) - w(x) + w(y)$$

This computation consists solely of additions and subtractions also.

Since the entering variable is not currently basic, its arc is not part of the basis spanning tree. Any arc not in a spanning tree creates a unique cycle when added to the tree. To preserve conservation of flow at all nodes, the flows along every arc in this cycle must reflect any increase in the flow of the entering arc. Consider traversing the cycle in the direction of the entering arc. As the entering arc flow is increased, we must also increase the flow on any arc in the cycle that is oriented in the same direction and decrease the flow on any arc that is oriented in the opposite direction (see Fig. 5.14). Since the flow on arcs being decreased will eventually hit zero and they all must change by the same amount, the increase in the entering arc flow is limited by the minimum of the arc flows being decreased. The arc on which this occurs will leave the basis [or if there is a tie, one will leave the basis and the others will remain basic at a zero (degenerate) level].

Thus, we have a simple network interpretation of the familiar *minimum ratio rule* of linear programming. Pivoting simply consists of adjusting the flows on the arcs of the cycle by the limiting value to preserve conservation of flow. Again, we point out that all computations are simple additions and subtractions. Thus, the entire simplex algorithm can be implemented on a network using elementary arithmetic operations. [In practice, sophisticated data structures are used to streamline the operations. The reader is referred to the paper by Bradley et al. (1977) for an excellent example of how this is done.]

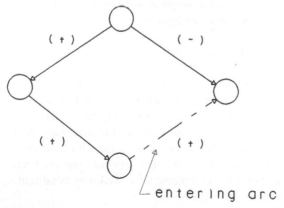

Fig. 5.14 Adjusting flows on cycles.

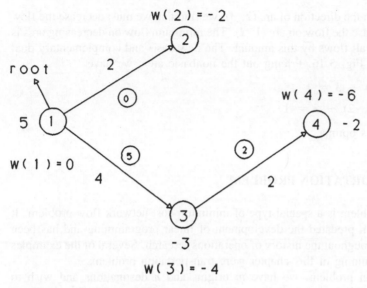

Fig. 5.15   Initial basic feasible solution.

EXAMPLE 6

We will illustrate the network simplex method using the example in Fig. 5.13. Figure 5.15 gives an initial basic feasible solution and the complementary dual solution. We price out the nonbasic arcs as follows:

$$\bar{c}(2, 3) = 1 - (-2) - 4 = -1$$
$$\bar{c}(2, 4) = 4 - (-2) - 6 = 0$$

Since $\bar{c}(2, 3) < 0$, we bring arc $(2, 3)$ into the basis. Adding this arc to the current basis spanning tree creates the following cycle:

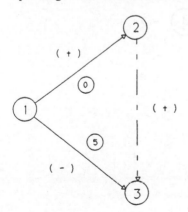

Traversing this cycle in the direction of arc (2, 3), we find that we must decrease the flow on arc (1, 3) and increase the flow on arc (1, 2). The minimum flow on decreasing arcs is 5; therefore we adjust all flows by this amount. The new basis and complementary dual solution are shown in Fig. 5.16. Pricing out the nonbasic arcs, we have

$$\bar{c}(1, 3) = 4 - 0 - 3 = 1$$
$$\bar{c}(2, 4) = 4 - (-2) - 5 = 1$$

Therefore, this basis is optimal.

5.5   THE TRANSPORTATION PROBLEM

The transportation problem is a special type of minimum-cost network flow problem. It has a rich history, as it predated the development of linear programming and has been studied extensively throughout the history of operations research. Several of the examples formulated at the beginning of this chapter were transportation problems.

In a transportation problem, we have m origins and n destinations and wish to transport some commodity from the origins to the destinations. Each origin x has a supply $s(x)$ and each destination y has a demand $d(y)$. A unit cost $c(x, y)$ is known for each origin-destination pair. We assume that the total supply equals the total demand. [If not, add a fictitious destination having demand equal to $\Sigma\ s(x) - \Sigma\ d(y)$.] We seek to determine the flow between origin x and destination y, $f(x, y)$, to minimize the total transportation cost.

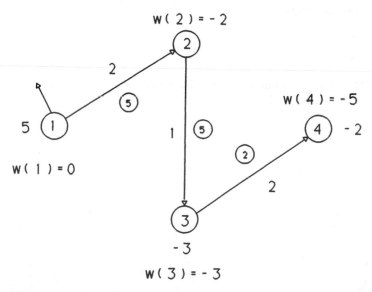

Fig. 5.16   Optimal basis.

As a linear program, the transportation problem traditionally is formulated as

$$\min \sum_x \sum_y c(x, y)f(x, y) \tag{14}$$

$$\sum_y f(x, y) = s(x), \text{ for all } x = 1, 2, \ldots, m \tag{15}$$

$$\sum_x f(x, y) = d(y), \text{ for all } y = 1, 2, \ldots, n \tag{16}$$

$$f(x, y) \geq 0, \text{ for all } x, y \tag{17}$$

This formulation is slightly different from the general minimum-cost network flow problem (6)–(8). If we stick to the convention that "flow out − flow in equals net supply," then multiplying constraints (16) by − 1 will put the transportation problem in the same form as (6)–(8). Hence, any algorithm presented in this chapter can be used to solve the transportation problem.

Any network flow problem on a graph $G = (X, A)$ can be converted to an equivalent transportation problem. This is done as follows. Let S be the set of source nodes and T be the set of sink nodes in G. Find the shortest path between each $s \in S$ and $t \in T$ using the unit arc costs as lengths. Let $c(s, t)$ be the length of the shortest path. Next, construct a bipartite graph using S as the set of sources and T as the set of sinks, joining each $s \in S$ and $t \in T$ with an arc. Assign the cost of each arc as $c(s, t)$. The supply of each source node and demand of each sink node are the same as in G. This network is the equivalent transportation problem. If $f(s, t)$ is the optimal flow on arc (s, t) in this transportation problem, sending $f(s, t)$ units of flow along the shortest path from s to t in G provides the optimal solution.

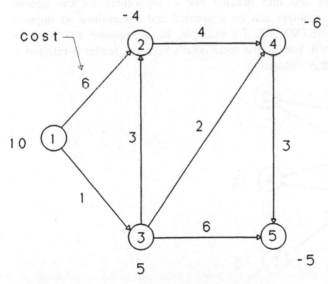

Fig. 5.17 A minimum-cost network flow problem.

To illustrate this, consider the network in Fig. 5.17. Using NETSOLVE, the shortest paths from source nodes 1 and 3 to the sink nodes are given below:

```
SHORTEST PATH TREE FROM 1

     NODE              DISTANCE      PREDECESSOR
     ----              --------      -----------
     1                    0.00       1
     2                    4.00       3
     3                    1.00       1
     4                    3.00       3
     5                    6.00       4

SHORTEST PATH TREE FROM 3

     NODE              DISTANCE      PREDECESSOR
     ----              --------      -----------
     1              INACCESSIBLE
     2                    3.00       3
     3                    0.00       3
     4                    2.00       3
     5                    5.00       4
```

Figure 5.18 shows the equivalent transportation problem that is constructed. We leave it as an exercise to show that the optimal cost of 59 is achieved by solving either problem.

Because of its special structure—such as having a bipartite network with all arcs directed out of supply nodes and into demand nodes—algorithms for the general minimum-cost network flow problem can be simplified and streamlined to improve computational performance. NETSOLVE, for example, has a separate algorithm for transportation problems. We will not discuss these details here; the reader is referred to Bazaraa et al. (1990) for further discussion.

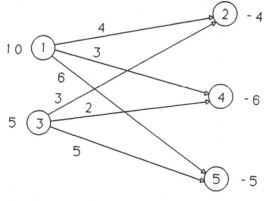

Fig. 5.18 Transportation problem formulation of Fig. 5.17.

5.6 FLOWS WITH GAINS

Previously, whenever a flow unit entered an arc, the same flow unit exited from the arc unchanged. Flow units were neither created nor destroyed as they traveled through an arc. In this section, we shall no longer assume that flow units are unchanged as they traverse an arc. Instead, we will allow the number of flow units traveling through an arc to increase or decrease. Specifically, we shall assume that if $f(x, y)$ flow units enter arc (x, y) at vertex x, then $k(x, y) f(x, y)$ flow units exit arc (x, y) at vertex y, for all arcs (x, y). Each flow unit traversing arc (x, y) can be regarded as being multiplied by $k(x, y)$. The quantity $k(x, y)$ is called the *gain* or *gain factor* of arc (x, y).

If $k(x, y) > 1$, then flow is *increased* along arc (x, y). If $k(x, y) = 1$, then flow remains *unchanged* along arc (x, y). If $0 < k(x, y) < 1$, then flow is *reduced* along arc (x, y). If $k(x, y) = 0$, then flow is *eradicated* by arc (x, y) and arc (x, y) can be regarded as a sink. In the discussion to follow we shall always assume that all arcs (x, y) with $k(x, y) = 0$ have been replaced by sinks. If $k(x, y) < 0$, then for each flow unit entering arc (x, y) at vertex x, $-k(x, y)$ flow units must arrive at vertex y; thus, arc (x, y) can be regarded as *creating a demand* for flow units.

The central problem of this section is the minimum-cost flow with gains problem. As you might expect, this problem is the problem of finding a minimum-cost way to dispatch a given number V of flow units from the source to the sink in a network in which the arcs have gain factors associated with them. Of course, if V flow units are discharged into the network at the source, then the number of flow units arriving at the sink need not equal V because of arc gains. (Note that in all previous flow problems, the number of units discharged into the network at the source always equaled the number that arrived at the sink.)

As before, arc (x, y) has associated with it an upper bound $u(x, y)$ and a lower bound $l(x, y)$ on the number of flow units that may enter it and a cost $c(x, y)$ for each unit that enters the arc. Again, let $f(x, y)$ denote the number of flow units that enter arc (x, y).

The minimum-cost flow with gains problem can be written as follows:
Minimize

$$\sum_{(x, y)} c(x, y) f(x, y) \tag{18}$$

such that

$$\sum_{y} f(x, y) - \sum_{y} k(y, x) f(y, x) = \begin{cases} V & \text{if } x = s \\ 0 & \text{if } x \neq s, x \neq t \end{cases} \tag{19}$$

$$l(x, y) \leq f(x, y) \leq u(x, y) \text{ for all } (x, y) \tag{20}$$

Expression (18) represents the total cost of the flow. Equation (19) states that the net flow out of vertex s must equal V and the net flow out of every other vertex, except t, must equal zero. Relation (20) states that each arc flow must lie within the upper and lower bounds on the arc's capacity.

The minimum-cost flow with gains problem, specified by relations (18)–(20), is a linear programming problem. Let $p(x)$ denote the dual variable corresponding to equation (19) for vertex x. Let $v_1(x, y)$ denote the dual variable corresponding to the upper capacity constraint in relation (20) for arc (x, y). Let $v_2(x, y)$ denote the dual variable corresponding to the lower capacity constraint in relation (20) for arc (x, y). With these definitions for the dual variables, the dual linear programming problem for the linear programming problem of relations (18)–(20) is
Maximize

$$Vp(s) - \sum_{(x,\ y)} u(x,\ y)\ v_1(x,\ y) + \sum_{(x,\ y)} l(x,\ y)v_2(x,\ y) \tag{21}$$

such that

$$p(x) - k(x,\ y)p(y) - v_1(x,\ y) + v_2(x,\ y) \le c(x,\ y) \text{ for all arcs } (x,\ y) \tag{22}$$

$$v_1(x,\ y) \ge 0 \text{ for all arcs } (x,\ y) \tag{23}$$

$$v_2(x,\ y) \ge 0 \text{ for all arcs } (x,\ y) \tag{24}$$

$$p(x) \text{ unrestricted (for all } x) \tag{25}$$

If the values for the dual variables $p(x)$, $x \in X$, are already given, then the remaining dual variables $v_1(x, y)$ and $v_2(x, y)$ must satisfy

$$-v_1(x,\ y) + v_2(x,\ y) \le c(x,\ y) - p(x) + k(x,\ y)p(y) \overset{\Delta}{=} w(x,\ y)$$
$$\text{for all arcs } (x,y) \tag{26}$$

For convenience, the right side of relation (26) is denoted by $w(x, y)$.

The dual objective function (21) is maximized if $v_1(x, y)$ and $v_2(x, y)$ are chosen as follows:
If

$$w(x,\ y) \ge 0$$

set

$$v_1(x,\ y) = 0, \qquad v_2(x,\ y) = w(x,\ y)$$

If

$$w(x,\ y) < 0$$

set

$$v_1(x,\ y) = -w(x,\ y), \quad v_2(x,\ y) = 0 \tag{27}$$

This follows since $v_1(x, y)$ and $v_2(x, y)$ appear in only one dual constraint (22).

Thus, the values of $v_1(x, y)$ and $v_2(x, y)$ for all arcs (x, y) are imputed by the values of the dual variables $p(x)$. Thus, we need only seek an optimal set of values for the vertex dual variables $p(x)$.

The complementary slackness conditions generated by this primal-dual pair of linear programming problems are

$$v_1(x, y) > 0 \Rightarrow f(x, y) = u(x, y) \qquad (28)$$

$$v_2(x, y) > 0 \Rightarrow f(x, y) = l(x, y) \qquad (29)$$

Since the values of $v_1(x, y)$ and $v_2(x, y)$ are determined by the value of $w(x, y)$, the complementary slackness conditions, (28) and (29), can be restated in terms of $w(x, y)$ as follows:

$$w(x, y) < 0 \Rightarrow f(x, y) = u(x, y) \qquad (30)$$

$$w(x, y) > 0 \Rightarrow f(x, y) = l(x, y) \qquad (31)$$

Hence, to solve the minimum-cost flow with gains problem, we need only find a set of feasible flow values $f(x, y)$ and a set of dual vertex variable values $p(x)$ such that the complementary slackness conditions, (30) and (31), are satisfied for all arcs (x, y).

Consider the cycle in Fig. 5.19 in isolation from the remainder of its graph. If one additional flow unit arrives at vertex x and enters arc (x, y), then during its passage across arc (x, y), it increases to two flow units at vertex y, and becomes two-thirds of a flow unit at vertex z, and finally becomes two flow units upon its return to vertex x. Thus, each additional flow unit that travels clockwise around this cycle becomes two flow units. This cycle is called a *clockwise generating cycle*. In general, a cycle C is called a clockwise generating cycle if

$$k_C \overset{\Delta}{=} \frac{\Pi_{(x, y)\text{forward arc}} \; k(x, y)}{\Pi_{(x, y)\text{backward arc}} \; k(x, y)} > 1 \qquad (32)$$

when cycle C is traversed in the clockwise direction. Similarly, we can define a *counter-clockwise generating cycle* as any cycle for which inequality (32) is satisfied when the cycle is traversed in the counterclockwise direction. The generating cycles (i.e., clockwise generating cycles and counterclockwise generating cycles) are important because they have the ability to create additional flow units.

Now, consider the cycle in Fig. 5.20. If an additional flow unit arrives at vertex x and enters arc (x, y) and traverses this cycle in the clockwise direction, then an additional ½

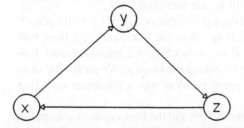

Fig. 5.19 Clockwise generating cycle; $k(x, y) = 2$, $k(y, z) = \frac{1}{3}$, $k(z, x) = 3$.

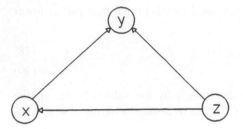

Fig. 5.20 Clockwise absorbing cycle; $k(x, y) = 1$, $k(z, y) = \frac{1}{2}$, $k(z, x) = \frac{1}{4}$.

flow unit will be required at vertex x. This follows from the fact that the unit traversing arc (x, y) creates an additional unit at vertex y. This additional flow unit arriving at vertex y requires the flow in arc (z, y) to increase by 2 units. Hence, because of the increase of 2 units in arc (z, y), an additional $\frac{1}{2}$ unit is required at vertex x. Thus, if $1\frac{1}{2}$ flow units arrive at vertex x, 1 flow unit can be sent clockwise around this cycle and the remaining $\frac{1}{2}$ flow unit can be used to supply the additional need at vertex x. Thus, this cycle absorbs flow units and is called a *clockwise absorbing cycle*. In general, a cycle C is called a clockwise absorbing cycle if

$$k_C = \frac{\Pi_{(x,y)\text{forward arc}} \, k(x, y)}{\Pi_{(x,y)\text{backward arc}} \, k(x, y)} < 1 \tag{33}$$

when the cycle is traversed in the clockwise direction. Similarly, we can define a *counterclockwise absorbing cycle* as any cycle for which inequality (33) is satisfied when the cycle is traversed in the counterclockwise direction.

Because a graph might contain generating and/or absorbing cycles, we cannot be certain that every flow unit leaving the source ultimately arrives at the sink (it could be absorbed by an absorbing cycle) or that every flow unit arriving at the sink initially came from the source (it could have been generated by a generating cycle).

Note that if k_C is calculated in the clockwise direction, then $1/k_C$ equals the value of k_C for the counterclockwise direction. This follows since changing the direction of travel around the cycle changes backward arcs to forward arcs and forward arcs to backward arcs. Hence, the numerator and denominator in k_C are switched.

If a graph contains no absorbing or generating cycles, that is, $k_C = 1$ for all cycles C for both clockwise and counterclockwise directions, then the minimum-cost flow with gains problem can be converted into a pure (i.e., without gains) minimum-cost flow problem which can be solved by any algorithm discussed previously. We shall now show how to convert the minimum-cost flow with gains problem into a minimum-cost flow problem.

Consider the conservation of flow requirements (19) and the flow capacity constraints (20) for the minimum-cost flow with gains problem. Let E denote the coefficient matrix of the left side of these constraints. If a row of E is multiplied by a nonzero constant, the set of feasible solutions to this set of simultaneous linear inequalities remains unchanged.

Suppose that the column of E corresponding to arc (x, y) is multiplied by a nonzero constant c. Then, in the original set of inequalities $cf(x, y)$ appears wherever $f(x, y)$ formerly appeared. If $l(x, y)$ and $u(x, y)$ are also multiplied by c, then $cf(x, y)$ can be replaced by a new variable $f'(x, y)$ and the constraints remain those of the original problem.

Can we find row and column multipliers such that after the redefinition of variables as defined above, the new flow problem will be a pure network flow problem? The following theorems due to Truemper (1976) provide an affirmative answer to this question:

Theorem 5.3. A flow with gains problem can be converted into a flow without gains problem if, and only if, there exists a vertex number $m(x)$ for each vertex x such that

$$\frac{m(x)k(x, y)}{m(y)} = 1 \tag{34}$$

for all arcs (x, y) in the graph.

Proof. Suppose there exist vertex numbers $m(x)$ such that equation (34) is satisfied for all arcs (x, y) in the graph. Multiply the conservation of flow equation (19) for vertex x by $1/m(x)$ for each vertex x. Let $f'(x, y) = f(x, y)/m(x)$ for all arcs (x, y). Rewrite equations (19) and (20) in terms of the f' variables. After doing this, the resulting equations in the f' variables are in the form of a minimum-cost flow (without gains) problem.

To prove the converse, suppose that the flow with gains problem can be converted into a flow without gains problem. Then there must exist a multiplier for each row in E that converts the corresponding equation into the constraint for a flow without gains problem. Let $1/m(x)$ denote the multiplier for constraint (19) for vertex x.

In a flow without gains problem, the coefficient of each variable $f'(x, y)$ in the conservation of flow constraint for vertex x is $+1$; consequently, $f'(x, y) = f(x, y)/m(x)$. In a flow without gains problem, the coefficient of each variable $f'(y, x)$ in the conservation of flow constraint for vertex x is -1: Consequently, $-f'(y, x) = -k(y, x)f(y, x)/m(x) = -k(y, x)m(y)f'(y, x)/m(x)$. Hence, $1 = k(y, x)m(y)/m(x)$ for all arcs (y, x), which is condition (34).   Q.E.D.

Thus, a flow with gains problem can be converted into a pure flow problem if there exist vertex numbers satisfying the conditions prescribed in (34). When do such vertex numbers exist? How can they be calculated? Read on.

Theorem 5.4. Vertex numbers satisfying Theorem 5.3 exist if, and only if, the graph contains no absorbing and no generating cycles, that is, $k_C = 1$ for all cycles C.

Proof. Suppose that a set of vertex numbers $m(x)$ exist such that $m(x, y) = k(x, y)m(x)/m(y) = 1$ for all arcs (x, y). Select any cycle C. It follows that

$$k_C = \frac{\Pi_{(x, y)\text{forward}} \ k(x, y)}{\Pi_{(x, y)\text{backward}} \ k(x, y)} = \frac{\Pi_{(x, y)\text{forward}} \ m(x)k(x, y)/m(y)}{\Pi_{(x, y)\text{backward}} \ m(x)k(x, y)/m(y)} \tag{35}$$

$$= \frac{\Pi_{(x, y)\text{forward}} \ m(x, y)}{\Pi_{(x, y)\text{backward}} \ m(x, y)} = 1$$

Hence, if all $m(x, y)$ equal 1, no absorbing or generating cycle can exist in the graph.

Conversely, suppose that no generating or absorbing cycle exists in the graph. Select any spanning tree T of the graph. Let $m(s) = 1$, and let $m(x, y) = 1$ for all arcs (x, y) in T. In the obvious way, fan out along the arcs of the tree calculating a value for $m(x)$ for each vertex x. These vertex numbers are uniquely determined.

Select any arc (x, y) not in T. This arc forms a unique cycle C with the arcs in T. By (35), it follows that $m(x, y) = 1$ since $m(i, j) = 1$ for all arcs (i, j) in T. Thus, $m(x, y) = 1$ for all arcs (x, y) not in T.   Q.E.D.

Thus, if a network possesses no absorbing or generating cycles, then the minimum-cost flow with gains problem can be transformed into a minimum-cost flow without gains problem, which can be solved optimally by a minimum-cost flow algorithm. This transformation is effected by multiplying the conservation of flow constraint (19) for vertex x by $1/m(x)$, where $m(x)$ is determined as in the proof of Theorem 5.4.

If the network possesses absorbing or generating cycles, the flow problem becomes more complicated. In this section, we shall present an algorithm for solving the minimum-cost flow with gains problem. This algorithm is called the *flow with gains algorithm*. It consists of three basic steps:

Step 1 (Initialization Step). This step finds flow values $f(x, y)$ and dual-variable vertex number values $p(x)$ such that the complementary slackness conditions (30), (31) are satisfied and the flow satisfies all feasibility requirements, (19), (20), except that possibly less than V flow units are dispatched from the source. Hence, step 1 finds a solution that satisfies all primal, dual, and complementary slackness requirements, except the required output at the source. (As will be seen later, step 1 will be expedited by constructing a graph that is a slight enlargement of the original graph.)

Step 2 (Flow Increment Step). In this step, the flow out of the source is increased while retaining all the primal, dual, and complementary slackness requirements described in step 1.

Step 3 (Dual-Variable Change Step). This step describes how to change the values of the dual variables $p(x)$ so that even more flow units can be discharged out of the source into the network.

The algorithm consists of performing step 1 to determine an initial flow. Step 2 is performed next to send more flow units out of the source, while maintaining all complementary slackness conditions. Once step 2 cannot discharge any more flow units from the source into the network while maintaining all complementary slackness conditions, step 3 is performed to change the dual-variable values so that even more flow units can be discharged into the network. The algorithm returns to step 2. Once step 2 ceases to discharge any more flow units into the network, step 3 is repeated, etc., until all V flow units have been discharged into the network from the source. If no feasible flow that discharges V units into the network exists, the algorithm will discover this during an iteration of step 3 and terminate.

With all this as motivation, we are now prepared to state formally the *flow with gains algorithm*.

Flow with Gains Algorithm

Step 1 (Initialization). This step shows how to find a set of flow values $f(x, y)$ and dual-variable values $p(x)$ for a network equivalent to our original network such that all feasibility conditions (19), (20) and complementary slackness conditions (30), (31) are satisfied except that the net flow out of source s is less than or equal to V, the required flow out of source s.

Arbitrarily select any values for the dual variables $p(x)$, $x \in X$. Consult the complementary slackness conditions

$$w(x, x) = c(x, y) - p(x) + k(x, y)p(y) < 0 \Rightarrow f(x, y) = u(x, y) \tag{36}$$

$$w(x, y) = c(x, y) - p(x) + k(x, y)p(y) > 0 \Rightarrow f(x, y) = l(x, y) \tag{37}$$

to determine which values for $f(x, y)$ are compatible with the complementary slackness conditions for each arc (x, y). Select any compatible value for (x, y) for each arc (x, y).

Next, determine the net surplus flow $V(x)$ out of each vertex x, where

$$V(s) = \sum_y f(x, y) - \sum_y k(y, x)f(y, x) - V$$

$$V(x) = \sum_y f(x, y) - \sum_y k(y, x)f(y, x) \tag{38}$$

If $V(x) = 0$ for all x, then the current solutions is an optimal solution since it is a feasible flow that satisfies all complementary slackness conditions for the chosen values of $p(x)$. Of course, we usually are not so fortunate.

Create a new network from the original network by adding a vertex s and an arc (s, x) from s to each vertex x with $V(x) \neq 0$. Let $u(s, x) = |V(x)|$ for each arc (s, x). If

$$V(x) < 0$$

let

$$k(s, x) = +1$$

If

$$V(x) > 0$$

let

$$k(s, x) = -1$$

Lastly, let $c(s, x) = 0$ for all arcs (s, x).

Let the flow in each newly created arc (s, x) equal zero and let the flow in all other arcs remain as before. Clearly, flow is not conserved at all intermediate vertices of the new network since at least one intermediate vertex x has $V(x) \neq 0$. However, if enough flow units can be sent out of vertex s and absorbed into the new network so that each arc (s, x) is saturated, then this new (nonfeasible) flow in the new network will correspond to a feasible flow in the original network. This follows since arc (s, x) will supply vertex x with exactly enough flow units to counteract the net surplus out of vertex x.

Moreover, a minimum-cost flow in the new network that saturates all arcs (s, x) corresponds to a minimum-cost flow in the original network since all $c(s, x) = 0$. Hence, we can find an optimal solution for the original network by searching for an optimal solution for the new network.

The linear programming formulation of this minimum-cost flow with gains problem on the new network is like the linear programming formulation of the minimum-cost flow with gains problem on the original network, relations (18)–(20), except that now vertex s is the source vertex and the right side of the conservation of flow equation (19) for vertex x should now be $V(x)$ instead of zero. The complementary slackness conditions for the new network are the same as for the original network.

If we retain the same values as above for $p(x)$ for each vertex x in the original network and let $p(S)$ be any large negative number, the complementary slackness conditions will be satisfied for the new network.

Hence, the original selection of flow values and dual-variable values generates a set of values for the flow variables and dual variables for the new network that satisfy feasibility and complementary slackness conditions. All is done, except more units must be sent into the network from vertex s. Proceed to step 2.

Step 2 (Flow Augmentation). This step shows how to increase the flow out of the source S as much as possible without losing any complementary slackness or changing any dual-variable value.

For each arc (x, y), determine whether the complementary slackness conditions (30), (31) will allow $f(x, y)$ to be increased or decreased. Let I denote the set of increasable arcs, and let R denote the set of reducible arcs. Only the arcs in I and R will be considered in this step.

To determine whether additional flow units can be dispatched from s using only arcs in I and R, we shall successively grow a tree of arcs rooted at the source. These arcs will be called labeled arcs. If a vertex is reached by this tree, then additional flow units can be dispatched from the source to this vertex via the arcs of the tree. Whenever an arc incident to a vertex x is added to the tree (labeled), vertex x will receive a label $f(x)$ denoting the number of flow units that would arrive at vertex x for each flow unit sent to it from the source.

If the sink is labeled, then a path from the source to the sink over which flow can be increased (called a *flow-augmenting* path) has been discovered. As much flow as possible is dispatched from the source along this flow-augmenting path to the sink.

If an arc that forms a cycle with the labeled arcs can also be labeled, then we must check if this cycle can absorb flow units sent out from the source. If so, we dispatch as much flow as possible from the source to be absorbed by this cycle. If this cycle cannot absorb flow units dispatched from the source, then some of the labeled arcs in this cycle are unlabeled, and the labeling process continues.

Now for the details of the labeling procedures: Initially all arcs are unlabeled and all vertices are unlabeled. Label the source s with the label $f(s) = +1$. Perform the following labeling operations:

Unlabeled arc (x, y) can be labeled if one of the following conditions is met:

1. x is labeled, $f(x) > 0$, and $(x, y) \in I$.
2. x is labeled, $f(x) < 0$, and $(x, y) \in R$.
3. y is labeled, $f(y) > 0$, and either $(x, y) \in R$, $k(x, y) > 0$ or $(x, y) \in I$, $k(x, y) < 0$.
4. y is labeled, $f(y) < 0$, and either $(x, y) \in I$, $k(x, y) > 0$ or $(x, y) \in R$, $k(x, y) < 0$.

If arc (x, y) is labeled because of item 1 or 2, then we say that arc (x, y) was labeled from vertex x. In this case, label y with $f(y) - f(x)k(x, y)$. If arc (x, y) is labeled because of item 3 or 4, then we say that arc (x, y) was labeled from vertex y. In this case, label x with $f(x) = f(y)/k(x, y)$. Note that a vertex label $f(x)$ denotes the number of flow units that arrive at vertex x for each unit discharged at s along the path of labeled arcs that generated the label $f(x)$.

Continue this labeling procedure until

1. The sink is labeled.
2. Some vertex receives two distinct labels.
3. No more labeling is possible.

If item 3 occurs, go to step 3. If item 1 occurs, then a unique path of arcs from the source to the sink has been labeled. This is a flow-augmenting path from the source to the sink. Send as many units as possible along this path, and return to the beginning of step 2.

If item 2 occurs, then some vertex x has received two distinct labels, $f_1(x)$ and $f_2(x)$. Without loss of generality, suppose that $f_1(x) < f_2(x)$. Each label corresponds to a distinct (possibly partially overlapping) flow-augmenting path from s to x. If the labels $f_1(x)$ and

$f_2(x)$ are of opposite signs, then a flow unit sent from s along the flow-augmenting path corresponding to the label $f_1(x)$ results in a demand of $f_1(x)$ flow units at vertex x. A flow unit sent from s along the flow-augmenting path corresponding to the label $f_2(x)$ results in $f_2(x)$ flow units arriving at x. Consequently, these two flow-augmenting paths contain an absorbing cycle. Send as many flow units as possible from s along these two paths so that the flow units required at x by the first path are supplied by the flow units arriving at x along the second path. Return to the beginning of step 2.

If $f_1(x)$ and $f_2(x)$ have the same sign, we cannot be certain that we have discovered an absorbing cycle that can accept flow units discharged at s. Without loss of generality, suppose that $|f_1(x)| < |f_2(x)|$. If the label $f_2(x)$ is descendent from the label $f_1(x)$ [i.e., involves $f_1(x)$ in its computation], then a cycle C has been labeled from vertex x back to itself. Since $|f_1(x)| < |f_2(x)|$, it follows that $k_C = f_1(x)/f_2(x) < 1$, and hence cycle C is an absorbing cycle. Send as many flow units as possible from the source to be absorbed by cycle C, and return to the beginning of step 2.

If label $f_2(x)$ is not descendent from label $f_1(x)$, we have discovered two alternate flow-augmenting paths from s to x but have not discovered any absorbing cycle. In this case, erase label $f_2(x)$ and all arc labels and node labels descendent from label $f_2(x)$. Continue the labeling procedure.

Step 3 (Dual-Variable Change). This step shows how to define new values $p'(x)$ for the dual variables $p(x)$ that will satisfy the complementary slackness conditions. After the values of the dual variables are redefined, return to step 2 to try to discharge more flow units from the source.

The details of redefining the dual variables are as follows: Let T denote the tree of arcs labeled during the last iteration of step 2. Let L denote the set of vertices labeled during the last iteration of step 2. Set L consists of the endpoints of all arcs in T.

For each vertex x, define a variable $q(x)$ as follows:

$$q(x) = \begin{cases} \dfrac{1}{f(x)} & \text{if } x \in L \\[2mm] 0 & \text{if } x \notin L \end{cases} \tag{39}$$

Define

$$\delta = \min \left\{ \frac{w(x, y)}{q(x) - k(x, y)q(y)} \right\} \tag{40}$$

where the minimization is taken over all arcs (x, y) for which the numerator and denominator have the same sign. If the denominator is zero for all arcs, then stop the algorithm since no feasible solution exists for the minimum-cost flow with gains problem. Otherwise, define the new values $p'(x)$ for the dual variables as follows:

$$p'(x) = p(x) + \delta q(x) \quad \text{(for all } x) \tag{40}$$

Return to step 2.

Proof of the Flow with Gains Algorithm. We must show that the flow with gains algorithm terminates with an optimal solution or that no solution exists.

Step 1 transforms the original minimum-cost flow with gains problem to a new network in which each source arc must be saturated. A minimum-cost solution in the new network in which every source arc is saturated is equivalent to an optimal solution to the minimum-cost flow with gains problem in the original network. Hence, we need only show that (a) the algorithm finds an optimal solution to the minimum-cost flow with gains solution in the new network that saturates all source arcs, or (b) no feasible solution exists.

(a) To show that the algorithm terminates with an optimal solution in the new network, we need only show that complementary slackness is maintained at all times, since the algorithm can only terminate with a flow that saturates all source arcs or else the algorithm claims no feasible solution exists.

The algorithm is initialized in step 1 with a solution in which all complementary slackness conditions are satisfied. Moreover, all complementary slackness conditions are maintained throughout all flow changes in step 2 since the flow changes in an arc are never permitted to violate any complementary slackness conditions.

Are all complementary slackness conditions maintained by step 3? Consider any arc $(x, y) \in A$. After the vertex number change of step 3, the new value $w'(x, y)$ of $w(x, y)$ becomes

$$
\begin{aligned}
c(x, y) &- p'(x) + k(x, y)p'(y) \\
&- c(x, y) - p(x) - \delta q(x) + k(x, y)[p(y) + \delta q(y)] \\
&= w(x, y) + \delta[-q(x) + k(x, y)q(y)]
\end{aligned}
\tag{41}
$$

Case 1. If arc $(x, y) \in T$, then $q(x) = k(x, y)q(y)$ and $w'(x, y) = w(x, y) + \delta[-q(x) + q(x)] = w(x, y)$. Since the dual-variable change of step 3 causes no change in $w(x, y)$, it follows that the complementary slackness conditions remain satisfied for arc (x, y).

Case 2. If arc $(x, y) \notin T$, $x \notin L$, $y \notin L$, then $q(x) = q(y) = 0$ and $w'(x, y) = w(x, y) + 0\delta = w(x, y)$. Since the dual-variable change of step 3 causes no change in $w(x, y)$, it follows that the complementary slackness conditions remain satisfied for arc (x, y).

Three more cases remain:

$(x, y) \notin T, x \in L, y \notin L$

$(x, y) \notin T, x \notin L, y \in L$

$(x, y) \notin T, x \in L, y \in L$

It is left to the reader to verify that for each of these cases the dual-variable change maintains all complementary slackness conditions for arc (x, y). These verifications follow the same lines as above.

(b) Suppose that the algorithm terminates in step 3 with δ undefined. We shall now show that no feasible flow exists.

Additional flow from the source can reach only the vertices in set L without violating any complementary slackness conditions imposed by the current choice of the dual variables $p(x)$. Can the dual-variable values be changed so that set L can be increased by at least one member, thereby bringing us closer to labeling the sink or coloring an absorbing cycle? Disregarding the complementary slackness conditions, there are five ways an arc (x, y) can be colored:

1. $x \in L$, $y \notin L$, $f(x) > 0$, $f(x, y) < u(x, y)$
2. $x \in L$, $y \notin L$, $f(x) < 0$, $f(x, y) > l(x, y)$
3. $x \notin L$, $y \in L$, $f(y) > 0$, $f(x, y) > l(x, y)$
4. $x \notin L$, $y \in L$, $f(y) < 0$, $f(x, y) < u(x, y)$
5. $x \in L$, $y \in L$, $(x, y) \notin T$, and (x, y) forms an absorbing cycle with the labeled arcs.

Careful examination of each of these cases shows that for each case $w(x, y)$ and $q(x) - k(x, y)q(y)$ must have the same sign. If δ is undefined, it follows that there is no arc (x, y) that is unlabeled and for which $w(x, y)$ and $q(x) - k(x, y)q(y)$ have the same sign. Hence, no arc is a candidate for labeling and flow units cannot reach any further out of the source. Hence, no feasible flow exists.   Q.E.D.

In the flow with gains algorithm and its proof as stated here, we avoided any mention of the number of steps required before termination. In fact, the algorithm as stated need not terminate in a finite number of steps. We shall now describe how to modify the flow with gains algorithm to ensure that it will terminate finitely. First some definitions are needed.

A flow with gains is called *canonical* if no connected component of the set of intermediate arcs (arcs with flow strictly between their lower and upper capacity) contains any of the following configurations:

1. A cycle C with $k_C = 1$
2. Two distinct but possibly overlapping cycles
3. The source and a cycle
4. The sink and a cycle
5. The source and the sink

Notice that if a connected component of intermediate arcs contains any of the above configurations, the flow in the arcs in the component can be altered so that

(a)   One or more arcs in the component become nonintermediate.
(b)   The net flow out of each vertex (except possibly the source and the sink) remains unchanged.

For example, if the intermediate arcs contain two cycles that overlap or are connected to each other via intermediate arcs (configuration 2), flow could be increased around one of the cycles and the surplus (or deficiency) created by the flow change around this cycle could be absorbed into the second cycle. It is left to the reader to verify that this flow change can always be made so that the net flow out of the source remains unchanged or is increased.

Hence, if a flow with gains is not canonical, it contains one of the configurations listed above and a flow change can be made to decrease the number of intermediate arcs without decreasing the net flow out of the source. If the new flow resulting from this flow change is also not canonical, this process can be repeated. After successive repetitions of this process, a canonical flow will be generated without decreasing the net flow out of the source. Hence, any noncanonical flow can be converted into a canonical flow without decreasing the net flow out of the source.

Theorem 5.5.   There exists only a finite number of distinct canonical flows in graph G.

Proof.   The number of distinct possibilities for the set M of intermediate arcs is finite since M is a subset of the finite arc set of graph G. The flow in each arc $(x, y) \notin M$ is either $l(x, y)$ or $u(x, y)$. Hence, there is only a finite number of possible sets of values for the flows of the arcs not in M.

For a given set M and a given set of flow values for the arcs not in M, we shall show that if an infinite number of distinct flows exists, then none of them are canonical. This will prove the theorem.

Some of the flow values of the arcs in M are uniquely determined by the conservation of flow requirements. Let M' denote the subset of M whose flow values are not uniquely determined by the conservation of flow requirements. Set M', if it is not empty, must contain a cycle or a path from s to t. (Otherwise, the flow values of all members of M' could be successively determined starting at some vertex incident to only one member of M'.)

If set M' contains a path from s to t, then none of the possible flows is canonical. If set M' contains a cycle C with $k_C = 1$, then none of the possible flows is canonical. Hence, if set M' is not empty, it must contain a cycle C with $k_C \neq 1$.

If infinitely many different flows are possible, then infinitely many different flows are possible around cycle C. Since $k_C \neq 1$, it follows that the surplus or deficiency caused by each of these flows in cycle C must be compensated for either by flow from the source or flow into the sink or by flow in some other cycle. Hence, the set I' must contain configuration 2, 3, or 4. Hence, none of the possible flows is canonical.   Q.E.D.

With these results in mind, we are now able to state formally the *finite termination modification* for the flow with gains algorithm.

Finite Termination Modification

Convert the initial flow generated by step 1 of the flow with gains algorithm to a canonical flow without decreasing the net flow out of the source. In step 2 of the flow with gains algorithm, never label a nonintermediate arc if you can instead label an intermediate arc.

Proof. First we shall show that if step 2 is started with a canonical flow, then the flow resulting from a flow change in step 2 is also a canonical flow. Suppose that the new flow is not canonical. Then the intermediate arcs of the new flow would contain at least one of the five configurations. Denote the (intermediate) arcs in this configuration by J. Let J' denote the arcs in J that were intermediate in the original flow. Let J'' denote the arcs in J that were not intermediate in the original flow. When the original flow was changed to the new flow, step 2 must have labeled the arcs in J''. This implies that, if the finite termination modification was used, no arcs in J' were available for labeling each time an arc in J'' was labeled. However, careful examination of each of the five possible configurations will show that this nonavailability of the arcs in J' for labeling is impossible. This contradicts the instructions of the finite termination modification.

Hence, step 2 of the algorithm will produce only a canonical flow from a canonical flow. Since each successive flow produced by step 2 increases the net flow out of the source, and since there is by Theorem 5.5 only a finite number of distinct canonical flows, step 2 will never generate the same canonical flow twice. Hence, step 2 cannot be performed an infinite number of times, and the only possible way for the flow with gains algorithm not to terminate finitely is for step 3 to be performed an infinite number of times.

If step 3 were performed an infinite number of times, then there must be an infinite number of successive iterations of step 3 between two flow changes since there is only a finite number of flow changes. However, after each iteration of step 3 at least one more arc is labeled, or δ is undefined and the algorithm terminates with the discovery that no feasible flow exists. Since there is only a finite number of arcs in the graph, step 3 can be performed only a finite number of successive times on the same flow. Hence, the algorithm must terminate after a finite number of steps.   Q.E.D.

Other Flows with Gains

So far, we have considered only the problem of finding a minimum-cost flow with gains that discharges a given number of flow units from the source. Suppose that instead we were interested in finding a minimum-cost flow that conveyed a given number of flow units to the sink. Could the flow with gains algorithm be used to solve this problem? The answer is yes. This is accomplished by using the inverse graph.

The *inverse graph* G^{-1} of graph $G = (X, A)$ is defined as the graph with vertex set X and arc set

$$A^{-1} = \{(y, x): (x, y) \in A\}$$

Let the cost of arc $(y, x) \in A^{-1}$ equal $c(x, y)/k(x, y)$. Let the gain of arc $(x, y) \in A^{-1}$ equal $1/k(x, y)$. Let $l(y, x) = l(x, y)/|k(y, x)|$ and $u(y, x) = u(x, y)/|k(x, y)|$. Note that the inverse of G^{-1} is G.

Given any flow $f(x, y)$ for graph G, we can define the inverse flow $f^{-1}(y, x)$, $(y, x) \in A^{-1}$, as follows:

If $k(y, x) > 0$, then $f^{-1}(y, x) = f(x, y)k(x, y)$.

If $k(y, x) < 0$, let $u(x, y)$ units be supplied at vertex x,

let $k(x, y)u(x, y)$ units be supplied at vertex y,

and let $f^{-1}(y, x) = [u(x, y) - f(x, y)](-k(x, y))$.

Let $b(x)$ denote the total number of flow units supplied at vertex x.

Why this particular definition of the inverse flow? If in graph G, $f(x, y)$ flow units are removed from vertex x by arc (x, y) and $k(x, y)f(x, y)$ flow units are delivered to vertex y by arc (x, y), then in the inverse graph, $f(x, y)$ flow units are delivered to vertex x and $k(x, y)f(x, y)$ flow units are removed from vertex y by arc (y, x). (Verify this for yourself using the definitions of the inverse flow.) Thus, the inverse flow has exactly the opposite effect to the original flow. Moreover, the source s in graph G becomes a sink in graph G^{-1}, and the sink t in graph G becomes a source in graph G^{-1}.

Suppose a flow $f(x, y)$ in graph G satisfies all arc capacities and supplies $b(x)$ flow units at each vertex x. Then the corresponding flow $f^{-1}(y, x)$ in graph G^{-1} satisfies all arc capacities and supplies $-b(x)$ flow units at each vertex x. Moreover, these two flows have the same total cost.

Hence, a minimum-cost flow of V units from the source in graph G^{-1} corresponds to a minimum-cost flow of V units into the sink in graph G. Thus the problem of finding a minimum-cost flow with gains that delivers a given number of flow units into the sink can be solved by using the flow with gains algorithm on the inverse graph to dispatch the same number of flow units from the source of the inverse graph.

When the flow with gains algorithm is used to find a minimum-cost flow that discharges V flow units from the source, there is no way to predict how many flow units ultimately arrive at the sink, since these V flow units may either arrive at the sink or be absorbed into the network. Is there any way to ensure that the algorithm generates a minimum-cost flow with gains in which at least W flow units arrive at the sink? Yes, simply append the original network with a new vertex T and an arc (t, T) from the original sink to vertex T. Let $l(t, T) = W, u(t, T) = \infty, k(t, T) = 1$, and $c(t, T) = 0$. Let vertex T be the sink in the appended network. Every feasible flow in the appended network corresponds to a flow in the original network that sends at least W flow units into the sink. Of course, this requirement that at least W flow units enter the sink may possibly increase the total cost of the minimum-cost flow.

Suppose that no requirement is placed on how many flow units arrive at the sink. There is possibly more than one minimum-cost flow that discharges V flow units from the source. Can we find the minimum-cost flow that discharges V flow units from the source and simultaneously sends as many flow units as possible into the sink? This flow is generated by appending the network, as above, with a vertex T and an arc (t, T). Let $l(t, T) = 0, u(t, T) = \infty, k(t, T) = 1$, and $c(t, T) = \epsilon$, a very small negative number.

If ϵ is very, very small, the minimum-cost flow generated by the flow with gains algorithm for the appended network corresponds to a minimum-cost flow for the original network. If there is more than one minimum-cost flow, the flow with gains algorithm when applied to the appended network will choose the minimum-cost flow that sends as many flow units as possible into the sink in order to incur the small negative cost ϵ for each unit flowing along arc (t, T).

Conversely, to minimize the number of flow units arriving at the sink, let ϵ be a very, very small positive number.

For the special case when all arc traverse costs are zero, see Grinold (1973).

Flows with Gains and the Simplex Method

The flow with gains problem can also be solved using the network simplex method; however, the structure of the problem is significantly different and several modifications must be made. Consider the problem shown in Fig. 5.21. The linear programming formulation is

$$\min c(1, 2) f(1, 2) + c(1, 3)f(1, 3) + c(2, 3)f(2, 3) + c(2, 4)f(2, 4)$$
$$+ c(3, 4) f(3, 4)$$

$$f(1, 2) + f(1, 3) = V$$

$$-k(1, 2)f(1, 2) + f(2, 3) + f(2, 4) = 0$$

$$-k(1, 3)f(1, 3) - k(2, 3)f(2, 3) + f(3, 4) = 0$$

A basis for a flow with gains problem will consist of $n - 1$ arcs just as in an ordinary network flow problem. For example, consider the set of arcs $(1, 2)$, $(2, 3)$, and $(3, 4)$. These arcs form a spanning tree in the network. The basis matrix is

$$\begin{bmatrix} -k(1, 2) & 1 & 1 \\ 0 & -k(2, 3) & 1 \\ 0 & 0 & -k(2, 4) \end{bmatrix}$$

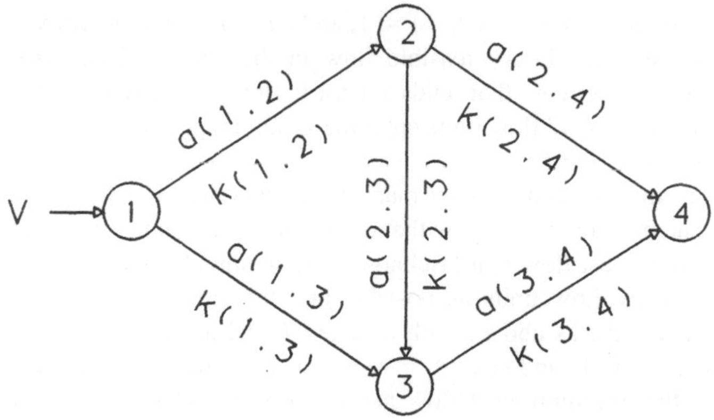

Fig. 5.21 Flow with gains network.

Notice that the determinant of this matrix is the product of the diagonal elements and consequently is nonzero. Thus, this matrix represents a basis.

Now, however, consider the set of arcs (2, 3), (2, 4), and (3, 4) which form a cycle in the network. The matrix corresponding to these arcs is

$$\begin{bmatrix} 1 & 1 & 0 \\ -k(2, 3) & 0 & 1 \\ 0 & -k(2, 4) & -k(3, 4) \end{bmatrix}$$

The determinant of this matrix is $k(2, 4) - k(2, 3)k(3, 4)$. If this value is nonzero, then these arcs represent a basis. Thus, for a flow with gains problem, a basis may include a cycle. In general, a basis consists of a forest in which each tree may contain a cycle.

The simplex method can be performed on the network in a manner similar to that for ordinary network flow problems. We shall not present the details here. The interested reader is referred to Brown and McBride (1984) for details of the computational aspects of the algorithm.

APPENDIX: USING NETSOLVE TO FIND MINIMUM-COST FLOWS

NETSOLVE has two algorithms for minimum-cost network flow problems: MINFLOW, for general networks; and TRANS, for transportation problems.

The minimum-cost flow algorithm determines a minimum-cost (or maximum-value) flow from supply nodes to demand nodes in a capacitated, directed network. Supply nodes have a positive value in the SUPPLY data field. Demand nodes have a negative value and transshipment nodes have a zero value. Total supply must be greater than or equal to total demand (that is, the sum of the SUPPLY values over all nodes must be nonnegative). Edge flows must lie between a nonnegative lower bound and an upper bound specified by the LOWER and UPPER data fields, respectively. The total cost (or value) of a flow is the sum over all edges of the edge flow multiplied by the unit edge weight given in the COST data field.

The algorithm is executed by entering the command line

    MINFLOW

or

    MINFLOW MIN

An alternate form with the keyword MAX instead of MIN is used to find a maximum-value flow in the network. If an optimal solution is found, the first line of output consists of the cost (value) for the optimal flow. If the terminal is ON, a listing of the edge data fields and edge flows for edges having nonzero flow is produced. If an optimal solution does not exist, then an indication is given of whether the problem is infeasible or the

Edge data:

FROM	TO	LB	UB	COST
II	M1	0.00	999999.00	0.00
M1	M2	0.00	20.00	0.25
M1	P1	14.00	18.00	0.00
M2	M3	0.00	20.00	0.25
M2	P2	14.00	18.00	0.00
M3	M4	0.00	20.00	0.25
M3	P3	14.00	18.00	0.00
M4	FI	0.00	999999.00	0.00
M4	P4	14.00	18.00	0.00
P1	R1	0.00	999999.00	0.00
P1	R2	0.00	999999.00	0.00
P1	R3	0.00	999999.00	0.00
P1	R4	0.00	999999.00	0.00
P2	R2	0.00	999999.00	0.00
P2	R3	0.00	999999.00	0.00
P2	R4	0.00	999999.00	0.00
P3	R3	0.00	999999.00	0.00
P3	R4	0.00	999999.00	0.00
P4	R4	0.00	999999.00	0.00
S	M1	0.00	999999.00	6.00
S	M2	0.00	999999.00	7.00
S	M3	0.00	999999.00	8.00
S	M4	0.00	999999.00	9.00

Node data:

NAME	SUPPLY
FI	-8.00
II	10.00
M1	0.00
M2	0.00
M3	0.00
M4	0.00
P1	0.00
P2	0.00
P3	0.00
P4	0.00
R1	-12.00
R2	-19.00
R3	-15.00
R4	-20.00
S	99.00

Fig. 5.22 NETSOLVE data for Fig. 5.4.   **Fig. 5.23** Optimal solution to Fig. 5.4.

MINIMUM COST FLOW PROBLEM: MINIMUM COST IS   455.00

FROM	TO	LB	FLOW	UB	COST
M4	FI	0.00	8.00	999999.00	0.00
II	M1	0.00	10.00	999999.00	0.00
S	M1	0.00	28.00	999999.00	6.00
M1	M2	0.00	20.00	20.00	0.25
S	M2	0.00	18.00	999999.00	7.00
M2	M3	0.00	20.00	20.00	0.25
S	M3	0.00	16.00	999999.00	8.00
M3	M4	0.00	20.00	20.00	0.25
S	M4	0.00	2.00	999999.00	9.00
M1	P1	14.00	18.00	18.00	0.00
M2	P2	14.00	18.00	18.00	0.00
M3	P3	14.00	16.00	18.00	0.00
M4	P4	14.00	14.00	18.00	0.00
P1	R1	0.00	12.00	999999.00	0.00
P1	R2	0.00	6.00	999999.00	0.00
P2	R2	0.00	13.00	999999.00	0.00
P2	R3	0.00	5.00	999999.00	0.00
P3	R3	0.00	10.00	999999.00	0.00
P3	R4	0.00	6.00	999999.00	0.00
P4	R4	0.00	14.00	999999.00	0.00

Do you want sensitivity analysis? (Y/N): y

SENSITIVITY ANALYSIS FOR EDGE COSTS

NODE	DUAL
FI	9.00
II	6.00
M1	6.00
M2	7.00
M3	8.00
M4	9.00
P1	8.00
P2	8.00
P3	8.00
P4	8.00
R1	8.00
R2	8.00
R3	8.00
R4	8.00
S	0.00

FROM	TO	EDGE STATE	REDUCED COST	COST RANGE LOWER	CURRENT	UPPER
II	M1	BASIC	0.00	-999999.00	0.00	999999.00
M1	M2	UPPER	-0.75	-999999.00	0.25	1.00
M1	P1	UPPER	-2.00	-999999.00	0.00	2.00
M2	M3	UPPER	-0.75	-999999.00	0.25	1.00
M2	P2	UPPER	-1.00	-999999.00	0.00	1.00
M3	M4	UPPER	-0.75	-999999.00	0.25	1.00
M3	P3	BASIC	0.00	-1.00	0.00	1.00
M4	FI	BASIC	0.00	-999999.00	0.00	999999.00
M4	P4	LOWER	1.00	-1.00	0.00	999999.00
P1	R1	BASIC	0.00	-999999.00	0.00	999999.00
P1	R2	BASIC	0.00	-999999.00	0.00	0.00
P1	R3	LOWER	0.00	0.00	0.00	999999.00
P1	R4	LOWER	0.00	0.00	0.00	999999.00
P2	R2	BASIC	0.00	0.00	0.00	999999.00
P2	R3	BASIC	0.00	-999999.00	0.00	0.00
P2	R4	LOWER	0.00	0.00	0.00	999999.00
P3	R3	BASIC	0.00	0.00	0.00	999999.00
P3	R4	BASIC	0.00	-999999.00	0.00	0.00
P4	R4	BASIC	0.00	-1.00	0.00	999999.00
S	M1	BASIC	0.00	-999999.00	6.00	6.75
S	M2	BASIC	0.00	6.25	7.00	7.75
S	M3	BASIC	0.00	7.25	8.00	8.75
S	M4	BASIC	0.00	8.25	9.00	999999.00

solution is unbounded. Additional information is produced if the terminal is ON. In the case of an infeasible formulation, a list of nodes where flow conservation cannot be attained is displayed. For an unbounded solution, a list of edges having a negative unit cost (positive unit value) and infinite flow is exhibited for a minimum-cost (maximum-value) problem.

In addition, the user will be asked whether a sensitivity analysis should be conducted on the problem. This analysis produces dual variables for each node and reduced costs for each edge. Also, tolerance intervals are produced for each relevant edge cost, such that varying the individual edge cost within that range will not change the optimality of the current solution.

We will illustrate the use of this algorithm for the production scheduling example in Fig. 5.4. Figure 5.22 shows a listing of the node and edge data for the network created with NETSOLVE. The optimal solution is given in Fig. 5.23.

The transportation algorithm determines a minimum-cost (or maximum-value) shipping pattern from supply nodes to demand nodes in a directed network. All edges must be directed from supply nodes (having a positive value in the SUPPLY data field) to demand nodes (having a negative SUPPLY value). All nodes must be either supply or demand nodes, and total supply must meet or exceed total demand. Flows between supply and demand nodes must lie between a nonnegative lower bound (specified by the LOWER data field) and an upper bound (specified by the UPPER data field).

The algorithm is executed by issuing the command line

        TRANS

or

        TRANS MIN

A maximum-value shipment can be determined by using the keyword MAX in the second form above. If an optimal solution is determined, the cost (value) of the associated shipping pattern is displayed. If the terminal is ON, an additional listing of all edges used in the optimal solution is displayed along with the associated edge data fields. If no optimal solution exists, then the system reports an infeasible formulation and (if the terminal is ON) a list is produced of nodes at which supply is insufficient or demand is excessive.

In addition, the user will be asked whether a sensitivity analysis should be conducted on the problem. This analysis produces dual variables for each node and reduced costs for each edge. Also, tolerance intervals are produced for each relevant edge cost, such that varying the individual edge cost within that range will not change the optimality of the current solution.

We will illustrate the use of the transportation algorithm using the distribution network in Fig. 5.2. Figure 5.24 shows a list of the node and edge data for the network created with NETSOLVE. Figure 5.25 provides the optimal solution.

Edge data:

FROM	TO	COST
A	1	5.00
A	2	4.00
A	3	10.00
B	1	4.00
B	3	5.00
C	2	6.00
C	3	6.00

Node data:

NAME	SUPPLY
1	-8.00
2	-8.00
3	-6.00
A	10.00
B	5.00
C	7.00

Fig. 5.24 NETSOLVE data for transportation problem.

EXERCISES

1. Airplanes are built at a factory at certain rates. Current contracts require delivery at specified future dates:

Month	Maximum Production	Contracted Amounts	Unit Cost ($1,000,000)
1	25	10	1.08
2	35	15	1.11
3	30	25	1.10
4	10	20	1.13

Storage and insurance costs are $15,000 per plane per month if they are produced but delivered in a later month. Formulate this problem as a minimum-cost network flow problem in which flow variables represent production in a given month or inventory holding from one month to the next.

TRANSPORTATION PROBLEM: MINIMUM COST IS   105.00

FROM	TO	LOWER	FLOW	UPPER	COST
A	1	0.00	3.00	999999.00	5.00
B	1	0.00	5.00	999999.00	4.00
A	2	0.00	7.00	999999.00	4.00
C	2	0.00	1.00	999999.00	6.00
C	3	0.00	6.00	999999.00	6.00

Do you want sensitivity analysis? (Y/N): y

SENSITIVITY ANALYSIS FOR EDGE COSTS

NODE	DUAL
1	7.00
2	6.00
3	6.00
A	2.00
B	3.00
C	0.00

FROM	TO	REDUCED COST	EDGE STATE	COST RANGE LOWER	CURRENT	UPPER
A	1	0.00	BASIC	3.00	5.00	999999.00
A	2	0.00	BASIC	4.00	4.00	6.00
A	3	6.00	LOWER	4.00	10.00	999999.00
B	1	0.00	BASIC	-999999.00	4.00	6.00
B	3	2.00	LOWER	3.00	5.00	999999.00
C	2	0.00	LOWER	4.00	6.00	999999.00
C	3	0.00	BASIC	-999999.00	6.00	8.00

Fig. 5.25 Optimal solution to transportation problem.

2. Formulate the problem in Exercise 1 as a transportation problem. (Hint: define flow variables to be the number of units produced in month x and delivered in month y, where $y \geq x$.)

3. Consider the network shown in Fig. 5.26. Write down the linear programming formulation, its dual problem, and the complementary slackness conditions.

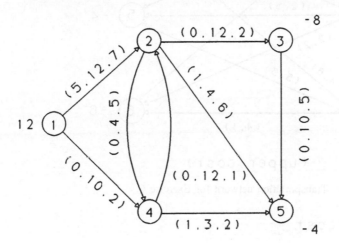

(lower . upper . cost)

Fig. 5.26 Network for Exercise 3.

4. In the network in Fig. 5.26, apply transformations to remove lower and upper bounds on arc flows and construct the resulting network.

5. Write the linear program and dual of the transformed problem in Exercise 4. Is the dual equivalent to the one developed in Exercise 3? If so, show the equivalence.

6. Convert the transportation problem in Fig. 5.27 to one without upper bounds on arc flows.

7. Consider the network in Fig. 5.28. Suppose we have the basis spanning tree consisting of arcs (1, 2), (2, 4), (4, 3), (4, 5), and (4, 6). Put the matrix corresponding to this basis in triangular form.

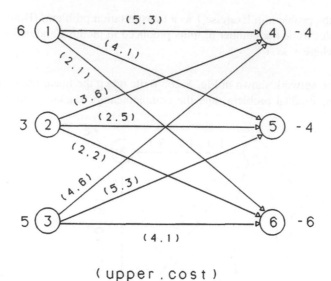

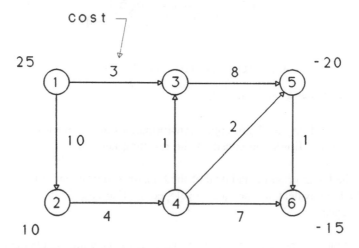

Fig. 5.27 Transportation network for Exercise 6.

Fig. 5.28 Network for Exercise 7.

8. Solve the problem in Fig. 5.28 using the negative-cycle algorithm.

9. Solve the problem in Fig. 5.28 using the successive shortest-path algorithm.

10. Solve the problem in Fig. 5.28 using the network simplex method.

11. Consider the transportation problem in Fig. 5.29. Solve this using the negative-cycle algorithm.

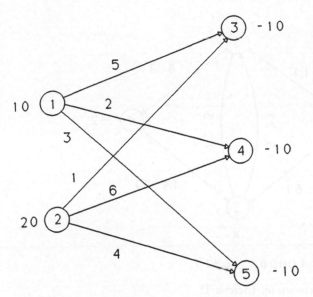

Fig. 5.29 Transportation problem for Exercise 11.

12. Solve the transportation problem in Fig. 5.29 using the successive shortest-path algorithm.

13. Solve the transportation problem in Fig. 5.29 using the network simplex method.

14. Convert the network in Fig. 5.28 to a transportation problem and solve it using the network simplex method.

15. Extend the negative-cycle and successive shortest-path algorithms to problems with finite upper bounds on the arcs (and zero lower bounds). Apply your modified algorithms to the network in Fig. 5.30.

16. Suppose that we have already calculated a minimum-cost flow for a very large graph. When reviewing the final results, we notice that: (1) an arc was omitted from the graph; (2) the capacity of arc (x, y) was overstated by 5 units; (3) the cost of arc (m, n) was understated by 2 units.

    (a) How can we determine which of these errors individually had an effect on the optimal solution?

    (b) Is it possible to correct individually for each of these errors without scrapping all our results? Is it possible to correct for all these errors without scrapping all our results?

17. Explain how a flow with gains network can be used to model the following situations:

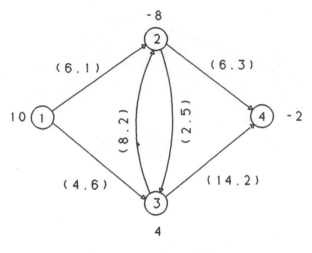

(upper, cost)

Fig. 5.30 Network for Exercise 15.

(a) A plant nursery must dispatch a shipment of plants to its distributors. On some of these routes, the plants experience a high fatality rate due to improper climatic conditions. On other routes, with more suitable climatic conditions, the plants generally experience significant growth during shipment.

(b) A corporate financial analyst must decide how to ration the corporation's investment funds between competing investments. How can the analyst develop a network with gains to aid the investment decision problem?

18. List all absorbing and generating cycles in the following graph. All lower arc capacities equal zero, all upper arc capacities equal one. The arc gain and arc cost are indicated next to each arc.

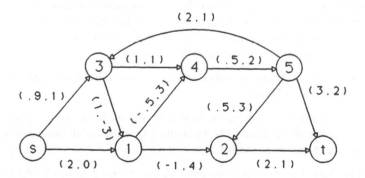

19. Use the flow with gains algorithm to find a minimum-cost way to discharge 5 units from the source in the graph in Exercise 18.

20. How can the flow with gains algorithm be used to find generating cycles that can transmit flow units to the sink?

21. Prove in detail that the complementary slackness conditions remain satisfied after the vertex number change in step 3 of the flow with gains algorithm.

REFERENCES

Barr, R. S., and J. S. Turner, 1981. Microdata File Merging Through Large Scale Network Technology. In *Network Models and Associated Applications* (eds. D. Klingman and J. M. Mulvey), North-Holland, Amsterdam, pp. 1–22.

Bazaraa, M. S., J. J. Jarvis, and H. Sherali, 1990. *Linear Programming and Network Flows, 2nd ed.*, Wiley & Sons, New York.

Bowers, M. R., and J. P. Jarvis, 1989. A Hierarchical Production Planning and Scheduling Model. Report 248, University of Tennessee, College of Business Administration, Knoxville.

Bradley, G. H., G. G. Brown, and G. W. Graves, 1977. Design and Implementation of Large Scale Primal Transshipment Algorithms, *Man. Sci.*, *24*, pp. 1–34.

Brown, G., and R. D. McBride, 1984. Solving Generalized Networks, *Man. Sci.*, *30*, pp. 1497–1523.

Busaker, R. G., and P. J. Gowen, 1961. A Procedure for Determining a Family of Minimal-Cost Network Flow Patterns, O.R.O. Technical Report No. 15, Operational Research Office, Johns Hopkins University, Baltimore.

Cooper, R. B., 1972. *Introduction to Queueing Theory*, Macmillan, New York.

Grinold, R. C., 1973. Calculating Maximal Flows in a Network with Positive Gains, *Oper. Res.*, *21*, pp. 528–541.

Iri, M. A New Method of Solving Transportation-Network Problems, *Journal of the Operations Research Society of Japan*, *3*, pp. 27–87.

Jewell, W. S., 1958. Optimal Flow Through Networks, Interim Technical Report No. 8, Operations Research Center, MIT, Cambridge, Massachusetts.

Jewell, W. S., 1962. Optimal Flow Through a Network with Gains, *Oper. Res.*, *10*, pp. 476–499.

Johnson, E. L., 1966. Networks and Basic Solutions, *Oper. Res.*, *14*, pp. 619–623.

Klein, M., 1967. A Primal Method for Minimal Cost Flows, *Man. Sci.*, *14*, pp. 205–220.

Larson, R. C., and A. R. Odoni, 1981. *Urban Operations Research*, Prentice-Hall, Englewood Cliffs, New Jersey.

Maurras, J.F., 1972. Optimization of the Flow Through Networks with Gains, *Math. Program.*, *3*, pp. 135–144.

Minieka, E. T., 1972. Optimal Flow in a Network with Gains, *INFOR*, *10*, pp. 171–178.

Truemper, K., 1973, Optimum Flow in Networks with Positive Gains, Ph.D. Thesis, Case Western Reserve University, Cleveland.

Truemper, K., 1976. An Efficient Scaling Procedure for Gains Networks, *Networks*, *6*, pp. 151–160.

6
MAXIMUM-FLOW ALGORITHMS

6.1 INTRODUCTION AND EXAMPLES

Flow problems are characterized by conservation of flow equations and possibly capacity constraints.In Chapter 5 we studied algorithms for minimum–cost flow problems. Several other types of common problems arise in flow situations. For example, one might wish to maximize the amount of flow from a source to a sink, maximize the flow through a network over multiple time periods, or to find the quickest way to deliver a shipment through a system. In this chapter we will present algorithms for maximum-flow and related network problems. First, we present some examples.

Maximum Flow Through a Pipeline Network

Figure 6.1 shows a pipeline network. Oil must be shipped from the refinery (the source, designated as node s) to a storage facility (the sink, designated as node t) along arcs of the network. Each arc has a capacity which limits the amount of flow along that arc. For example, at most 6 units can flow from node s to node 1. The problem is to determine what the maximum flow rate is between the refinery and the storage facility with the restriction that no arc capacity can be exceeded.

We may formulate this problem as a linear program by letting v denote the total flow from s to t. Applying the conservation of flow principle at the nodes of the network, we have

$$f(s, 1) + f(s, 2) - v = 0$$
$$f(1, 3) + f(1, 2) - f(s, 1) = 0$$
$$f(2, 3) + f(2, 4) - f(s, 2) - f(1, 2) = 0$$
$$f(3, t) - f(4, 3) - f(2, 3) - f(1, 3) = 0$$
$$f(4, 3) + f(4, t) - f(2, 4) = 0$$
$$v - f(3, t) - f(4, t) = 0$$

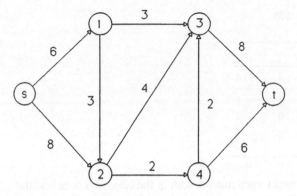

Fig. 6.1 A pipeline network.

Arc capacities are represented by the following constraints:

$$f(s, 1) \leq 6$$
$$f(s, 2) \leq 8$$
$$f(1, 2) \leq 3$$
$$f(1, 3) \leq 3$$
$$f(2, 3) \leq 4$$
$$f(2, 4) \leq 2$$
$$f(4, 3) \leq 2$$
$$f(3, t) \leq 8$$
$$f(4, t) \leq 6$$

The objective is to maximize the total flow, or

$$\max v$$

Even though we have formulated this problem as a linear program, we shall see that there are much more efficient solution procedures.

Site Selection (Balinski, 1970)

Consider the problem of selecting sites for an electronic message transmission system. Any number of sites can be chosen from a finite set of potential locations. We know the costs c_i of establishing site i and the revenue r_{ij} generated between sites i and j, if they are both selected. What configuration of sites will maximize the net profit?

Suppose that we have four potential sites, 1, 2, 3, and 4. Cost and revenue data are given in Table 6.1. Figure 6.2 shows a network representation. From this network, we

Table 6.1 Data for the Site Selection Problem

i	c_i	j:	1	2	3	4
				r_{ij}		
1	6		—	7	5	1
2	5			—	6	2
3	4				—	1
4	5					—

wish to choose sites forming a subnetwork such that the sum of the edge revenues less the node costs is as large as possible. Clearly, one way to solve this problem would be to enumerate all possible combinations of subsets of sites. For moderate-size problems, this becomes computationally infeasible. We will convert this problem into a "logical network" that addresses our objective. Note that if edge (i, j) is included yielding the revenue r_{ij}, then sites i and j must be included at a cost of $c_i + c_j$. We represent each edge in the physical network by the logical network component shown in Fig. 6.3. We then create a source node s and an arc from s to each revenue node ij with a capacity of r_{ij}, and a sink node t with an arc from each cost node i to it with a capacity of c_i. All other arcs have infinite capacity. The complete network for this example is shown in Fig. 6.4.

The solution to this problem involves finding a particular type of cutset. Recall that cutsets were introduced in Chapter 1. We shall be interested in simple cutsets that separate a network into two components such that s and t are in different components. Such a cutset will be called an *s–t cut*, or simply a *cut*. The sum of the capacities of all arcs directed from the component containing s to the component containing t is called the *capacity of the cut*. We seek a cut having minimum capacity. For the network in Fig. 6.4, any potential minimum-capacity cut must include the cost nodes associated with the chosen

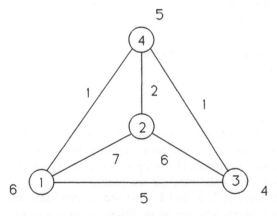

Fig. 6.2 Network representation of the site selection problem.

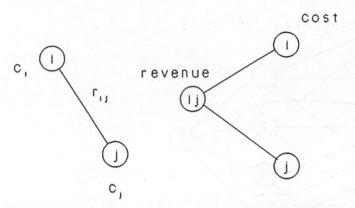

Fig. 6.3 Construction of a logical network for site selection.

revenue nodes since the connecting arcs have infinite capacity. For instance, in Fig. 6.5, if $R = \{23, 24, 34\}$ lies on the same side of a minimum-capacity cut as s, then $C = \{2, 3, 4\}$ must also lie on the same side of the cut. Observe that the capacity of the cut is

$$r_{12} + r_{13} + r_{14} + c_2 + c_3 + c_4$$
$$= \sum r_{ij} - (r_{23} + r_{24} + r_{34} - c_2 - c_3 - c_4)$$
$$= \text{constant} - (\text{revenue} - \text{cost})$$

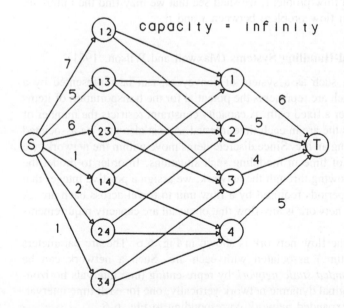

Fig. 6.4 Logical network for site selection example.

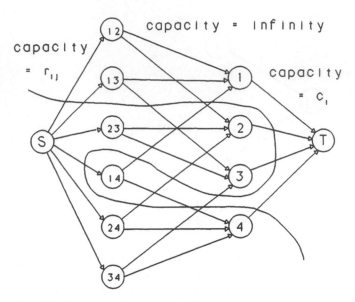

Fig. 6.5 Example of a cut in the site selection network.

Thus to maximize the net profit, we need only minimize the capacity of a cut between S and T. Although this is not a flow problem, we shall see that we may find the minimum cut by solving the maximum flow problem between s and t!

Dynamic Flows in Material-Handling Systems (Maxwell and Wilson, 1981)

A material-handling system, such as a system of conveyors, can be represented by a directed network in which each arc represents the potential for the transportation of items from one point to another over a fixed path. A capacity constraint restricts the number of items which may flow across the arc in each time interval. A cost also may be associated with each item that flows along an arc. Since discrete items move within the network, we must include the dimension of time in modeling such situations. In order to model the dynamic behavior of items flowing through the network, we assign a positive integer that denotes the number of time periods required by a flow unit to travel across each arc. A *dynamic flow* from s to t in a network is any flow that obeys all arc capacity requirements at all times.

An example of a dynamic flow network is shown in Fig. 6.6. The arc parameters represent (capacity, travel time) associated with each arc. Such a network can be transformed into a *time-expanded static network* by representing time intervals horizontally and the nodes of the original dynamic network vertically, one for each time interval. Figure 6.7 shows the time-expanded network corresponding to Fig. 6.6.

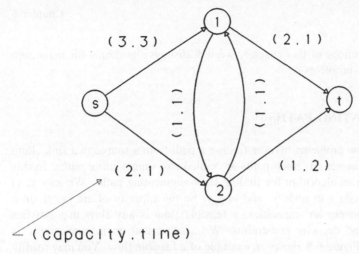

Fig. 6.6 Dynamic network flow example.

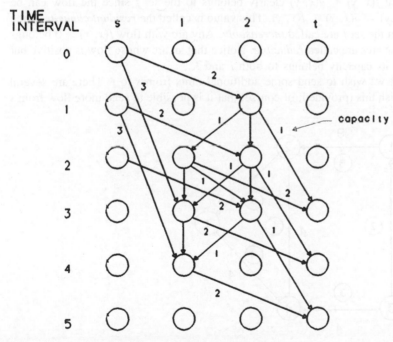

Fig. 6.7 Time-expanded static network for Fig. 6.6.

In the remaining sections of this chapter, we will discuss algorithms for maximum flow and dynamic flow problems.

6.2 FLOW-AUGMENTING PATHS

Many algorithms for flow problems involve finding a path from a source to a sink along which the flow can be increased. Such paths are called *flow-augmenting paths*. In this section we will develop an algorithm for finding flow-augmenting paths. We let $f(x, y)$ denote the flow from node x to node y, and $u(x, y)$ be the capacity of arc (x, y). In a network with finite, nonzero arc capacities, a feasible flow is any flow that satisfies conservation of flow and capacity constraints. We assume that an infinite supply is available at the source. Figure 6.8 shows an example of a feasible flow. You may readily verify that no arc capacity is exceeded and that conservation of flow holds at each node. The net flow from s to t is 2 units.

The arcs in a flow network can be put into two categories:

I, the set of arcs whose flow can be increased
R, the set of arcs whose flow can be reduced

Any arc such that $f(x, y) < u(x, y)$ clearly belongs to the set I since the flow can be increased by $i(x, y) = u(x, y) - f(x, y)$. This value is called the *residual capacity* of the arc (x, y). Arcs in the set I are called *increasable*. Any arc with flow $f(x, y) > 0$ belongs to the set R. These arcs are called *reducible*. Notice that an arc whose flow is positive but strictly less than its capacity belongs to both I and R.

Suppose that we wish to send some additional units from s to t. There are several ways to accomplish this (provided, of course, that it is possible to send more flow from s

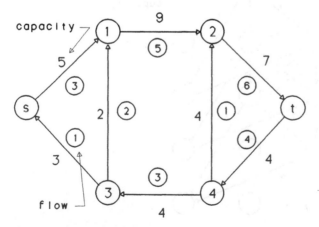

Fig. 6.8 A feasible flow.

to t). First, you might find a directed path P from s to t consisting entirely of increasable arcs. The path $s - 1 - 2 - t$ in Fig. 6.8 is one example. How many additional flow units can be sent along this path? Since $i(x, y)$ is the maximum amount by which the flow in arc (x, y) can be increased, then at most

$$\min_{(x, y) \in P} \{i(x, y)\}$$

additional flow units can be sent from s to t. For the path $s - 1 - 2 - t$ in Fig. 6.8 we can send

$$\min\{i(s, 1), i(1, 2), i(2, t)\} = \min\{2, 4, 1\} = 1$$

additional unit of flow. This would increase the net flow from s to t by one unit and preserve conservation of flow. Therefore, this path is a flow-augmenting path.

Similarly, if we could find a directed path from t to s consisting entirely of decreasable arcs, the flow could be decreased in each arc of this path, resulting in a greater net flow from s to t. In Fig. 6.8, all arcs on the path $P = t - 4 - 3 - s$ have a nonzero flow; thus each is reducible. The maximum amount by which the flow along the entire path can be decreased is

$$\min_{(x, y) \in P} \{f(x, y)\}$$

Thus, for the path $t - 4 - 3 - s$, we see that the net flow from s to t can be increased by one unit since

$$\min\{f(4, t), f(3, 4), f(s, 3)\} = \min\{4, 3, 1\} = 1$$

This path would also be called a flow-augmenting path. Observe that flow is conserved at all nodes along the path.

A third way to increase the net flow from s to t is to combine the two ideas described above. Let P be any path from s to t with the following properties:

1. If arc (x, y) is on P and is directed from s to t (called a *forward arc*), then (x, y) is a member of I.
2. If arc (x, y) is on P and is directed from t to s (called a *backward arc*), then (x, y) is a member of R.

For example, consider the path $s - 1 - 3 - 4 - t$ in Fig. 6.8. Arc $(s, 1)$ is a forward arc and is a member of I. Arcs $(3, 1)$, $(4, 3)$, and $(t, 4)$ are backward arcs and are members of R. If each forward arc belongs to I and each backward arc belongs to R, then additional flow can be sent from s to t along the path by increasing the flow on the forward arcs and decreasing the flow along the backward arcs. The maximum amount of additional flow that can be sent along such a path from s to t is the minimum of the following two quantities:

$\min\{i(x, y): (x, y) \text{ is a forward arc}\}$

$\min\{f(x, y): (x, y) \text{ is a backward arc}\}$

Thus, for the path $s - 1 - 3 - 4 - t$, the net increase in flow from s to t is the minimum of

$\min\{i(s, 1)\} = \min\{2\} = 2$

and

$\min\{f(3, 1), f(4, 3), f(t, 4)\} = \min\{2, 3, 4\} = 2$

or 2 units. We send these 2 units from s to t by increasing the flow along all forward arcs by 2 units and decreasing the flow along all backward arcs by 2 units. You can verify that conservation of flow is maintained since if a forward arc and a backward arc are incident to a node, both must be directed into the node. Since the flow on the forward arc is increased and the flow on the backward arc is decreased, the net change in flow at the node is zero.

In any of these three situations, the amount by which the net flow from s to t can be increased is called the *maximum flow augmentation of the path*. Algorithms that use flow-augmenting paths often use a related network, called a *residual network*, for finding flow-augmenting paths more conveniently. The residual network for any feasible flow is constructed as follows. If arc (x, y) is a member of I, then construct an arc in the residual network from x to y and label it with a residual capacity $r(x, y) = u(x, y) - f(x, y)$. If arc (x, y) is a member of R, then construct an arc in the residual network from y to x, labeling it with a residual capacity $r(y, x) = f(x, y)$. The residual network corresponding to the flow in Fig. 6.8 is shown in Fig. 6.9.

The advantage of using residual networks is that every flow-augmenting path from s to t in the original network will be a *directed path* from s to t in the residual network.

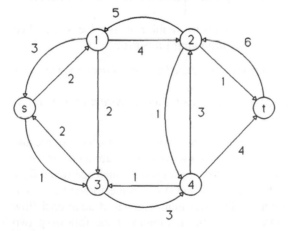

Fig. 6.9   Residual network for flow in Fig. 6.8.

Thus, we could use the path-finding algorithm described in Chapter 2 directly on the residual network to locate flow-augmenting paths. By keeping track of the minimum value of the residual capacity along the path, we maintain current information about the maximum flow augmentation along the path. These ideas are incorporated in the following *flow-augmenting algorithm*. The essential idea of this algorithm is to grow from the source s an arborescence along which additional units of flow can be sent from s. If t is included in this arborescence, then the unique path from s to t will be a flow-augmenting path. If t cannot be included in the arborescence, then no flow-augmenting path exists. Nodes are labeled with two values: $e(x)$ denotes the maximum amount of additional flow that can be sent from s to x; $p(x)$ is the predecessor node of x in the arborescence. The predecessor labels are used to find the actual path from s to t, if it exists, just as in shortest-path algorithms studied in Chapter 4. The value of $e(t)$ will represent the maximum flow augmentation of the path.

Flow-Augmenting Algorithm

Step 1. Construct the residual network corresponding to the current feasible flow. Label node s with $e(s) = \infty$ and $p(s) = 0$. All other nodes are initially unlabeled. All arcs are unmarked.

Step 2. Select a labeled node x which has not yet been considered. If none exist, then stop; no flow-augmenting path from s to t exists.

Step 3. If (x, y) is a member of the forward star of x, then label node y with $e(y) = \min\{e(x), r(x, y)\}$ and $p(y) = x$, if y is currently unlabeled. Mark arc (x, y). If t is labeled, then stop; a flow-augmenting path from s to t has been found. Otherwise, return to step 2.

Proof of the Flow-Augmenting Algorithm. To prove that the algorithm locates a flow-augmenting path if one exists, we must demonstrate three facts:

1. If t is labeled by the algorithm, then there does exist a flow-augmenting path from s to t.
2. If t cannot be labeled by the algorithm, then there does not exist any flow-augmenting path from s to t.
3. The algorithm terminates after a finite number of steps.

Proof of 1. When node y is labeled, arc $(p(y), y)$ is marked. The algorithm will never mark an arc if both of its endpoints are labeled. Thus the marked arcs will never form a cycle. Since only the forward star is scanned and node s is labeled initially, the marked arcs must form an arborescence that contains node s. Therefore, if vertex y is labeled by the algorithm, there must exist a unique directed path of marked arcs from s to y. Specifically, if t is labeled, there exists a unique directed path of marked arcs from s to t. This path must be a flow-augmenting path by construction of the residual network.

Proof of 2. If there exists a flow-augmenting path P from s to t, then there exists at least one flow-augmenting path from s to each node in P. Thus, each node in P can be labeled by the algorithm, and hence node t must also be labeled. Conversely, if t cannot be labeled by the algorithm, then no flow-augmenting path from s to t exists.

Proof of 3. The algorithm must terminate after a finite number of labelings since only a finite number of nodes and arcs can be labeled or marked, and no node or arc can be labeled or marked more than once.

EXAMPLE 1

We shall apply the flow-augmenting algorithm to the network in Fig. 6.8. The residual network has already been constructed in Fig. 6.9, so we label node s with $e(s) = \infty$, $p(s) = 0$ and begin with step 2.

    Step 2. Let $x = s$.
    Step 3. Label node 1 with $e(1) = \min\{e(s), r(s, 1)\} = \min\{\infty, 2\} = 2$, and $p(1) = s$.
        Label node 3 with $e(3) = \min\{e(s), r(s, 3)\} = \min\{\infty, 1\} = 1$, and $p(3) = s$.
    Step 2. Let $x = 1$.
    Step 3. Label node 2 with $e(2) = \min\{e(1), r(1, 2)\} = \min\{2, 4\} = 2$, and $p(2) = 1$.
        Do not attempt to label node 3 since it is already labeled.
    Step 2. Select either node 3 or 2. We will select node 2.
    Step 3. Label node 4 with $e(4) = \min\{e(2), r(2, 4)\} = \min\{2, 1\} = 1$ and $p(4) = 2$.
        Label node t with $e(t) = \min\{e(2), r(2, t)\} = \min\{2, 1\} = 1$, and $p(t) = 2$.
        Since t is labeled, stop.

The arborescence grown by the algorithm is shown in Fig. 6.10, and the flow-augmenting path from s to t is found by tracing the predecessor labels back from t: $p(t) = 2$; $p(2) = 1$; $p(1) = s$. Thus the flow-augmenting path found by the algorithm is the path

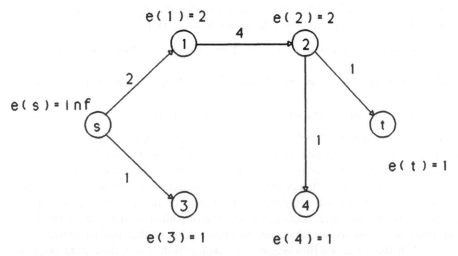

Fig. 6.10 Arborescence grown by the flow-augmenting algorithm.

$s-1-2-t$ with maximum flow augmentation $e(t) = 1$. Since all arcs along this path are forward arcs, we can increase the flow along arcs by 1 unit.

As can be seen from Example 1, the flow-augmenting algorithm does not specify completely which vertex is to be labeled next, nor does it necessarily find the flow-augmenting path with the largest augmentation. The best decision at each step is a matter of research and often depends on the larger problem in which the flow-augmenting algorithm is being used.

6.3 MAXIMUM-FLOW ALGORITHM

The pipeline problem described in Section 6.1 is an example of a maximum-flow problem in which we are interested in determining the largest possible amount of flow that can be sent from a source s to a sink t. In that example, we formulated the maximum-flow problem as a linear program. This example can be generalized. The maximum-flow problem on a network $G = (X, A)$ can be stated as

$$\max v$$

$$\sum_{y \in X} f(x, y) - \sum_{y \in X} f(y, x) = 0 \quad \text{for all } x \neq s, t \tag{1}$$

$$\sum_{y \in X} f(s, y) - \sum_{y \in X} f(y, s) = v \tag{2}$$

$$\sum_{y \in X} f(x, t) - \sum_{y \in X} f(t, x) = v \tag{3}$$

$$0 \leq f(x, y) \leq u(x, y) \quad \text{for all } (x, y) \in A \tag{4}$$

However, using the simplex method to solve this problem is like killing a mouse with a cannon. A more elegant and far more intuitive approach is available. This is the Ford and Fulkerson (1962) maximum-flow algorithm.

The idea underlying the maximum-flow algorithm is quite simple: Start with any flow from s to t and look for a flow-augmenting path using the flow-augmenting algorithm. If a flow-augmenting path from s to t is found, then send as many flow units as possible along this path. Then, start again to look for another flow-augmenting path, etc. If no flow-augmenting path is found, then stop because the current flow from s to t is shown to be a maximum flow from s to t.

With these ideas in mind, we can state formally the maximum-flow algorithm for finding a maximum flow from s to t in a network.

Maximum-Flow Algorithm

Step 1. Let s denote the source vertex, and let t denote the sink vertex. Select any initial flow from s to t, i.e., any set of values for $f(x, y)$ that satisfy relations (1)–

(4). If no such initial flow is known, use as the initial flow $f(x, y) = 0$ for all (x, y).

Step 2. Construct the residual network relative to the current flow.

Step 3. Perform the flow-augmenting algorithm. If no flow-augmenting path is discovered by the flow-augmenting algorithm, stop; the current flow is a maximum flow. Otherwise, make the maximum possible flow augmentation along the flow-augmenting path discovered by the flow-augmenting algorithm. Return to step 2.

Proof of the Maximum-Flow Algorithm: To show that the maximum-flow algorithm constructs a maximum flow from s to t, we must show that

1. The algorithm constructs a flow.
2. This flow is a maximum flow.
3. The algorithm terminates after a finite number of steps.

Proof of 1. To show that the algorithm constructs a flow, we need only note that the algorithm is initialized in step 1 with a flow and that during all steps, the algorithm maintains a flow since each flow augmentation maintains a flow. Hence, the algorithm must terminate with a flow.

Proof of 2. In section 6.1 we defined a $(s\text{--}t)$ cut. In Fig. 6.11, arcs $(s, 1)$ and $(s, 2)$ form a cut since their removal would disconnect the graph into two components. This cut is a simple cut since neither arc is a cut by itself.

Consider any simple cut that separates the source into one component and the sink into another component. Let the component containing the source be denoted by X_s and let the component containing the sink be denoted by X_t. Each arc in this cut must either (a) have its head in X_s and its tail in X_t, or (b) have its head in X_t and its tail in X_s. The sum of the capacities of the arcs of type (b) is called the *capacity of the cut*. Since the cut

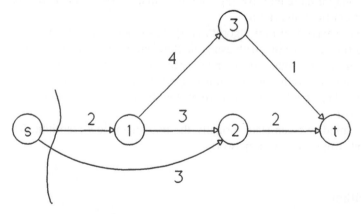

Fig. 6.11 Example of a cut.

separates s from t, clearly it is impossible to send more flow units from s to t than the capacity of this cut, since every flow unit must cross the cut. In general, the maximum flow possible from s to t must surely be less than or equal to the smallest cut capacity of a cut separating s from t. Simply stated,

max flow $\leq$ min cut

In order to demonstrate that the maximum-flow algorithm has indeed produced a maximum flow, we need only produce a cut separating s from t whose capacity equals the value of the terminal flow produced by the algorithm. This is done as follows.

The algorithm terminates when no flow-augmenting path can be found from s to t. After the final application of the flow-augmenting algorithm, node t could not be labeled. Consider the set of all arcs that upon termination of the last application of the flow-augmenting algorithm have one endpoint labeled and the other endpoint unlabeled. Call the set of labeled nodes N and the set of unlabeled nodes N'. Clearly $N \cup N' = X$. Since N and N' represent a partition of the node set X, the arcs joining these two sets is a cut. Since s is labeled and t is not labeled, this cut separates s from t (see Fig. 6.11). The capacity of this cut is defined as the sum of the capacities of the arcs with labeled tails and unlabeled heads, that is, the arcs directed from N to N'.

Upon termination of the maximum-flow algorithm, each arc with an unlabeled head and labeled tail carries a flow equal to its capacity; otherwise the head of this arc could be labeled by the flow-augmenting algorithm. Also, each arc with an unlabeled tail and labeled head carries no flow units; otherwise the tail of this arc could be labeled by the flow-augmenting algorithm.

Clearly, every flow unit must traverse an arc of this cut at least once. Since all arcs from the sink side to the source side of the cut carry no flow, no flow unit can traverse this cut more than once. Thus, the total flow from s to t equals the capacity of this cut because every arc from the source side to the sink side carries a flow equal to its capacity.

Proof of 3. To show that the algorithm terminates in a finite number of steps requires the assumption that all arc capacities are integers and all initial flow values are integers. Practically speaking, this is not a drastic assumption since most arc capacities can usually be rounded off to an integer without affecting the underlying physical problem.

The only possible way for the maximum-flow algorithm not to terminate in a finite number of steps would be for the algorithm to encounter an infinite number of flow-augmenting paths. However, each time a flow-augmenting path is found, the total flow v from s to t is increased by a positive integer because all flow values and arc capacities are always integers. Since v is bounded above by the capacity of any cut separating s from t, there cannot be an infinite number of flow augmentation.   Q.E.D.

EXAMPLE 2

Consider the pipeline network in Fig. 6.1.

Step 1. Initialize the algorithm with zero flow; that is, $f(x, y) = 0$ for all arcs (x, y).

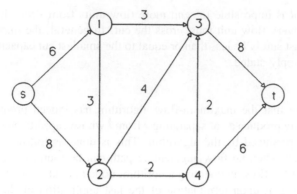

Steps 2 and 3. The residual network and labeling from the flow-augmenting algorithm are shown below (labels are $[e(x), p(x)]$).

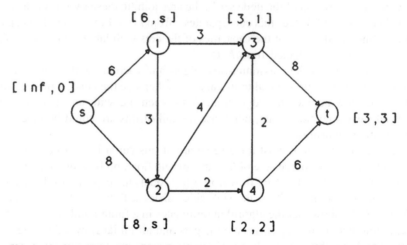

We have found the flow-augmenting paths $s - 1 - 3 - t$. The flow along this path can be increased by 3 units. The new flow on the original network is therefore

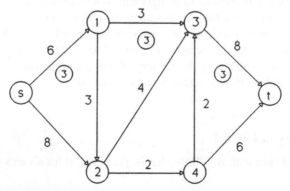

We return to step 2.

Steps 2 and 3. We construct the residual network relative to this new flow and apply the flow-augmenting algorithm.

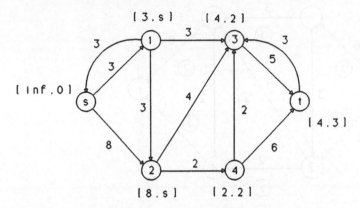

The flow-augmenting path $s - 2 - 3 - t$ has been identified. We increase the flow along this path by 4 units. The new flow is shown below.

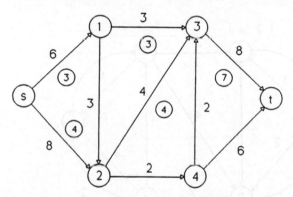

Steps 2 and 3.

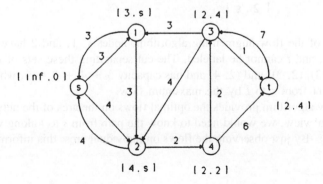

The flow-augmenting path is $s-2-4-t$ with a maximum increase in flow of 2 units. The resulting flow is shown below.

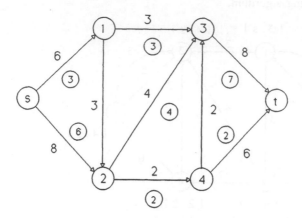

Steps 2 and 3. At this point, we cannot find a directed path from s to t in the residual network. Thus, the maximum flow is found.

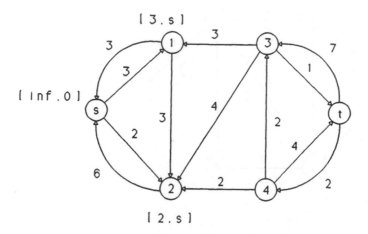

At the completion of the flow-augmenting algorithm, nodes s, 1, and 2 have been labeled and nodes 3, 4, and t cannot be labeled. The cut separating these sets of nodes consists of the arcs (1, 3), (2, 3), and (2, 4) and has capacity $3 + 4 + 2 = 9$, which is the number of units sent from s to t by the maximum flow.

The maximum-flow algorithm provides the optimal flows on the arcs of the network. From a practical point of view, we would need to know the path from s to t along which each unit of flow travels. By just observing the flows on individual arcs, this information is not apparent. Fortunately, it is very easy to transform flows on arcs to flows on paths using the path-finding algorithm with some appropriate modifications. This algorithm is called the *path decomposition algorithm*.

Path Decomposition Algorithm

Step 1. Begin with the solution to the maximum-flow problem in which each arc is labeled with its optimum flow $f(x, y)$. Set $k = 1$.

Step 2. Using the path-finding algorithm, find a directed path P_k from s to t along arcs with $f(x, y) > 0$. If no path exists, then stop; all flows have been allocated to some path. Let $v(P_k)$ be the minimum flow on any arc along the path P_k.

Step 3. For each arc (x, y) on P_k, set $f(x, y) = f(x, y) - v(P_k)$. Set $k = k + 1$ and return to step 2.

We will illustrate this algorithm using the solution to Fig. 6.1 (Example 2).

EXAMPLE 3

The network and its optimum arc flows are shown in Fig. 6.12a. Suppose that the first path we find is $s - 1 - 3 - t$. The minimum flow on any arc on this path is 3; therefore we subtract 3 from each arc on this path. The result is shown in Fig. 6.12b. Next, assume that we identify the path $s - 2 - 4 - t$. The minimum flow on this path is 2. This leaves the network shown in Fig. 6.12c, which itself is the final path remaining. Therefore, we have decomposed the arc flows into the set of path flows shown in Fig. 6.13.

Finite Termination Modification

The proof of the maximum-flow algorithm required that all initial arc flow values were integers and that all arc capacities were also integers. If some arc capacities are not integers, there is no guarantee that the maximum-flow algorithm will terminate finitely. An example of a graph with noninteger arc capacities for which the maximum-flow algorithm requires an infinite number of flow augmentations can be found in Ford and Fulkerson (1962). Fortunately, Johnson (1966) and Edmonds and Karp (1972) have provided two different ways to modify the maximum-flow algorithm to ensure that it terminates after only a finite number of flow augmentations. The modification due to Edmonds and Karp is presented now.

In step 2 of the flow-augmenting algorithm, we have a choice of labeled nodes to consider. The finite termination modification specifies the next node to choose. Number the nodes as they are labeled (clearly, node s will receive number one). First, attempt to label all nodes incident to node number one. Next, attempt to label all nodes incident to node number two, and so on. If the labeling process is performed in this fashion (called *first labeled, first scanned*), the path of marked arcs connecting any node to the source

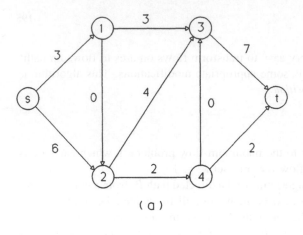

(a)

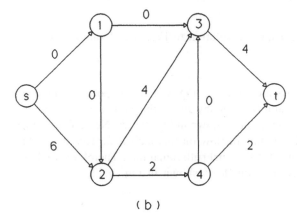

(b)

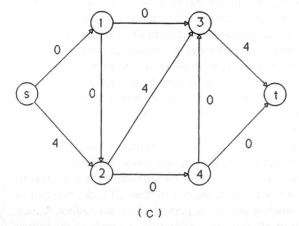

(c)

Fig. 6.12 Steps in the path decomposition algorithm.

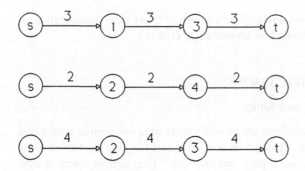

Fig. 6.13 Path decomposition.

will contain as few arcs as possible. Hence, each flow-augmenting path generated by this modified labeling method will contain the minimum number of arcs possible.

Arc (x, y) is called a *bottleneck arc* whenever arc (x, y) limits the amount of the flow augmentation. If arc (x, y) is a bottleneck arc, $f(x, y)$ either increases to $u(x, y)$ or reduces to zero.

Suppose that arc (x, y) is a bottleneck arc in both flow-augmenting paths C_1 and C_2 but not in any flow-augmenting path occurring between C_1 and C_2. Without loss of generality, suppose that $f(x, y) = u(x, y)$ after the flow augmentation along path C_1. Thus, (x, y) must be a forward arc in C_1 and a backward arc in C_2. For $i = 1, 2$, denote the number of arcs from vertex p to vertex q in flow-augmenting path C_i by $C_i(p, q)$. From the observation that the modified labeling method always labels the shortest possible flow-augmenting path from s to any vertex, it follows that

$$C_1(s, y) \leq C_2(s, y)$$
$$C_1(x, t) \leq C_2(x, t)$$
$$C_1(s, t) = C_1(s, y) + C_1(x, t) - 1$$
$$\leq C_2(s, y) + C_2(x, t) - 1$$
$$= C_2(s, t) - 2$$

Thus, $C_1(s, t) \leq C_2(s, t) - 2$, and each time arc (x, y) is a bottleneck arc, the minimum number of arcs in a shortest flow-augmenting path has increased by at least two. Since no flow-augmenting path from s to t can contain more than $m - 1$ arcs (recall that m is the number of vertices in the graph), it follows that arc (x, y) cannot be a bottleneck arc more than $m/2$ times. Since each flow-augmenting path has at least one bottleneck arc, there can be at most $mn/2$ flow augmentations.

A similar result follows if the flow in arc (x, y) is reduced to zero by the flow augmentation in path C_1. The labeling in Example 2 was performed in this fashion.

Thus, for a network with m nodes and n arcs, the number of flow augmentations is of order $O(nm)$. Since each augmentation will require a search of $O(n)$ arcs, the overall time complexity of this modified maximum-flow algorithm is $O(n^2m)$.

6.4 EXTENSIONS AND MODIFICATIONS

Modification for Several Sources and Sinks

Next, let us consider a graph in which there are possibly more than one source vertex and more than one sink vertex. Can this situation be accommodated by the maximum-flow algorithm, which works with only one source and one sink? Yes, simply create a new source vertex S, called the *supersource*, and a new sink vertex T, called the *supersink*. Join the supersource S to each original source s_1, s_2, . . . by an arc (S, s_1), (S,s_2), . . . with infinite capacity. Join each original sink t_1, t_2, . . . to the supersink T by an arc (t_1, T), (t_2, T), . . . with infinite capacity.

   Clearly, any flow on the new, enlarged graph from S to T corresponds to a flow on the original graph from the original sources to the original sinks, and vice versa. Moreover, a maximum flow in the enlarged graph corresponds to a maximum flow in the original graph. Thus, the maximum-flow algorithm can be applied to the enlarged graph, and the maximum flow generated by the algorithm yields a maximum flow in the original graph. See Fig. 6.14.

Modification for Node Capacities

In some situations, capacities may also be associated with nodes in the network. For instance, in a chemical processing plant, arcs may correspond to flow through pipes while nodes correspond to processing operations, which may also have restricted flow rates. A transformation that allows the maximum-flow algorithm to be used is to split the node into two nodes joined by an arc having the node capacity. This transformation is shown in Fig.

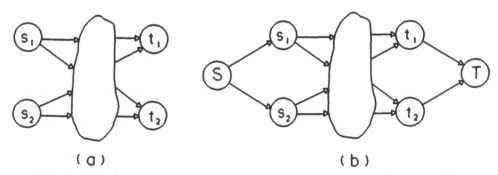

(a)   (b)

Fig. 6.14 Graph with several sources and sinks: (a) original graph, (b) expanded graph.

6.15. Notice that all arcs entering the original node k enter into the new node k', while those leaving k leave from node k''.

Lower Bounds

In some situations, lower bounds exist on arc flows; that is, at least l_{ij} units must flow from node i to node j. This complicates the situation since we do not even know if a feasible flow exists in the network. Also, the maximum-flow algorithm requires a feasible flow to start. If we have a feasible flow, a simple modification of the maximum-flow algorithm will allow us to solve the problem (see Exercise 11).

To find an initial feasible flow, if it exists, we will create an extended network and solve a maximum-flow problem on it. This is constructed as follows. Add two new nodes, say u and w. Add an arc from t to s having infinite capacity. Assume that arc (i, j) has a nonzero lower bound $l(i, j)$. Add an arc from u to j with capacity $l(i, j)$, and an arc from i to w with the same capacity. Replace the original capacity by $u(i, j) - l(i, j)$. Notice that this new network has all zero lower bounds, so the maximum-flow algorithm can be applied. Solve the maximum flow from u to w.

Theorem 6.1. A feasible flow exists in G if and only if the maximum flow in G^* is $\Sigma\, l(i, j)$.

We can recover a feasible flow in G by setting $f(i, j) = f^*(i, j) + l(i, j)$, where $f^*(i, j)$ is the optimal flow in G^*.

EXAMPLE 4

Consider the maximum-flow problem given in Fig. 6.16. The extended network is shown in Fig. 6.17 along with its optimum solution. Figure 6.18 shows the initial feasible flow in G. We can then apply the maximum-flow algorithm to this starting solution.

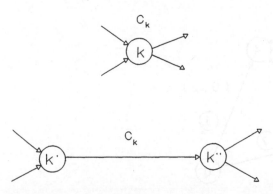

Fig. 6.15 Transformation for node capacities.

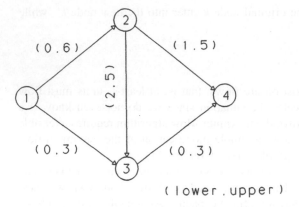

Fig. 6.16 Maximum-flow problem with lower bounds.

Maximum Flow in Planar Networks

In this section we will show how to solve for the maximum flow in undirected planar networks by solving an equivalent shortest-path problem. A network is *source-sink planar* if it is planar after the source and sink are connected by an edge. Given a source-sink planar network with edge capacities, assign the capacity of infinity to the edge connecting

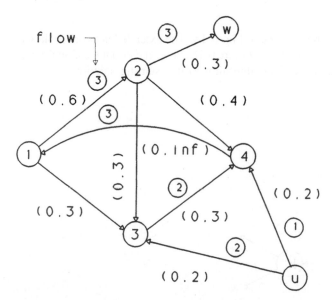

Fig. 6.17 Extended network for Example 4.

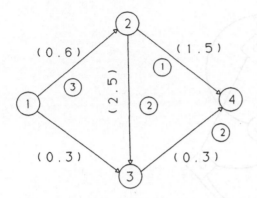

Fig. 6.18   Initial feasible flow for Example 4.

the source and the sink. Create the dual graph and assign to each edge in the dual graph a length corresponding to the capacity of the corresponding edge in the original graph. Find the shortest path from the dual vertex in the region bounded by the source-sink edge to the vertex in the outer (infinite) region. The length of the shortest path is equal to the value of the maximum flow, and the corresponding edges in the original graph comprise the minimum cut.

EXAMPLE 5

Consider the graph shown in Fig. 6.19. The dual graph is shown in Fig. 6.20. The shortest path from node a to node d has length 3 and corresponds to the cut $\{(1, 2), (1, 4)\}$ in the original graph.

6.5   PREFLOW-PUSH ALGORITHMS FOR MAXIMUM FLOW

A different class of algorithms for solving the maximum-flow problem is based on the concept of a *preflow*. A preflow is any flow satisfying

$$\sum_y f(y, x) - \sum_y f(x, y) \ge 0 \quad \text{for all } x \ne s, t$$

That is, the flow into a node is greater than or equal to the flow out of the node. We no longer maintain conservation of flow. Preflow-push algorithms send flows to nodes along one arc at a time (this is called a *push*), not along paths from the source to sink. Hence, they are quite different from the algorithms that we have discussed. These algorithms outperform path-augmenting algorithms in terms of both theoretical time complexity and actual computation time.

We define $e(x) = \Sigma f(y, x) - \Sigma f(x, y)$ to be the excess flow available at node x. If $e(x) > 0$, the node is called *active*, and we seek to push this flow closer toward the sink.

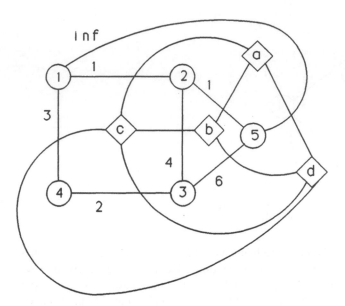

Fig. 6.19 Maximum-flow problem on an undirected planar network.

The algorithm maintains at least one active node at each iteration until the maximum flow is found.

The algorithm begins by setting *distance labels* $d(x)$ for each node x having the following properties:

1. $d(t) = 0$.
2. $d(x) \leq d(y) + 1$ for every arc (x, y) with $r(x, y) > 0$ in the residual network.

These distance labels actually represent a lower bound on the length of the shortest path from x to t in the residual network. We call an arc (x, y) in the residual network *admissible* if $d(x) = d(y) + 1$. As we shall see, we push flow only along admissible arcs.

Step 1. Set initial distance labels by searching the backward star of each node, beginning with the sink t and working backward, so that the properties stated above hold. We set initial flows $f(s, x) = u(s, x)$ for all arcs in the forward star of s, and set $d(s) = m$, the number of nodes.

Step 2. Select an active node x. If none exist, then stop; the maximum flow has been found.

Step 3. If the network contains an admissible arc (x, y), then push δ units of flow from x to y, where $\delta = \min\{e(x), r(x, y)\}$. Otherwise, replace $d(x)$ by $\min\{d(y) + 1\}$ for all y is in the forward star of x and $r(x, y) > 0$ in the residual network. Return to step 2.

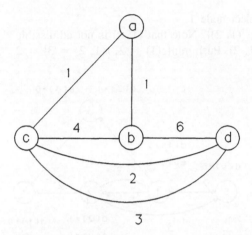

Fig. 6.20 Dual graph.

This algorithm is due to Karzanov (1974). The time complexity of this algorithm can be shown to be $O(n^2m)$, the same time bound as for the Edmonds and Karp modification of the Ford-Fulkerson algorithm. Many variations of this algorithm can be developed by specifying different rules for node selection. In particular, if we select the active node with the highest distance label, the time complexity becomes $O(m^3)$.

EXAMPLE 6

We shall illustrate the preflow-push algorithm using the example in Fig. 6.11. We will show both the flow network and the residual network at each iteration. Flows on arcs will be encircled on the flow network.

Step 1.

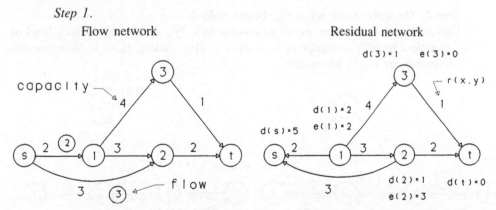

Note that the distance labels satisfy the two properties described above.

Step 2. The active nodes are {1, 2}. Select node 1.

Step 3. The admissible arcs are {(1, 3), {(1, 2)}. Note that (1, *s*) is not admissible since $d(l) \neq d(s) + 1$. Select arc (1, 2). Push min$\{e(1) = 2, r(1, 2) = 3\} = 2$ units of flow to node 2.

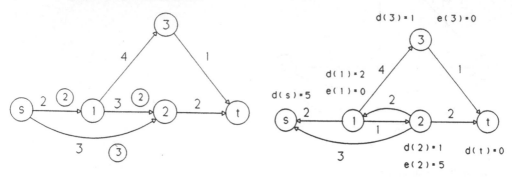

Step 2. The active node set is {2}. Select node 2.

Step 3. Admissible arcs = {(2, *t*)}. Push min{5, 2} = 2 units of flow to node *t*.

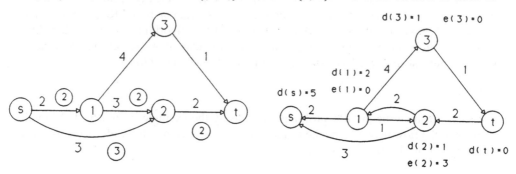

Step 2. The active node set is {2}. Select node 2.

Step 3. At this point there are no admissible arcs. We change the distance label of node 2 by $d(2) = \min\{d(s) + 1, d(1) + 1\} = \min\{6, 3\} = 3$. Note that this makes arc (2, 1) admissible.

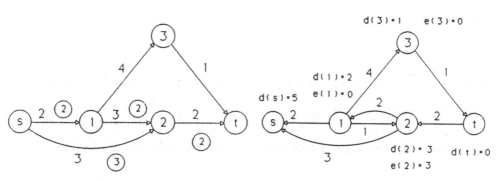

Step 2. Active node set = {2}. Select node 2.

Step 3. Admissible arcs = {(2, 1)}. Push min{3, 2} = 2 units of flow from 2 to 1. Observe that since the flow from 2 to *t* saturates the arc, we must send the excess flow at node 2 back toward the source. Relabeling the distance label for node 2 makes this possible.

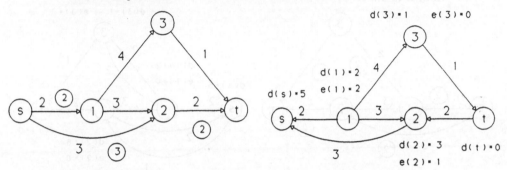

Step 2. Active node set = {1, 2}. Select node 1.

Step 3. Admissible arcs = {(1, 3)}. Note that (1, 2) is not admissible since the distance label of node 2 is 3. Push min{2, 4} = 2 units of flow to node 3.

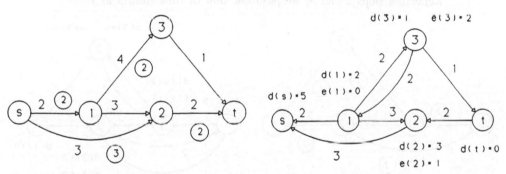

Step 2. Active node set = {3, 2}. Select node 3.

Step 3. Admissible arcs = {(3, *t*)}. Push min{2, 1} = 1 unit of flow to node *t*.

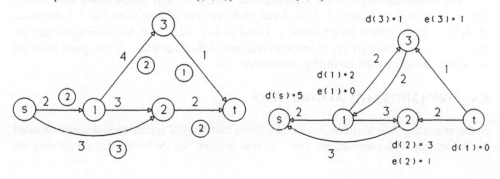

Step 2. Active node set = {2, 3}. Select node 2.
Step 3. There are no admissible arcs. Change the distance label of node 2 to $d(2) =$
   $\min\{d(s) + 1\} = 6$. This makes arc $(2, 1)$ admissible.

Repeating steps 2 and 3, we select node 2 and push $\min\{1, 3\} = 1$ unit of flow
from 2 to s.

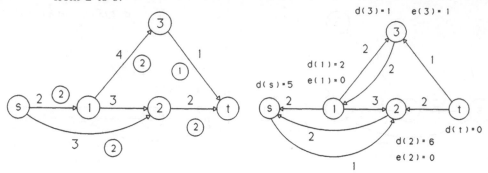

Step 2. Active node set = {3}. Select node 3.
Step 3. No admissible arcs are available. Set $d(3) = \min\{d(1) + 1\} = 3$.
   Repeating steps 2 and 3, we push one unit of flow from 3 to 1.

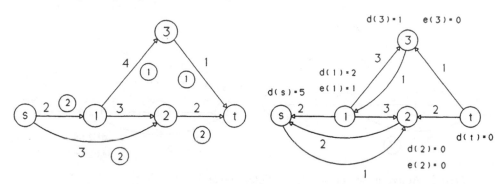

The remaining steps of the algorithm are summarized as follows. We now relabel
$d(1) = \min\{d(3) + 1, d(s) + 1\} = 4$ and push one unit of flow from 1 to 3. Label node
3, $d(3) = 5$. Push flow back to node 1. Label node 1, $d(1) = 6$. the admissible arcs are
$\{(1, 3), (1, s)\}$. Select arc $(1, s)$ and push one unit of flow back to s. At this point there are
no active nodes, so the algorithm terminates.

6.6 DYNAMIC FLOW ALGORITHMS

In the preceding four sections, we studied flows that obeyed certain requirements dictated
by the arc capacities and the arc costs. In this section, we shall consider yet another arc

requirement, namely arc traverse time, and we shall study flows in which all the flow units must make the trip from the source to the sink within a given amount of time.

Associate with each arc (x, y) in graph $G = (X, A)$ a positive integer $c(x, y)$ that denotes the number of time periods required by a flow unit to travel across arc (x, y) from x to y. The quantity $c(x, y)$ is called the *traverse time* of arc (x, y). [Yes, previously $c(x, y)$ was used to denote the cost of arc (x, y). As will be seen later, both arc cost and traverse time will fulfill the same role in algorithms, and hence we have chosen to identify them with the same symbol. This will become clearer later.] Let $u(x, y, T)$ denote the maximum number of flow units that can enter arc (x, y) at the start of time period T, for $T = 0, 1, \ldots$. A dynamic flow from s to t is any flow from s to t in which not more than $u(x, y, T)$ flow units enter arc (x, y) at the start of time period T, for all arcs (x, y) and all T. Note that in a dynamic flow, units may be departing from the source at time $0, 1, 2, \ldots$.

A *maximum dynamic flow for p time periods* from s to t is any dynamic flow from s to t in which the maximum possible number of flow units arrive at the sink t during the first p time periods. Obviously, the problem of finding a maximum dynamic flow is more complex than the problem of finding a maximum flow, since the dynamic flow problem requires that we keep track of when each unit travels through an arc so that no arc's entrance capacity is violated at any time. Happily, this additional complication can be resolved by rephrasing the dynamic flow problem into a static (nondynamic) flow problem on a new graph called the time-expanded replica of the original graph.

The *time-expanded replica of graph* $G = (X, A)$ *for p time periods* is a graph G_p whose vertex set is

$$X_p = \{x_i : x \in X, i = 0, ., \ldots, p\}$$

and whose arc set is

$$A_p = \{(x_i, y_j) : (x, y) \in A, i = 0, 1, \ldots, p - c(x, y), j = i + c(x, y)\}$$

Let

$$u(x_i, y_j) = u(x, y, i)$$

Note that the vertex set X_p of graph G_p consists of duplicating each vertex in X once for each time period. An arc joins vertices x_i and y_j only if it takes $j - i$ time periods to travel from vertex x to vertex y. Thus, a flow unit leaving vertex x along arc (x, y) in graph G at time 5 and taking 8 time periods to arrive at vertex y is represented in graph G_p as a unit flowing along the arc (x_5, y_{13}). Figure 6.21 shows the time-expanded replica of a graph for 6 time periods.

Clearly, any dynamic flow from s to t in graph G is equivalent to a flow from the sources to the sinks in G_p, and vice versa. Figure 6.22 shows a dynamic flow and its static equivalent for the graph in Fig. 6.21.

Since each dynamic flow is equivalent to a static flow in the time-expanded replica

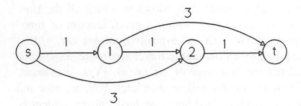

Fig. 6.21a Original graph. All arc capacities equal one. Traverse times are given next to each arc.

graph, a maximum dynamic flow for p time periods can be found simply by finding a maximum flow in the time-expanded replica for p time periods using the maximum-flow algorithm. Thus, no additional algorithms are required to solve the maximum dynamic flow problem. However, if p is very large, then the graph G_p becomes very large and the number of calculations required to find a maximum flow of graph G_p becomes prohibitively large.

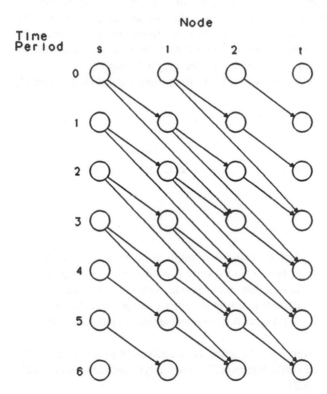

Fig. 6.21b Time-expanded replica graph (all arc capacities are one).

Dynamic Flow				Static Flow	
Path	Departure Time	Amount		Path	Amount
$s,1,2,t$	0	1 unit	$\leftrightarrow$	$s_0,1_1,2_2,t_3$	1 unit
$s,1,2,t$	1	1 unit	$\leftrightarrow$	$s_1,1_2,2_3,t_4$	1 unit
$s,1,t$	2	1 unit	$\leftrightarrow$	$s_2,1_3,t_6$	1 unit

Fig. 6.22 Equivalence between dynamic and static flows.

Happily, Ford and Fulkerson (1962, p. 142) have devised an algorithm, called the *maximum dynamic flow algorithm*, that generates a maximum dynamic flow much more efficiently than the method suggested above. However, the maximum dynamic flow algorithm works only when the entrance capacity of each arc remains unchanged through time, i.e., when

$$u(x, y, T) = u(x, y) \quad \text{for all } T = 0, 1, \ldots, p$$

for all arcs (x, y).

The maximum dynamic flow algorithm uses a simple minimum-cost flow algorithm as a subroutine. This algorithm is based on solving a sequence of maximum flow problems; thus, we could not present it in the previous chapter independently. We will first present this algorithm, and then present the maximum dynamic flow algorithm which employs it.

A Minimum-Cost Flow Algorithm Based on Maximum Flows

Suppose that the minimum cost objective function (see equation 5.1) were replaced by

$$\max \left\{ pv - \sum_{(x, y)} c(x,y) f(x,y) \right\}$$

where p is any large number, e.g., larger than the maximum total cost a unit could incur by traveling from s to t. If p is interpreted as the profit received for each unit sent from s to t, then this expression can be interpreted as the best possible net profit after shipping costs have been deducted. From this interpretation, it follows that any flow that maximizes this expression will also minimize the original objective, and vice-versa.

The minimum-cost flow algorithm first sends as many units as possible from s to t that incur a total cost of 0 each for the entire journey from s to t. Next, the minimum-cost flow algorithm sends as many units as possible from s to t that incur a total incremental cost of 1 each for the entire journey from s to t, etc. The algorithm stops when a total of v units have been sent from s to t, or when no more units can be sent from s to t, whichever happens first. In other words, the algorithm solves the problem first for $p = 0$, then for $p = 1$, then for $p = 2$, etc.

Suppose that as many units as possible with a total incremental cost of $p - 1$ or less have been sent by the algorithm from s to t. How does the algorithm determine how to send flow units from s to t with a total incremental cost of p each? To do this, the algorithm must locate a flow augmenting path from s to t with the property that the total incremental cost of sending a unit "along this path" equals p. Thus, the algorithm must find a flow augmenting path with the property that the sum of the costs of the forward arcs in the path less the arc of the costs of the backward arcs in the path equals p.

The algorithm accomplishes this by assigning an integer $p(x)$ to each vertex x in the graph. These vertex numbers $p(x)$ have the properties that $p(s) = 0$, $p(t) = p$, $0 \leq p(x) \leq p$ for all vertices $x \neq s$, $x \neq t$. The algorithm makes flow changes only along arcs (x, y) for which

$$p(y) - p(x) = c(x, y)$$

If the algorithm finds a flow-augmenting path from s to t consisting entirely of arcs that satisfy this equality, then it follows that the total incremental cost for each unit dispatched from s to t along this path equals p.

With this as motivation, we are now prepared to state formally the algorithm.

Step 1 (Initialization): Initially, let the flow $f(x, y)$ in each arc (x, y) equal zero. Initially, let $p(x) = 0$ for all vertices x.

Step 2 (Deciding Which Arcs Can Have Flow Changes): Let I be the set of all arcs (x, y) for which

$$p(y) - p(x) = c(x, y)$$

and

$$p(x, y) < u(x, y)$$

Let R be the set of all arcs for which

$$p(y) - p(x) = c(x, y)$$

and

$$0 < f(x, y)$$

Let N be the set of all arcs not in $I \cup R$. [The arcs in I and R will be the only arcs considered for possible flow changes. Hence, flow changes are only possible on arcs that satisfy $p(y) - p(x) = c(x, y)$.]

Step 3 (Flow Change): Perform the maximum flow algorithm with I, R, and N as defined above in Step 2. Stop when a total of v flow units have been sent from s to t or when no more flow from s to t is possible for the current composition of sets I, R, and N. If the former occurs first, stop because the terminal flow is a minimum cost flow that sends v units from s to t.

If the latter occurs first, check to see if the current flow is a maximum flow from s to t. (This is done by verifying if the cut generated by the last labeling of the flow augmenting algorithm is saturated.) If so, then stop because no more flow units can be sent from s to t and the terminal flow is a minimum cost flow. If not, then go to Step 4.

Step 4 (Vertex Number Change): Consider the last labeling done by the flow augmenting algorithm. (Recall that the flow augmenting algorithm is a sub-routine in the maximum flow algorithm which is used a a subroutine in this algorithm.) Increase by $+1$ the vertex number $p(x)$ of each unlabeled vertex x. (Note that $p(t)$ increases by $+1$ since t is unlabeled since otherwise a flow augmenting path would have been discovered.) Return to Step 2.

Proof of the Minimum-Cost Flow Algorithm. We shall prove that the minimum-cost flow algorithm does in fact produce a minimum-cost flow of v units from s to t by using the complementary slackness conditions of linear programming.

As mentioned earlier, the minimum-cost flow problem can be stated as the linear programming problem. Let $p(x)$ denote the dual variable associated with conservation of flow equation for vertex x. Let $p(s)$ denote the dual variable for the conservation of flow equation for the source s. Let $p(t)$ denote the dual variable for the conservation of flow equation for the sink t. (Later on, we will show that the dual variables $p(x)$ are the same as the vertx numbers $p(x)$ generated by the algorithm. So, don't worry about this duplicate notation.) Let $w(x, y)$ denote the dual variable for the capacity constraint for arc (x, y). Lastly, let us regard v as a variable rather than a constant.

The dual linear programming problem for the minimum-cost flow problem is
Minimize

$$\sum_{(x, y)} u(x, y)\, w(x, y) \tag{5}$$

such that

$$-p(s) + p(t) = p \tag{6}$$

$$p(x) - p(y) + w(x, y) > -c(x, y) \text{ [for all } (x, y)] \tag{7}$$

$$w(x, y) > 0 \text{ [for all } (x, y)] \tag{8}$$

$$p(x) \text{ unconstrained (for all } x) \tag{9}$$

Equation (6) is the dual constraint corresponding to the unconstrained primal variable v. Each relation (7) is a dual equation corresponding to a primal variable $f(x, y)$.

The complementary slackness conditions for the primal-dual pair linear programming problems become

$$p(x) - p(y) + w(x, y) > -c(x, y) \Rightarrow f(x, y) = 0 \tag{10}$$

and

$$w(x, y) > 0 \Rightarrow f(x, y) = u(x, y) \tag{11}$$

If we let

$$p(s) = 0, \; p(t) = p \tag{12}$$

and

$$w(x, y) = \max\{0, \, p(y) - p(x) - c(x, y)\} \tag{13}$$

for all arcs (x, y), then complementary slackness condition (10) becomes

$$p(y) - p(x) < c(x, y) \Rightarrow f(x, y) = 0 \tag{14}$$

since equation (12) implies that $w(x, y) = 0$. Complementary slackness condition (11) becomes

$$p(y) - p(x) > c(x, y) \Rightarrow f(x, y) = u(x, y) \tag{15}$$

Hence from the complementary slackness conditions of linear programming, we need only construct values for $p(x)$ for all vertices x and values for $f(x, y)$ for all arcs (x, y) that satisfy conditions (12), (14), and (15). Of course, the values chosen for $f(x, y)$ must form a feasible solution to the minimum cost flow problem.[†]

When the algorithm is initialized, $p(x) = 0$ for all x and $p = 0 = p(t)$. Hence, the complementary slackness conditions are satisfied. We shall now show that the complementary slackness conditions remain satisfied throughout all iterations of this minimum cost flow algorithm. This is accomplished in two parts:

(a)   By showing that all flow changes made by the algorithm maintain the complementary slackness conditions

(b)   By showing that all vertex number changes maintain the complementary slackness conditions for $p = p(t)$.

[Note that condition (12) is always satisfied by the algorithm and that the values for $f(x, y)$ always yield a feasible flow.]

The algorithm allows a flow change in arc (x, y) only when $p(y) - p(x) = c(x, y)$. Hence, flow changes cannot disturb the complementary slackness conditions which pertain only to arcs for which $p(y) - p(x) \neq c(x, y)$. Thus, part (a) is verified.

It remains to show that the vertex number changes made by the algorithm do not destroy complementary slackness. (Recall that the algorithm increases a vertex number by $+1$ only when the vertex could not be labeled. If a vertex cannot be labeled, then no additional flow units can arrive at that vertex from the source.) If both endpoints of arc (x, y) are labeled or if both endpoints of arc (x, y) are unlabeled, then $p(y) - p(x)$ remains unchanged and complementary slackness is preserved.

If vertex x is labeled and vertex y is unlabeled, then we know from the algorithm that one of the following occurs:

(a) $p(y) - p(x) < c(x, y)$
(b) $p(y) - p(x) > c(x, y)$
(c) $p(y) - p(x) = c(x, y)$ and $f(x, y) = u(x, y)$.

If (a) occurs, then after increasing $p(y)$ by $+1$, $p(y) - p(x) \leq c(x, y)$ and (14) remains satisfied. If (b) occurs, then after increasing $p(y)$ by $+1$, $p(y) - p(x)$ remains $> c(x, y)$ and (15) remains satisfied. If (c) occurs, then after increasing $p(y)$ by $+1$, $p(y) - p(x) > c(x, y)$ and (15) is satisfied.

If vertex x is unlabeled and vertex y is labeled, then we know from the algorithm that one of the following occurs:

[†] Note that the values for $p(x)$, $x \in X$, determine feasible values for all $w(x, y)$ by condition (13).

(a) $p(y) - p(x) < c(x, y)$
(b) $p(y) - p(x) > c(x, y)$
(c) $p(y) - p(x) = c(x, y)$ and $f(x, y) = 0$

If (a) occurs, then after increasing $p(x)$ by $+1$, $p(y) - p(x)$ remains $< c(x, y)$, and (14) remains satisfied. If (b) occurs, then after increasing $p(x)$ by $+1$, $p(y) - p(x) \geq c(x, y)$, and (15) remains satisfied. If (c) occurs, then after increasing $p(x)$ by $+1$, $p(y) - p(x) < c(x, y)$ and (14) is satisfied. This completes the verification of part (b).

Hence, this minimum-cost flow algorithm constructs a feasible flow that satisfies all complementary slackness conditions for $p = p(t)$. When the algorithm terminates after dispatching the last flow unit from s to t, the terminal value of $p(t)$ will equal the total incremental cost incurred by the last flow unit. Hence, $p = p(t)$ will be suitably large enough to insure that

$$pv - \sum_{(x, y)} c(x, y) f(x, y)$$

can replace the original objective function of the minimum cost flow linear programming problem. Hence, the flow constructed by the algorithm is a minimum cost flow.

The proof is complete except we must show that the minimum-cost flow algorithm terminates in a finite number of steps. The minimum-cost flow algorithm terminates after the last flow unit has been dispatched from the source to the sink. When this occurs, $p(t)$ equals the total incremental cost incurred by this last flow unit. If all arc costs are finite, positive integers, then at termination $p(t)$ must be a finite positive integer. Hence only $p(t) + 1$ applications of the maximum flow algorithm are required by the minimum-cost flow algorithm. As we know from before, the maximum-flow algorithm can be modified to terminate in a finite number of steps. Hence, the minimum-cost flow algorithm also terminates in a finite number of steps. Q.E.D.

To apply this minimum-cost flow algorithm to the maximal dynamic flow problem, let $F_1, F_2, \ldots, F_p$ denote the resulting sequence of flows generated by the minimum-cost flow algorithm. Each flow F_i consists of a set of paths $f_{i,1}, f_{i,2}, \ldots, f_{i,r_i}$ from s to t taken by the units sent from s to t by flow F_i. Let $n_{i,j}$ denote the number of flow units that would follow path $f_{i,j}$ in flow F_i. Clearly, no flow path $f_{i,j}$ from s to t has a total cost greater than i, since otherwise the incremental cost of using this path would be greater than i and this path could not have been generated by the minimum-cost flow algorithm. Let $c(f_{i,j})$ denote the total cost of flow path $f_{i,j}$.

With these definitions in mind, we can now state the maximum dynamic flow algorithm.

Maximum Dynamic Flow Algorithm

This algorithm generates a maximum dynamic flow for p time units from the source s to the sink t in a graph $G = (X, A)$ for which $u(x, y, T) = u(x, y)$ for all $T = 0, 1, \ldots, p$ and for all $(x, y) \in A$.

Step 1. Let the cost of using arc (x, y) equal the transit time $c(x, y)$ for all arcs (x, y).
Perform the minimum-cost flow algorithm on graph G, stopping after the flow F_p has been generated.

Flow F_p consists of flow paths $f_{p,1}, f_{p,2}, \ldots, f_{p,r_p}$ from s to t, respectively, carrying $n_{p,1}, n_{p,2}, \ldots, n_{p,r_p}$ flow units from s to t. The decomposition of F_p into flow paths is a by-product of the minimum-cost flow algorithm.

Step 2. For $j = 1, 2, \ldots, r_p$, send $n_{p,j}$ flow units from s to t along flow path $f_{p,j}$ starting out from s at time periods $0, 1, \ldots, p - c(f_{p,j})$. The resulting flow is a maximum dynamic flow for p time periods.

Thus, the maximum dynamic flow algorithm consists of performing the minimum-cost flow algorithm using the arc traverse times as the arc costs. The terminal flow generated by the minimum-cost flow algorithm is decomposed into paths from s to t. Lastly, flow is sent along each of these paths starting at time period $0, 1, \ldots$ until flow can no longer reach the sink by time period p. Figure 6.23 shows a maximum dynamic flow for $p = 4, 5, 6$ for the graph in Fig. 6.21.

The flow generated by the maximum dynamic flow algorithm is called a *temporally repeated* flow for the obvious reason that it consists of repeating shipments along the same flow paths from s to t. Of course, the maximum dynamic flow for p time periods generated by the maximum dynamic flow algorithm need not be the only possible maximum dynamic flow for p time periods. However, as shown by the algorithm, there is always one maximum dynamic flow for p time periods that is a temporally repeated flow.

Proof.   To prove the maximum dynamic flow algorithm, we must show

(a)   That the algorithm constructs a flow
(b)   That this flow is a maximum dynamic flow for p time periods
(c)   That the algorithm terminates in a finite number of steps

Proof of (a). The algorithm sends flow units from s to t along the paths in F_p. Except for the source and the sink, the flow into a vertex equals the flow out of that vertex. Moreover, no flow units are dispatched unless they can reach the sink before time period p.

How many flow units enter arc (x, y) during any given time period? Flow F_p cannot send more than $u(x, y)$ flow units across arc (x, y); hence the flow paths $f_{p,1}, f_{p,2}, \ldots,$ f_{p,r_p} cannot send more than a total of $u(x, y)$ flow units across arc (x, y). Hence, not more than $u(x, y)$ flow units can enter arc (x, y) during any given time period. Hence, the dynamic flow constructed by the maximum dynamic flow algorithm obeys all arc capacities.

Since all flow units start at s and arrive at t before time period p, violate no arc capacities, and do not hold over at any intermediate vertex, the algorithm produces a dynamic flow.

Proof of (b). To show that the flow generated by the algorithm is a maximum dynamic flow, we shall consider its equivalent in the time-expanded replica graph G_p. We shall accomplish the proof by showing that this static flow saturates a cut of arcs separating the sources from the sinks in graph G_p.

Results of the Minimum-Cost Flow Algorithm

p	F_p
0	0
1	0
2	0
3	1 unit along path s, 1, 2, t
4	1 unit along path s, 1, 2, t
5	1 unit along path s, 2, t
	1 unit along path s, 1, t
6	1 unit along path s, 2, t
	1 unit along path s, 1, t

Maximum Dynamic Flow for p Time Periods

$p = 0$ No flow.

$p = 1$ No flow.

$p = 2$ No flow.

$p = 3$ Send one unit along path s, 1, 2, t starting at time 0 to arrive at the sink at time 3.
Total flow = 1 unit.

$p = 4$ Send one unit along path s, 1, 2, t starting at time 0 to arrive at the sink at time 3.
Send one unit along path s, 1, 2, t starting at time 1 to arrive at the sink at time 4.
Total flow = 2 units.

$p = 5$ Send one unit along path s, 2, t starting at time 0 to arrive at the sink at time 4.
Send one unit along path s, 2, t starting at time 1 to arrive at the sink at time 5.
Send one unit along path s, 1, t starting at time 0 to arrive at the sink at time 4.
Send one unit along path s, 1, t starting at time 1 to arrive at the sink at time 5.
Total flow = 4 units.

$p = 6$ Send one unit along path s, 2, t starting at times 0, 1, 2 to arrive at the sink at times 4, 5, 6.
Send one unit along path s, 1, t starting at times 0, 1, 2 to arrive at the sink at times 4, 5, 6.
Total flow = 6 units.

Fig. 6.23 Example of maximum dynamic flow algorithm.

Let $p(x)$ denote the value of the dual variable for vertex x just before the $(p + 1)$st iteration of the minimum-cost flow algorithm begins. Thus, $p(t) = p + 1$. Let

$$C = \{x_i: x_i \in X_p, \, p(x) \leq i\}$$

Note that since $p(s) = 0$, all source vertices $s_0, s_1, \ldots, s_p$ are members of set C. Also, since $p(t) = p + 1$, no sink vertex $t_0, t_1, \ldots, t_p$ is a member of set C. The set of all arcs with one endpoint in C and the other endpoint not in C forms a cut K that separates the sources from the sinks.

Suppose that arc $(x_i, y_j) \in K$, $x_i \in C$, $y_j \notin C$; then $c(x, y) = j - i < p(y) - p(x)$ from the definition of cut K. It follows from the complementary slackness conditions of the minimum-cost flow algorithm that $f(x, y) = u(x, y)$. Hence, the flow paths that comprise F_p must saturate arc (x, y). Each of these flow paths that uses arc (x, y) has a total traverse time (equivalently total cost) not greater than p time periods. Moreover, each of these paths is short enough so that flow units traveling along them can reach vertex x at time i and still arrive at the sink before or during the pth time period. Thus, arc (x_i, y_j) is saturated in graph G_p. By a similar argument, if $x_i \notin C$ and $y_j \in C$, then arc (x_i, y_j) is empty.

Thus, each cut arc in graph G_p from the source side of the cut to the sink side of the cut is saturated and each arc from the sink side of the cut to the source side of the cut is empty. Hence, this cut K is saturated and the flow produced by the algorithm corresponds to a maximum flow in graph G_p. Thus, the dynamic flow produced by the algorithm is a maximum dynamic flow for p time periods.

Proof of (c). The minimum-cost flow algorithm terminates in a finite number of steps. Hence, the maximum dynamic flow algorithm must terminate in a finite number of steps since it consists of one performance of the minimum-cost flow algorithm and a temporal repetition of flow along a finite number of flow paths from s to t.   Q.E.D.

The preceding discussion of dynamic flows did not consider the possibility of a flow unit stopping to rest or holding over at a vertex for one or more time periods before continuing its journey to the sink. If holdovers are permitted, then graph G_p must be revised to contain arcs of the form (x_i, x_{i+1}) so that flow units holding over at vertex x can depart from vertex x at a later time period.

If holdovers are permitted, how does the maximum dynamic flow for p time periods change? Obviously, the possibility of having holdovers cannot decrease the maximum number of flow units that can be sent from s to t in p time periods. In fact, it is easy to show that the maximum dynamic flow for p time periods remains unchanged: Suppose holdovers are permitted and graph G_p is augmented with additional arcs as described above. Consider the flow generated by the maximum dynamic flow algorithm and the cut K that this flow saturates in the original unaugmented graph G_p. This cut K is also a cut of the augmented graph G_p and is still saturated by the maximum dynamic flow produced by the maximum dynamic flow algorithm, since each holdover arc in K is directed from $\overline{C}$ to C and carries no flow. Thus, the possibility of holdovers does not change the maximum dynamic flow.

Lexicographic Dynamic Flows

Let $V(p)$ denote the maximum number of flow units that can be sent from the source to the sink in graph G within p time periods. Obviously,

$$V(1) \leq V(x) \leq \cdots \leq V(p) \leq V(p + 1)$$

For the special case when arc transit times remained stationary (i.e., were the same during all time periods), the maximum dynamic flow algorithm constructed a maximum dynamic flow for p time periods by temporally repeating the flow F_p obtained after the pth iteration of the minimum-cost flow algorithm. Thus,

$$V(p) = \sum_{i=1}^{r_R} [p - c(f_{p,i})]n_{p,i}$$

since each flow path $f_{p,i}$ in F_p is repeated $p - c(f_{p,i})$ times and carries $n_{p,i}$ flow units. Let V_p denote the number of flow units sent from s to t at the end of the pth iteration of the minimum-cost flow algorithm. Then, rewriting the preceding equation yields

$$V(p) = pV_p - \sum_{i=1}^{r_R} c(f_{p,i})n_{p,i} = pV_p - \sum_{(x, y)} c(x, y)f_p(x, y)$$

Thus, we can conclude that

$$V(p+1) - V(p) = (p+1)V_{p+1} - pV_p - \sum_{(x, y)} c(x, y)[f_{p+1}(x, y) - f_p(x, y)]$$

$$= V_{p+1} \geq V_p$$

Note that $V(p + 1) - V(p) = V_p$ if, and only if, flow F_{p+1} is identical to flow F_p. Consequently, *as p increases the maximum dynamic flow increases by at least V_p units.*

An *earliest arrival flow for p time periods* is any maximum dynamic flow for p time periods in which $V(i)$ flow units arrive at the sink during the first i time periods for $i = 0$, $1, \ldots, p$. Thus, an earliest arrival flow (if it exists) is a flow that is simultaneously a maximum dynamic flow for $0, 1, \ldots, p$ time periods. It is logical to call such a flow an earliest arrival flow since it would be impossible for any flow units to arrive any earlier at the sink.

Theorem 6.2. An earliest arrival flow for p time periods always exists for graph G.

Proof. Use the maximum-flow algorithm to construct a maximum flow from the sources to the sink in graph G_p for zero time periods. Call this flow MDF_0 since it is a maximum dynamic flow for zero time periods.

Next, starting with MDF_0 in graph G_p, generate a maximum flow from the sources into sinks t_0 and t_1. Clearly, this flow is a maximum dynamic flow for one time period. Call this flow MDF_1. Note that when the maximum flow algorithm is performed to generate MDF_1 from MDF_0, no flow unit entering sink t_0 is ever rerouted to sink t_1 since the algorithm will never reroute a flow unit already at a sink. However, the route taken by a unit arriving at sink t_0 might be altered when MDF_1 is constructed. Consequently, MDF_1 will be a maximum dynamic flow for zero time periods as well as for one time period.

In a similar way, MDF_2 can be generated from MDF_1 and MDF_2 will be a maximum dynamic flow for zero, one, and two time periods. Repeat this process to generate MDF_p, which is an earliest arrival flow.   Q.E.D.

This proof not only demonstrates that an earliest arrival flow always exists for any graph G and for any $p = 0, 1, \ldots$ but also shows how to construct an earliest arrival flow. For large graphs or for large values of p, this construction procedure requires an excessive and perhaps prohibitive number of calculations. Fortunately, there is an efficient algorithm to construct an earliest arrival flow for the special case when arc capacities are stationary through time, i.e., when $u(x, y, T) = u(x, y)$ for all T and all arcs (x, y). This algorithm is called the earliest flow algorithm (Minieka, 1973; Wilkinson, 1971).

Before discussing the earliest arrival flow algorithm, let us explore conditions under which the maximum dynamic flow algorithm will and will not construct a maximum dynamic flow that is also an earliest arrival flow.

Consider the graph in Fig. 6.21 and the maximum dynamic flow for $p = 6$ for this graph given in Fig. 6.23. This flow consists of 6 units. One flow unit is dispatched at time 0, 1, 2 along each of the two flow paths $(s, 1, t)$ and $(s, 2, t)$. Thus, two flow units arrive at the sink at time 4, 5, 6. This flow cannot be an earliest arrival flow because no flow units arrive at the sink at time 3. Note that during the fifth iteration ($p = 5$) of the minimum-cost flow algorithm performed on this graph, the flow-augmenting path $(s, 2, 1, t)$ was discovered. This path has total cost (traverse time) equal to $3 - 1 + 3 = 5$. This augmenting path transformed flow path $(s, 1, 2, t)$ of total cost $1 + 1 + 1 = 3$ into path $(s, 1, t)$ with total cost $1 + 3 = 4$. Hence, the units traveling this path will now be detoured to a path that costs one unit more.

The key idea behind the earliest arrival flow algorithm is to channel as many flow units as possible along shorter paths, like $(s, 1, 2, t)$, before these paths are superseded by longer paths, like $(s, 1, t)$.

To facilitate the presentation of the earliest arrival algorithm, we shall assume that all $u(x, y)$ are integers and replace each arc (x, y) by $u(x, y)$ replicas of itself, each with a capacity of one unit. Thus, since all arc capacities are one, all flow paths will have a capacity of one, and each arc will be either saturated or empty.

As before, let F_i denote the flow produced after the ith iteration of the minimum-cost flow algorithm. Let $f_{i,1}, f_{i,2}, \ldots, f_{i,r_i}$ denote the flow paths that constitute flow F_i. By the above assumption, each of these flow paths carries one flow unit. (Thus, $n_{i,j} = 1$ for all i and all j.) Let $P_{i,j}$ denote the jth flow-augmenting path from s to t discovered during the ith iteration of the minimum-cost flow algorithm. In the example in Fig. 6.23, $P_{3,1} = (s, 1, 2, t)$, and $P_{5,1} = (s, 2, 1, t)$.

With these definitions and motivation, we are now prepared to state the earliest arrival flow algorithm.

Earliest Arrival Flow Algorithm

Step 1 (Initialization). Initially, there is no flow in any arc. Perform the minimum-cost flow algorithm for p iterations. Record each flow F_i, each set of flow paths

$f_{i,1}, f_{i,2}, \ldots, f_{i,r_i}$, and each breakthrough path $P_{i,j}$ for $i = 0, 1, \ldots, p$ and $j = 1, 2, \ldots, w_i$.

Step 2 (Flow Construction). Repeat the following procedure for $\delta = 0, 1, \ldots, p$: Consider the sequence

$$P_{1,1}, P_{1,2}, \ldots, P_{1,w_1}, \ldots, P_{\delta,1}, \ldots, P_{\delta,r_\delta}$$

At time period $\delta - i$ dispatch one flow unit from s to t along each path $P_{i,j}$ in the above sequence. Let the flow unit spend $c(x, y)$ time periods in each forward arc (x, y) in path $P_{i,j}$ and $-c(x, y)$ time periods in each backward arc in path $P_{i,j}$. Label each forward arc in path $P_{i,j}$ with the time that this flow unit enters the arc, and remove for each backward arc the label (if it exists) that denotes the time that the flow unit would leave the backward arc.

Upon termination of the above procedure, the labels represent the times that a flow unit enters an arc and correspond to an earliest arrival flow for p time periods.

EXAMPLE 7

We shall now use the earliest arrival flow algorithm to construct an earliest arrival flow for the graph in Fig. 6.21 for $p = 6$. Recall that when the minimum-cost flow algorithm was performed on this graph only two flow-augmenting paths were discovered, namely $P_{3,1} = (s, 1, 2, t)$ and $P_{5,1} = (s, 2, 1, t)$. The labeling results of the earliest arrival flow algorithm are

For $\delta = 0$, no labeling
For $\delta = 1$, no labeling
For $\delta = 2$, no labeling,
For $\delta = 3$, dispatch one unit along $P_{3,1}$ at time $\delta - 3 = 0$

This creates the following labels (see Fig. 6.24):

0 on arc $(s, 1)$
1 on arc $(1, 2)$
2 on arc $(2, t)$

For $\delta = 4$, dispatch one unit along path $P_{3,1}$ at time $\delta - 3 = 1$. This creates the following labels:

1 on arc $(s, 1)$
2 on arc $(1, 2)$
3 on arc $(2, t)$

For $\delta = 5$, dispatch one unit along path $P_{3,1}$ at time $\delta - 3 = 2$. This creates the following labels:

2 on arc $(s, 1)$
3 on arc $(1, 2)$
4 on arc $(2, t)$

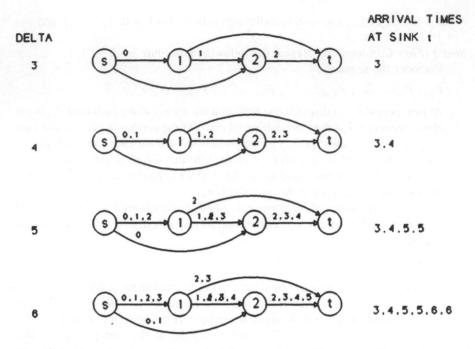

Fig. 6.24 Example of an earliest arrival flow algorithm. All arc capacities equal 1; $a(s, 1) = a(1, 2) = a(2, t) = 1$, $a(s, 2) = a(1, t) = 3$.

Dispatch one unit along path $P_{5,1}$ at time $\delta - 5 = 0$. This creates the following labels:

    0 on arc $(s, 2)$
    2 off arc $(1, 2)$
    2 on arc $(1, t)$

For $\delta = 6$, dispatch one unit along path $P_{3,1}$ at time $\delta - 3 = 3$. This creates the following labels:

    3 on arc $(s, 1)$
    4 on arc $(1, 2)$
    5 on arc $(2, t)$

Dispatch one unit along path $P_{5,1}$ at time $\delta - 5 = 1$. This creates the following labels:

    1 on arc $(s, 2)$
    3 off arc $(1, 2)$
    3 on arc $(1, t)$

The resulting set of labels corresponds to an earliest arrival flow for the 6 time periods.

Before presenting the proof of the earliest arrival flow algorithm, three lemmas are needed.

Lemma 6.1. Consider the sequence of all flow-augmenting paths generated by the minimum-cost flow algorithm. Suppose that two paths P' and P''' both contain arc (x, y) as a forward arc. Then there exists a flow-augmenting path P'' between P' and P''' in this sequence that contains arc (x, y) as a backward arc.

Proof. Recall that the capacity of arc (x, y) is one unit. Without loss of generality, assume that P' precedes P''' in the sequence of flow-augmenting paths generated by the minimum-cost flow algorithm. Since arc (x, y) is a forward arc in P', arc (x, y) is saturated when path P' augments flow. A similar situation occurs when P''' augments flow. Hence, between these two flow augmentations, a flow unit must be removed from arc (x, y). Hence, arc (x, y) must be a backward arc in some flow-augmenting path P''. Q.E.D.

Lemma 6.2. Suppose a flow unit travels along flow-augmenting path $P_{i,j}$ from s to t spending $c(x, y)$ time periods in each forward arc (x, y) and $-c(x, y)$ time periods in each backward arc (x, y). Then, the total travel time from s to t is i time periods.

Proof. When flow-augmenting path $P_{i,j}$ is generated by the minimum-cost flow algorithm during its ith iteration $p(s) = 0$, $p(t) = i$, and $p(y) - p(x) = c(x, y)$ for all arcs (x, y) in flow-augmenting path $P_{i,j}$, and the lemma follows. Q.E.D.

Lemma 6.3. Let $p_i(x)$ denote the value assigned to $p(x)$ during the ith iteration of the minimum-cost flow algorithm. Then, for $i \leq j$,

$$p_i(x) \leq p_j \leq p_i(x) + j - i$$

Proof. At the beginning of each iteration of the minimum-cost flow algorithm, each vertex number either increases by $+1$ or remains unchanged. Hence, $p_{i+1}(x) = p_i(x)$ or $p_{i+1}(x) = p_i(x) + 1$, and the lemma follows. Q.E.D.

Proof. To prove the algorithm, we must prove

    (a)   That the labels generated by the algorithm correspond to a flow
    (b)   That this flow is an earliest arrival flow for p time periods
    (c)   That the algorithm terminates after a finite number of steps.

Proof of (a). To show that the labels produced by the algorithm correspond to a flow, it suffices to prove two claims:

    1.   The algorithm never indicates the removal of a label that does not exist already
    2.   The algorithm never duplicates a label on an arc.

To prove the first claim, suppose that arc (x, y) is a backward arc in path $P_{m,n}$. Consequently, arc (x, y) must be a forward arc in some path that precedes $P_{m,n}$. Let the last such augmenting path preceding $P_{m,n}$ be denoted by $P_{i,j}$. It follows that $i \le m$.

According to the instructions of the earliest arrival flow algorithm, path $P_{i,j}$ labels arc (x, y) with $p_i(x)$, $p_i(x) + 1$, ..., $p_i(x) + m - i$, before path $P_{m,n}$ removes any labels from arc (x, y). The path $P_{m,n}$ removes the label $p_m(x)$ which is already present on arc (x, y) by Lemma 6.3. Moreover, before path $P_{m,n}$ is required to remove another label from arc (x, y), path $P_{i,j}$ has placed another label on arc (x, y). Since the additional labels are successive integers and since the labels to be removed are successive integers, it follows that any label to be removed from arc (x, y) has already been placed on arc (x, y). This proves the first claim.

To prove the second claim, suppose that arc (x, y) has received several identical labels. These labels must have been generated by distinct flow-augmenting paths that contain arc (x, y) as a forward arc. Let $P_{i,j}$ and $P_{m,n}$ denote any two such flow-augmenting paths. By Lemma 1, there exists a flow-augmenting path $P_{k,l}$ sequenced between $P_{i,j}$ and $P_{m,n}$ that contains arc (x, y) as a backward arc.

Path $P_{k,l}$ removes labels $p_k(x)$, $p_k(x) + 1$, ..., $p_k(x) + m - k$ from arc (x, y) before path $P_{m,n}$ adds any labels to arc (x, y). The first label that path $P_{m,n}$ generates on arc (x, y) is $p_m(x)$, which by Lemma 3 has already been removed. Similarly, during each successive iteration of the algorithm, path $P_{k,l}$ removes a label before path $P_{m,n}$ places the same label on arc (x, y). This completes the proof of the second claim.

Proof of (b). We shall now demonstrate that the algorithm does in fact produce an earliest arrival flow. This is accomplished by an inductive argument. Clearly, the algorithm constructs an earliest arrival flow for $p = 0$. Suppose that the algorithm constructs an earliest arrival flow for $p = p_0 - 1$. It remains to show that the algorithm constructs an earliest arrival flow for $p = p_0$.

The flow generated by the algorithm for p_0 time periods consists of the flow generated by the algorithm for $p_0 - 1$ time periods plus some additional flow units arriving at the sink at time period P_0. By Lemma 6.2, for each flow-augmenting path $P_{0,1}, ..., P_{0,r_0}$, $P_{1,1}, ..., P_{1,r_1}, ..., P_{p_0,1}, ..., P_{p_0,w_{p_0}}$ there is a flow unit arriving at the sink at time period p_0. This follows since each of these flow-augmenting paths contributes one flow unit to flow F_{p_0}. Hence, the total number of flow-augmenting chains in the above sequence equals V_{p_0}.

By the induction hypothesis, the flow units arriving at the sink by time period $p_0 - 1$ in the flow generated by the algorithm for $p = p_0$ constitute an earliest arrival flow for $p_0 - 1$ time periods and hence also a maximum dynamic flow for $p_0 - 1$ time periods. It follows that the flow generated by the algorithm is not only a maximum dynamic flow for p_0 time periods but also an earliest arrival flow for p_0 time periods since V_{p_0} flow units arrive at t at time period p_0.

Proof of (c). The algorithm terminates in a finite number of steps since labels have to be placed or removed along only a finite number of flow-augmenting paths. (There is only a finite number of flow-augmenting paths; otherwise the minimum-cost flow algorithm would not terminate after a finite number of steps.)   Q.E.D.

In some situations, it might be preferable to have flow units arrive at the sink as late as possible rather than as early as possible. For example, if each flow unit represents a check drawn on your account, you might prefer that they arrive at the payee (sink) as late as possible so that your checking balance would be as large as possible. Or the flow units might represent manufactured goods whose storage is very expensive, in which case you would prefer not to have them in storage (i.e., at the sink) until as late as possible. For situations such as these, we would be interested in a maximum dynamic flow in which the flow units arrive at the sink as late as possible rather than as early as possible.

A *latest arrival flow for p time periods* is any maximum dynamic flow for p time periods in which as many flow units as possible arrive during the last time period, as many flow units as possible arrive during the last two time periods, etc.

Does there always exist a latest arrival flow for p time periods? Yes. The existence of a latest arrival flow for p time periods can be proved by construction in the same way that the existence of an earliest arrival flow for p time periods was proved by construction. The only difference in the proof is that now units are first sent into sink t_p, then into sink $t_{p-1}, \ldots$

Clearly, the construction of the latest arrival flow for p time periods suggested above is hardly efficient when the graph has many arcs or vertices or when p is large. Even for the special case when all arc capacities are stationary, there does not appear to be an efficient algorithm for constructing a latest arrival flow, unlike the case of the earliest arrival flow, which could be generated efficiently by the earliest arrival flow algorithm. The reason for this is that a latest arrival flow necessitates that flow units leave the source and linger or hold over in the network as much as possible before arriving at the sink.

Up to now, we have discussed only the pattern of arrivals of flow units at the sink and ignored the pattern of departures of flow units from the source. There are, of course, many situations in which the departure pattern at the source is important. For example, if you are shipping manufactured items out of your warehouse, you might prefer to dispatch them into the distribution network as soon as possible in order to minimize the storage costs at your warehouse. On the other hand, if the flow units represent schoolchildren returning to school (sink) after a summer holiday at home (source), it might be preferable for the flow units to remain at the source as long as possible (if the children's sentiments are taken into account).

In a similar way, we can define an *earliest departure flow for p time periods* and a *latest departure flow for p time periods*. The existence of each of these flows can be proved constructively in a way similar to the proof for the earliest arrival flow by substituting sources for sinks in the proof.

As was the case for the earliest (and latest) arrival flow, the construction suggested by the existence proof is not efficient computationally but seems to be the best method available.

Fortunately, however, for the special case when arc capacities are stationary through time, there is an efficient way to construct a latest departure flow. This construction is shown below:

Let $G = (X, A)$ be any graph. Define the *inverse graph* G^{-1} of graph G as the graph with vertex set X and arc set

$$A^{-1} = \{(y, x): (x, y) \in A\}$$

Thus graph $G^{-1} = (X, A^{-1})$ is simply graph G with its arc directions reversed. Let the capacity and traverse time of each arc in the inverse of the time-expanded replica graph equal the capacity and traverse time for the corresponding arc in the time-expanded replica graph.

Any flow from s to t in graph G corresponds to a unique flow from t to s in graph G^{-1} and vice versa by simply reversing the route taken by each flow unit.

Lemma 6.4. Suppose there exists a maximum dynamic flow F for p time periods in graph G in which x_i flow units depart from the source during time period i and y_i flow units arrive at the sink during time period i, for $i = 0, 1, \ldots, p$. Then there exists a maximum dynamic flow for p time periods in graph G^{-1} in which x_i units arrive at the sink during time period $p - i$ and y_i units depart from the source during time period $p - i$, for $i = 0$, $1, \ldots, p$.

Proof. The flow in graph G^{-1} that corresponds to flow F in graph G has the desired properties. Note that the flow units in graph G^{-1} travel backward in time, and hence their departure and arrival times must be counted downward from p rather than from time zero.   Q.E.D.

It follows from Lemma 6.4 that an earliest arrival flow in G for p time periods corresponds to a latest departure flow in graph G^{-1} for p time periods. Moreover, since $(G^{-1})^{-1} = G$, an earliest arrival flow in graph G^{-1} for p time periods corresponds to a latest departure flow in graph $(G^{-1})^{-1} = G$ for p time periods.

A latest departure flow for p time periods for graph G can be constructed as follows:

1. Construct graph G^{-1} by reversing all arcs in G.
2. Determine an earliest arrival flow for p time periods for graph G^{-1} using the earliest arrival flow algorithm.
3. Determine the corresponding flow in graph G. This flow is a latest departure flow for p time periods for graph G.

Hence, the earliest arrival flow algorithm can also be used to generate a latest departure flow.

Now suppose that an earliest arrival flow for graph G for p time periods has already been found by the earliest arrival flow algorithm. Is it necessary to perform the earliest arrival flow algorithm again (this second time on graph G^{-1}) to determine the latest departure flow for graph G for p time periods? The following theorem answers this question negatively.

Theorem 6.3. Suppose that the earliest arrival flow for graph G for p time periods consists of y_i flow units arriving at the sink during time period i, for $i = 0, 1, \ldots, p$. Then the latest departure flow for graph G for p time periods consists of y_i units departing from the source during time period $p - i$, for $i = 0, 1, \ldots, p$.

Proof. We shall prove this result by contradiction. Suppose that in any latest departure flow for graph G for p time periods Z, $Z \neq \Sigma_{j=0}^{j=1} y_j$, flow units depart from the source during the last i time periods. (This contradicts the theorem.) Then, from Lemma 6.4, there exists a maximum dynamic flow in graph G^{-1} for p time periods in which Z flow units arrive during the first i time periods at the sink. However, the maximum number of flow units that can be sent from source to sink in the last i time periods in graph G is equal to the maximum number of flow units that can be sent from the source to the sink in the first i time periods in graph G^{-1}. Thus, $Z = \Sigma_{j=0}^{j=1} y_j$, which is a contradiction. Q.E.D.

Hence, we can conclude that the earliest arrival and latest departure schedules for p time periods are symmetric; i.e., the number of flow units arriving at the sink at time period i in an earliest arrival flow equals the number of flow units leaving the source at time period $p - i$ in a latest departure flow.

Happily, the earliest arrival flow constructed by the earliest arrival flow algorithm possesses another important property:

Theorem 6.4. The earliest arrival flow constructed by the earliest arrival flow algorithm is also a latest departure flow.

Proof. Let $P_{i,j}$ denote, as before, the jth flow-augmenting path found during the ith iteration of the minimum-cost flow algorithm. The earliest arrival flow algorithm for p time periods will send one flow unit along path $P_{i,j}$ from the source to the sink at times 0, $i, \ldots, p - i$ by Lemma 6.2. These flow units arrive at the sink at times $i, i + 1, \ldots, p$ by Lemma 6.2. Since this is true for every flow-augmenting path $P_{i,j}$, for each flow unit arriving at the sink at time $p - k$ there is a flow unit departing from the source at time k. Hence, the earliest arrival flow constructed by the earliest arrival flow algorithm has a latest departure schedule. Q.E.D.

For example, consider the earliest arrival flow generated by the earliest arrival flow algorithm in Fig. 6.24. The departure schedule of this flow is

Time period	0	1	2	3	4	5	6
Number of units	2	2	1	1	0	0	0

The arrival schedule for this flow is

Time period	0	1	2	3	4	5	6
Number of units	0	0	0	1	1	2	2

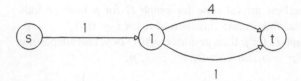

Fig. 6.25 Counterexample. All arc capacities equal 1. Arc transit times are indicated.

Notice the reverse symmetry between the arrival schedule and the departure schedule.

Unfortunately, the same reverse symmetry does not hold between latest arrival flows and earliest departure flows. To demonstrate this, consider the graph in Fig. 6.25. For p = 5, a maximum of 4 flow units can be sent from the source to the sink. The earliest departure schedule is by inspection:

Time period	0	1	2	3	4	5
Number of units	1	1	1	1	0	0

The latest arrival schedule is by inspection:

Time period	0	1	2	3	4	5
Number of units	0	0	0	1	1	2

Consider a set $S = \{a, b, c, \ldots\}$. We say that there is a *lexicographic preference a*, $b, c, \ldots$ on set S if one unit of item a is preferred to any number of units of the remaining items $b, c, \ldots$, and one unit of item b is preferred to any number of units of the remaining items $c, \ldots$, etc. An earliest arrival flow is called a *lexicographic flow* since it places a lexicographic preference $t_0, t_1, t_2, \ldots, t_p$ on the sinks to be used by arriving flow units. Similarly, the latest arrival flow, earliest departure flow, and latest departure flow are also lexicographic flows since for each of these flows there is some lexicographic preference on which sources or sinks the flow units are to use.

We shall conclude this section with a result which shows that for maximum flows any departure schedule at the source is compatible with any arrival schedule at the sink.

Theorem 6.5. Let F' and F'' be any two maximum flows on graph G. Then there exists a maximum flow on graph G that simultaneously has the same departure schedule at the source as flow F' and the same arrival schedule at the sink as flow F''.

Proof. Select any minimum cut separating the source from the sink in graph G. Since both F' and F'' are maximum flows, both saturate this cut. Construct a hybrid maximum flow from F' and F'' that is identical to flow F' on the source side of this cut and is identical to flow F'' on the sink side of this cut. This hybrid flow has the departure schedule identical to flow F' and an arrival schedule identical to flow F''. Thus, the hybrid flow satisfies the theorem. Q.E.D.

Lemma 6.5. The following dynamic flows exist:

1. Earliest departure and earliest arrival
2. Earliest departure and latest arrival
3. Latest departure and earliest arrival
4. Latest departure and latest arrival

Proof. The proof is achieved by applying Theorem 6.5 to graph G_p. Q.E.D.

As shown in Theorem 6.4, the earliest arrival flow algorithm constructs an earliest arrival and latest departure flow, i.e., item 3 in Lemma 6.5. The three remaining flows, items 1, 2, and 4, could be constructed by splicing together two flows as suggested by the proof of Theorem 6.5. Unfortunately, finding more efficient methods of constructing the other three flows seems to be an open question.

An excellent review of algorithms and applications for dynamic network flow is given by Aronson (1989).

APPENDIX: USING NETSOLVE TO FIND MAXIMUM FLOWS

The NETSOLVE maximum-flow algorithm determines a maximum flow from a source to a sink in a capacitated, directed network. Edge flows must lie between a nonnegative lower bound and an upper bound specified by the LOWER and UPPER data fields, respectively. The value of a flow is the net flow leaving the source (or equivalently, entering the sink).

The algorithm is executed by entering the command line

> MAXFLOW ORIGIN DEST

which would find the maximum flow from the node named ORIGIN to the node named DEST. If an optimal solution is found, the first line of output contains the value of the optimal flow. If the terminal is ON, a listing of the edge data fields and edge flows for edges having nonzero flow is produced. If an optimal solution does not exist, then an indication is given of whether the problem is infeasible or the solution is unbounded. Additional information is produced if the terminal is ON and the formulation is infeasible. In that case, a list of nodes where flow conservation cannot be attained is displayed.

We will illustrate the use of this procedure by solving the site selection example shown in Fig. 6.4. A listing of the input data is shown in Fig. 6.26. The LOWER and UPPER fields corresponding to arcs between the revenue nodes and cost nodes have been defaulted; thus, the capacity is set to 999999 by NETSOLVE. The solution to the maximum-flow problem between nodes s and t is given in Fig. 6.27. Figure 6.28 shows

FROM	TO	UB
1	T	6.
12	1	999999.
12	2	999999.
13	1	999999.
13	3	999999.
14	1	999999.
14	4	999999.
2	T	5.
23	2	999999.
23	3	999999.
24	2	999999.
24	4	999999.
3	T	4.
34	3	999999.
34	4	999999.
4	T	5.
S	12	7.
S	13	5.
S	14	1.
S	23	6.
S	24	2.
S	34	1.

Fig. 6.26 NETSOLVE edge data for maximum-flow example.

MAXIMUM FLOW PROBLEM: MAXIMUM FLOW IS   19.

FROM	TO	LOWER	FLOW	UB
12	1	0.	5.	999999.
13	1	0.	1.	999999.
S	12	0.	7.	7.
S	13	0.	5.	5.
S	14	0.	1.	1.
12	2	0.	2.	999999.
23	2	0.	3.	999999.
S	23	0.	3.	6.
S	24	0.	2.	2.
13	3	0.	4.	999999.
S	34	0.	1.	1.
14	4	0.	1.	999999.
24	4	0.	2.	999999.
34	4	0.	1.	999999.
1	T	0.	6.	6.
2	T	0.	5.	5.
3	T	0.	4.	4.
4	T	0.	4.	5.

Fig. 6.27 NETSOLVE output for maximum-flow problem.

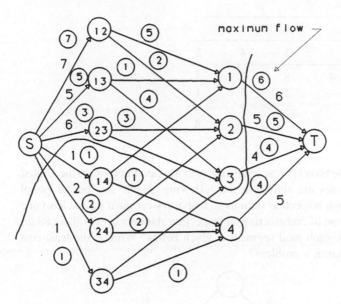

Fig. 6.28 Optimal solution (max flow and min cut).

this solution on the network. If you examine the residual network, you will find that nodes s, 12, 13, 23, 1, 2, and 3 are labeled at termination of the maximum-flow algorithm. This defines the cut shown in Fig. 6.28. The labeled "cost" nodes corresponding to the sites that should be built. Thus, we should build sites 1, 2, and 3. In the chapter we noted that

$$\text{Capacity of the minimum cut} = \Sigma \, r_{ij} - (\text{revenue} - \text{cost})$$

Since the maximum flow value $= 19 =$ capacity of the minimum cut and $\Sigma \, r_{ij} = 22$, we find that the net profit will be 3. This can be confirmed easily from the original data.

EXERCISES

1. A travel agent must arrange for the flights of a group of ten tourists from Chicago airport to Istanbul airport on a certain day. On that day, there are 7 seats left on the Chicago-Istanbul direct flight, there are 5 seats left on the Chicago-Paris flight, and there are 4 seats left on the connecting Paris-Istanbul flight. What should the agent do?

2. Formulate and solve the following site selection problem as a maximum-flow problem. What is the minimum cut?

i	c_i	1	2	3	4	5
				r_{ij}		
1	3	—	4	8	2	5
2	6		—	2	5	1
3	4			—	7	9
4	3				—	8
5	6					—

3. A large number of people travel by car from city A in Mexico to city B in the United States. The possible routes are shown in the following graph. The border patrol wishes to construct enough inspection stations so that every car must pass at least one inspection station. The cost of constructing an inspection station varies with location. The cost associated with each road segment is given below. What is the least-cost solution to the border patrol's problem?

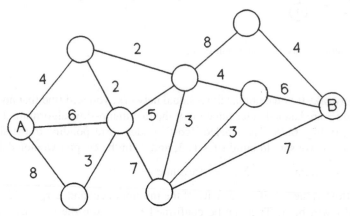

4. List all possible flow-augmenting paths from s to t in the following graph:

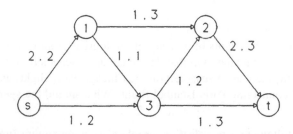

current flow, capacity

5. Formulate the maximum-flow problem for the network given in Exercise 4 as a linear program.

6. Use the Ford-Fulkerson maximum-flow algorithm to find the maximum flow for the network in Exercise 4. Begin with the current flow given. Determine the minimum cut.

7. Apply the path decomposition algorithm to the solution to Exercise 6.

8. If the maximum-flow algorithm discovers a flow-augmenting path that contains a backward arc, some flow units will be removed from this backward arc and rerouted in their journey to the sink.
   (a) Is it possible for the maximum-flow algorithm never to reroute any flow units? If so, what conditions generate such a situation?
   (b) Under what conditions can you be certain that the flow units assigned to a specific arc will not be rerouted during a subsequent iteration of the maximum-flow algorithm.

9. Find the maximum flow in the following network by solving a shortest-path problem on the dual graph.

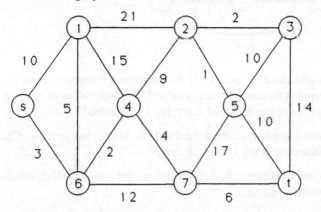

10. Apply the preflow-push maximum-flow algorithm to the following network.

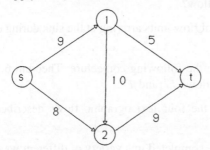

11. Suppose that the lower bound on arcs (s, 1) and (1, 2) of the network in Exercise 10 is 4 units. Construct a feasible flow using the maximum-flow algorithm and show how to modify the maximum flow algorithm to find the maximum flow from the initial feasible flow.

12. The travel agent in Exercise 1 wishes to send 75 passengers from Chicago to Istanbul within the next 48 hours. Explain how to model this problem as a dynamic flow problem.

13. Construct a maximum dynamic flow for 10 time periods for the graph given below. The number adjacent to each arc denotes the arc capacity and arc transit time.

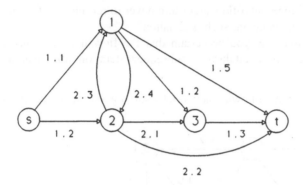

14. Suppose that the transit time of arc (1, 3) in Exercise 13 changes from 2 time periods to 1 time period at the end of time period x. For which values of x would this traverse time change leave the result of Exercise 13 unaltered?

15. Prove that a flow unit cannot travel backward in time in a dynamic flow. Can the time-expanded replica of a graph contain a cycle?

16. Generalize the maximum dynamic flow algorithm to the case in which arc transit times are not necessarily integers.

17. Under what conditions will the maximum dynamic flow algorithm produce an earliest arrival flow? Latest departure flow?

18. Under what conditions will the number of flow units arriving at the sink during each time period remain a constant?

19. Construct a counterexample to disprove the following conjecture: Theorem 6.5 of Section 6.5 is valid for nonmaximum flows F' and F''.

20. For the graph in Fig. 6.24, construct the four lexicographic flows described in Lemma 6.5.

21. A management training program may be completed in a variety of different ways as shown below. Each arc represents a subprogram. Each trainee must start at s and

pursue subprograms until he or she graduates at t. The number of trainees allowed in each subprogram is limited. Although each subprogram takes exactly one month to complete, the cost per student varies. Maximum enrollment and cost are shown in the following graph for each subprogram. What is the maximum number of students who may complete the program within the next five months so that not more than $1000 is spent on the training of any one student?

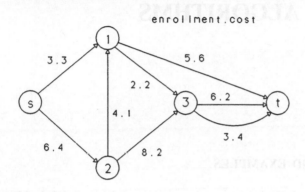

enrollment.cost

REFERENCES

Aronson, J. E., 1989. A Survey of Dynamic Network Flows, *Ann. Oper. Res.*

Edmonds, J., and R. M. Karp, 1972. Theoretical Improvements in Algorithm Efficiency for Network Flow Problems, *J. ACM*, *19*, no. 2, pp. 218–264.

Ford, L. R., and D. R. Fulkerson, 1962. *Flows in Networks*, Princeton University Press, Princeton, New Jersey.

Grinold, R. C., 1973. Calculating Maximal Flows in a Network with Positive Gains, *Oper. Res.*, *21*, pp. 528–541.

Jewell, W. S., 1962. Optimal Flow through a Network with Gains, *Oper. Res.*, *10*, pp. 476–499.

Johnson, E. L., 1966. Networks and Basic Solutions, *Oper. Res.*, *14*, pp. 619–623.

Karzanov, A. V., 1974. Determining the Maximal Flow in a Network by the Method of Preflows, *Sov. Math. Dokl.*, *15*, pp. 434–437.

Maurras, J. F., 1972. Optimization of the Flow through Networks with Gains, *Math. Program.*, *3*, pp. 135–144.

Maxwell, W. L., and R. C. Wilson, 1981. Dynamic Network Flow Modeling of Fixed Path Material Handling Systems, *AIIE Trans.*, *13*, no. 1, pp. 12–21.

Minieka, E. T., 1973. Maximum, Lexicographic and Dynamic Network Flows, *Oper. Res.*, *21*, pp. 517–527.

Minieka, E. T., 1972. Optimal Flow in a Network with Gains, *INFOR*, *10*, pp. 171–178.

Truemper, K., 1973. Optimum Flow in Networks with Positive Gains, Ph.D. Thesis, Case Western Reserve University, Cleveland.

Truemper, K., 1976. An Efficient Scaling Procedure for Gains Networks, *Networks*, *6*, pp. 151–160.

Wilkinson, W. L., 1971. An Algorithm for Universal Maximal Dynamic Flows in a Network, *Oper. Res.*, *19*, pp. 1602–1612.

7
MATCHING AND ASSIGNMENT ALGORITHMS

7.1 INTRODUCTION AND EXAMPLES

A *matching* in a graph is any set of edges such that each vertex of the graph is incident to *at most* one edge in this set. For example, in Fig. 7.1 the sets $\{(1, 2), (5, 3)\}$, $\{(2, 3), (4, 5)\}$, and $\{(1, 4)\}$ all are matchings. Clearly, any subset of a matching is also a matching.

A *bipartite* graph $G(S, T, E)$ is a graph $G(X, E)$ whose vertex set X can be partitioned into two subsets S and T such that no edge in the graph joins two vertices in the same subset. Thus, each edge in a bipartite graph has one end in S and the other in T. Figure 7.2 is an example of a bipartite graph with $S = \{1, 2, 3\}$ and $T = \{4, 5, 6\}$. Any matching in a bipartite graph must have one end of each edge in S and the other end in T. In Fig. 7.2, the set $\{(1, 5), (2, 4), (3, 6)\}$ is an example of a matching. A matching in a bipartite graph is commonly called an *assignment*. Bipartite graphs possess properties which general graphs do not possess. For instance, every cycle in a bipartite graph must contain an even number of edges. This follows because each edge joins vertices in different subsets and the cycle must return to the subset in which it originated. Also, every path in a bipartite graph alternates between vertices in S and vertices in T. Such properties often lead to more efficient algorithms than for general graphs.

Matching and assignment problems have many practical applications.

Crew Scheduling

Consider the problem of scheduling mass transit vehicles, for example, buses. A single vehicle's schedule is defined by the time of a crew change (T) and the location of the change (L) as shown in Fig. 7.3. Each segment in which the same crew is on the vehicle is called a piece. For example, a piece might typically be 4 hours.

1
2
3
4
5

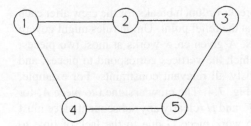

Fig. 7.1 An undirected graph.

1
4
2
5
3
6

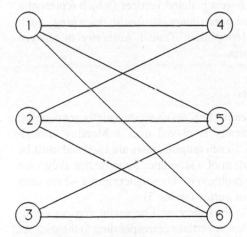

Fig. 7.2 A bipartite graph.

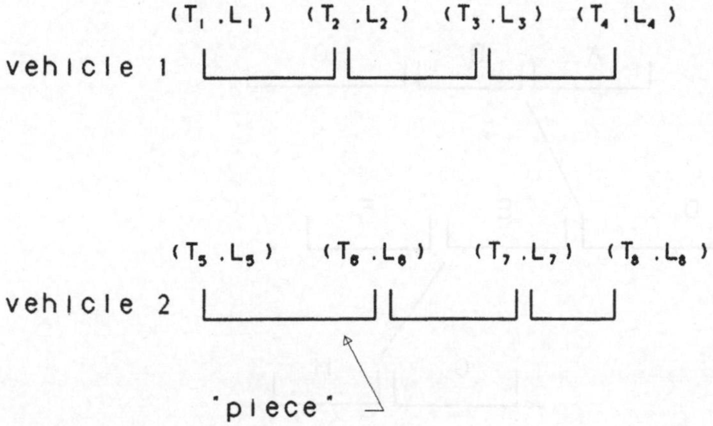

(T1 . L1) (T2 . L2) (T3 . L3) (T4 . L4)
vehicle 1
(T5 . L5) (T6 . L6) (T7 . L7) (T8 . L8)
vehicle 2
'piece'

Fig. 7.3 Crew scheduling structure.

Constraints usually take the form of rest breaks or lunch breaks for the crew after each piece. The vehicle continues with a new crew at the relief point. Union rules might govern the acceptable lengths and locations of breaks. A given crew works at most two pieces, called a *run*. We may construct a graph in which the vertices correspond to pieces, and there exists an edge between pieces that satisfy all relevant constraints. For example, consider the three-vehicle schedule shown in Fig. 7.4. The crew assigned to piece *A*, for instance, might be eligible to work pieces *C*, *F*, and *H* (clearly any subsequent piece must occur later in time) but might not be able to work piece *G* due to the lack of time to transfer locations.

The associated graph is shown in Fig. 7.5. We might seek a matching having the largest number of edges and consequently the fewest isolated vertices (which represent a one-piece run). This type of matching is called a *maximum-cardinality matching*. Such models have been used by the Baltimore Metropolitan Transit Authority to analyze alternative changes in timetables and work rules.

Major League Baseball Umpire Assignments

The American Baseball League consists of 14 teams; thus seven series usually are ongoing at any time during the season. A series typically is played over a Monday through Thursday, or over a weekend (Friday–Sunday). Seven umpire crews are used and must be assigned to each series. Figure 7.6 shows a portion of a schedule. Suppose that crews are assigned to each of the games in series 2. The problem involves determining where each crew should be assigned during the next set of series (series 3).

We may represent each series by a vertex; thus there will be seven vertices corresponding to the current series assignments and seven vertices corresponding to the next set of series. We draw an edge from vertex *x* to vertex *y* if the crew assigned to the series at vertex *x* can be assigned to the series at vertex *y*. Thus, the graph will be bipartite. Many assignments are infeasible due to travel restrictions. For instance, a crew on the west coast

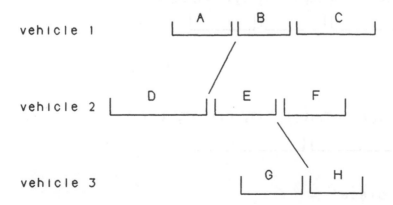

Fig. 7.4 Three-vehicle schedule.

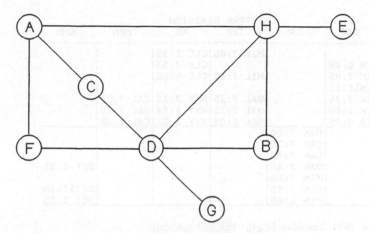

Fig. 7.5 Graph corresponding to crew scheduling example.

(SEA, CAL, OAK) requires a day off in order to travel to any city in the Eastern Division or Chicago. Similarly, a crew cannot travel to another city if it has just concluded a night game and has a day game scheduled on the next day. An example of the graph corresponding to Fig. 7.6 is shown in Fig. 7.7. The problem is to determine a matching or assignment of the two sets of series that will minimize the travel cost or some related measure of performance. This is an example of a *weighted matching* problem on a bipartite graph.

Basketball Conference Scheduling

A basketball conference consists of 10 teams, each playing one home and one away game with every other team in the conference. A team can play at most two games over a weekend. We seek to minimize the total distance traveled by all teams. Thus we must schedule $10 \times 9 = 90$ home games with these restrictions.

Suppose the teams $A, B, C, D, E, F, G, H, I,$ and J are placed into five groups of two each. Consider the following strategy for the pairing $\{X, Y\}$. On a given weekend, either team X plays at home versus team Y for one game, or some other team Z_1 travels to X and then to Y and returns home, and the second team of the pairing $\{Z_1, Z_2\}$ travels to Y and then to X and returns home. For example, if we group the ten teams as $\{A, B\}$, $\{C, E\}$, $\{D, H\}$, $\{F, J\}$, $\{G, I\}$, then one weekend schedule might look like the following:

Team A plays at C and E; team B plays at E and C; team D plays at F and J; team H plays at J and F; team G plays at I.

Schedules for other weeks can be constructed in a similar fashion by simply associating the groups in different ways.

We can construct a network model for this problem by letting each vertex correspond to a team and assigning a weight to each edge (X, Y) as the travel distance incurred if team

```
                                   WESTERN DIVISION
Series      SEA        OAK        CAL        TEX       KC        MIN       CWS
       --------------------------------------------------------------------
            |          |          |          |MIL 7:05|CLE 1:35|          |
            |CAL 7:35|MIN 8:05|          |         |CLE 7:35|          |
       1    |CAL 7:05|MIN 7:05|          |MIL 7:35|CLE 4:05|          |
            |CAL12:35|MIN12:15|          |         |         |          |
            |          |SEA 7:35|          |BAL 7:35|NYY 7:35|CAL 7:05|
       2    |          |SEA 1:05|          |BAL 7:35|NYY 1:35|CAL 7:05|
            |          |SEA 5:05|          |BAL 2:05|NYY 1:35|CAL 1:05|
            |MIN 7:05|          |OAK 7:00|         |         |          |
       3    |MIN 7:35|          |OAK 7:35|         |         |          |
            |MIN 7:05|          |OAK 7:35|         |         |          |
            |          |          |OAK 7:35|         |         |          |DET 1:35
            |OAK 7:35|          |MIN 7:35|         |         |          |
       4    |OAK 7:05|          |MIN 7:05|         |         |          |DET12:15
            |OAK 1:35|          |MIN 1:05|         |         |          |DET 1:35
```

Fig. 7.6   A portion of the 1991 American League baseball schedule.

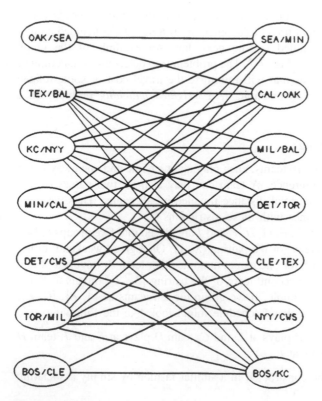

Fig. 7.7   Feasible umpire crew assignments.

```
                          EASTERN DIVISION
   1991    MIL      DET      CLE      TOR      BAL      NYY      BOS
  ----------------------------------------------------------------------
  |04/08|        |NYY 1:35|        |BOS 2:05|CWS 2:05|        |
  |04/09|        |        |        |BOS 7:35|        |        |
  |04/10|        |NYY 1:35|        |BOS 7:35|CWS 7:35|        |
  |04/11|        |NYY 1:35|        |MIL 7:35|        |        |CLE 1:05
  |04/12|        |CWS 7:35|        |MIL 7:35|        |        |
  |04/13|        |CWS 1:15|        |MIL 1:35|        |        |CLE 1:05
  |04/14|        |CWS 1:35|        |MIL 1:35|        |        |CLE 1:05
  |04/15|BAL 1:35|TOR 7:35|        |        |        |CWS 1:00|CLE11:05
  |04/16|        |TOR 7:35|TEX 1:35|        |        |CWS 7:30|KC  1:05
  |04/17|BAL 6:05|TOR 1:35|        |        |        |CWS 1:00|KC  1:05
  |04/18|BAL 1:35|        |TEX 7:35|        |        |        |KC  1:05
  |04/19|TOR 6:05|        |BOS 7:35|        |TEX 7:35|KC  7:30|
  |04/20|TOR 1:35|        |BOS 1:35|        |TEX 1:35|KC  1:30|
  |04/21|TOR 1:35|        |BOS 1:35|        |TEX 2:35|KC  1:30|
  ----------------------------------------------------------------------
```

Fig. 7.6   (*Continued*)

X visits Y. Figure 7.8 shows the interpretation of the pairing $\{A, B\}$, $\{C, D\}$, . . . , $\{I, J\}$ from the perspective of team A. Since travel to $B, C, . . . , J$ must be incurred some time, the sum of such distances is fixed. However, the sum of the distances between successive opponents will vary with the pairings. Thus to minimize the total distance traveled by team A, we need to minimize the sum $d(A, B) + d(C, D) + \cdots + d(I, J)$. Since the edges must not be adjacent, we require a matching of the vertices that minimizes the total weight. The intuitive idea is that we wish to group together teams that are in relatively close proximity. This will actually minimize the total distance traveled by all teams from a global perspective. From this, several schedules that satisfy other constraints can be constructed easily. This problem is a *weighted matching problem on a general graph*.

In this chapter we shall present algorithms for maximum-cardinality and maximum-weight matchings in both bipartite and general graphs.

7.2   MAXIMUM-CARDINALITY MATCHING IN A BIPARTITE GRAPH

When graph $G = (X, E)$ is a bipartite graph, the maximum-cardinality matching problem can be solved easily by any maximum-flow algorithm found in Chapter 6. This is accomplished as follows:

1. Direct all edges from S to T.
2. Create a source vertex s and an arc (s, x) from the source to each vertex $x \in S$.
3. Create a sink vertex t and an arc (x, t) from each $x \in T$ to the sink.
4. Let each arc capacity equal 1 (see Fig. 7.9).

Call this graph G'. Since all arc capacities in G' equal 1, all flow-augmenting paths will carry 1 flow unit if the maximum-flow algorithm is started with a zero flow. Thus, the maximum-flow algorithm will terminate with a maximum flow in which each arc carries

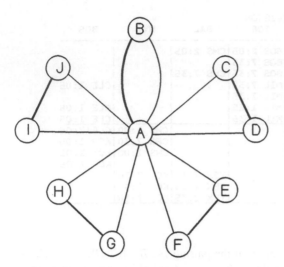

Fig. 7.8 An example of basketball team pairings.

either one flow unit or no flow units. The arcs from S to T in G' that carry one flow unit correspond to a matching in G. Moreover, each matching in G corresponds to a flow in G' in which the edges in the matching correspond to arcs in G' that carry one flow unit. Consequently, the matching in G corresponding to a maximum flow in G' must be a maximum-cardinality matching in G'. If this matching were not a maximum-cardinality

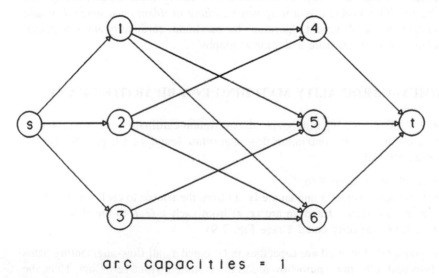

arc capacities = 1

Fig. 7.9 Flow network.

matching, there would be an even larger flow in G' corresponding to the maximum-cardinality matching, which is a contradiction. So, for bipartite graphs, the maximum-cardinality matching problem can be solved by using the maximum-flow algorithm.

7.3 MAXIMUM-CARDINALITY MATCHING IN A GENERAL GRAPH

If graph G is not bipartite, then G must contain a cycle with an odd number of edges in it (an odd cycle). Otherwise, the vertices of G could be partitioned into subsets S and T as described above. The presence of odd cycles complicates matters, since now there is no obvious way to convert the matching problem into a flow problem.

How about trying to solve the maximum-cardinality matching problem using linear programming? Consider the following linear programming problem:

Maximize

$$\sum_{(i,\,j)} x(i, j) \tag{1}$$

such that

$$\sum_{j} [x(j, i) + x(i, j)] \leq 1 \quad \text{for all vertices } i \tag{2}$$

$$0 \leq x(i, j) \leq 1 \quad \text{for all edges } (i, j) \tag{3}$$

where $x(i, j)$ denotes the number of times edge (i, j) is used in the matching. Each constraint (2) requires that the number of matching edges incident to vertex i not exceed one.[†] Each constraint (3) requires that each edge (i, j) not be used in the matching more than once.

Clearly, every matching satisfies constraints (2) and (3) when $x(i, j) = 1$ if (i, j) is in the matching and $x(i, j) = 0$ if (i, j) is not in the matching. However, noninteger values for the variables $x(i, j)$ may also satisfy constraints (2) and (3). For example, consider the graph in Fig. 7.10. The solution

$$x(1, 2) = x(2, 3) = x(3, 1) = \tfrac{1}{2}$$

satisfies constraints (2) and (3). For this solution the objective function (1) equals $\tfrac{1}{2} + \tfrac{1}{2} + \tfrac{1}{2} = 1\tfrac{1}{2}$. By inspection, the largest possible value for the objective function for a solution that is a matching is 1. For example, for the matching

$$x(1, 2) = 1, \quad x(2, 3) = x(3, 1) = 0$$

the objective function (1) equals $1 + 0 + 0 = 1$. Consequently, we cannot always be sure that an optimal solution to the linear programming problem (1)–(3) will be a matching. (In Section 7.5 additional constraints will be added to this linear programming

---

[†] Recall that an edge joining vertices i and j is denoted by either (i, j) or (j, i).

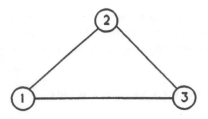

Fig. 7.10

problem so that the optimal solution will be a matching. However, so many additional constraints will be needed that the linear programming problem would become too large to solve efficiently.)

Alas, not being able to use either flow algorithms or linear programming to solve the maximum-cardinality matching problem for nonbipartite graphs, we must study an algorithm specifically designed for the maximum-cardinality matching problem. The remainder of this section presents the *maximum-cardinality matching algorithm* due to Edmonds (1965).

Given a matching M in graph G, an *alternating path* is a simple path in which the edges are alternately in and out of matching M. For example, for the matching in Fig. 7.11, the path $(1, 2)$, $(2, 3)$, $(3, 6)$ is an alternating path since the first and third edges $(1, 2)$ and $(3, 6)$ are not in M and the second edge $(2, 3)$ is in M. Also, the path $(4, 1)$, $(1, 2)$, $(2, 3)$, $(3, 6)$ is an alternating path since its first and third edges are in M and its second and fourth edges are not in M.

A vertex is called *exposed* if it is not incident to any matching edge. A vertex is called *matched* if it is incident to a matching edge. An *augmenting path* is an alternating path whose first and last vertices are exposed. For example, in Fig. 7.11 the path $(5, 2)$, $(2, 3)$, $(3, 6)$ is an augmenting path.

If the edges in an augmenting path have their roles in the matching reversed (i.e., matching edges are removed from the matching, edges not in the matching are placed in the matching), then the resulting matching contains one more edge than did the original matching. Consequently, if a matching possesses an augmenting path, then the matching cannot be a maximum-cardinality matching. Moreover, the converse of this result is also true.

Theorem 7.1. If matching M is not a maximum-cardinality matching, then matching M possesses an augmenting path.

Proof. Let M^* denote the maximum-cardinality matching that has the most edges in common with M. Let G' denote the subgraph consisting of all edges that are in exactly one of these two matchings M and M^*.

No vertex of G' can have more than two edges incident to it. Otherwise, two edges in the same matching would be incident to the same vertex, which is impossible. For this

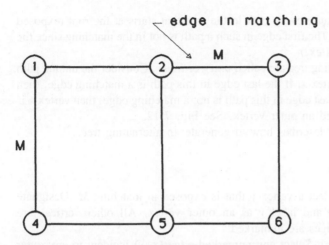

Fig. 7.11 Matching example.

reason, each connected component of G' must be either a simple path or a simple cycle.

A connected component of G' cannot be an odd cycle. Otherwise, one vertex in this odd cycle would be incident to two edges in the same matching, which is impossible. Moreover, a connected component of G' cannot be an even cycle. If a connected component of G' were an even cycle, then the role of each edge in this even cycle in M^* could be reversed. This new matching would contain the same number of edges as M^* but would have more edges in common with M, which is impossible. Consequently, no connected component of G' can be a cycle.

Thus, each connected component of G' is a simple path. Consider the first and last edges in such a path. If both of these edges are in M, then this path is an augmenting path in M^*, which contradicts the assumption that M^* is a maximum-cardinality matching. If one of these edges is in M and the other is in M^*, then the roles of the edges in this path can be reversed in matching M^*. This creates a new matching with the same number of edges as M^* and with more edges in common with M, which is a contradiction.

Consequently, both the first and last edges of the path must be in M^*, which implies that this path is an augmenting path in M. Q.E.D.

A matching is a maximum-cardinality matching if, and only if, it does not contain an augmenting path. The *maximum-cardinality matching algorithm* is based on this result. The algorithm selects an exposed vertex and searches for an augmenting path from this vertex to some other exposed vertex. If an augmenting path is found, the roles in the matching of the edges in this path are reversed. This creates a matching with greater cardinality. If no augmenting path is found, another exposed vertex is similarly examined. The algorithm stops when all exposed vertices have been examined.

The maximum-cardinality matching algorithm searches for an augmenting path rooted at an exposed vertex by growing a tree rooted at this exposed vertex. This tree is called an *alternating tree* because every path in the tree that starts at the root (exposed vertex) is an alternating path. (The first edge in such a path is not in the matching since the path begins at an exposed vertex.)

The vertices in an alternating tree are called *outer* or *inner*. Consider the unique path in the tree from the root to vertex x. If the last edge in this path is a matching edge, then vertex x is called outer. If the last edge in this path is not a matching edge, then vertex x is called inner. The root is called an outer vertex. See Fig. 7.12.

The following procedure describes how to generate an alternating tree.

Alternating-Tree Subroutine

Step 1 (Initialization). Select a vertex v that is exposed in matching M. Designate vertex v as the root and label v as an outer vertex. All other vertices are unlabeled, and all edges are unmarked.

Step 2 (Growing the Tree). Select any unmarked edge (x, y) incident to any outer vertex x. (If no such edge exists, go to step 4.) Three cases are possible:

(a)   Vertex y is an inner vertex.
(b)   Vertex y is an outer vertex.
(c)   Vertex y is not labeled.

If item (a) occurs, then mark (x, y) as *not* in the tree and repeat step 2.

If item (b) occurs, then mark (x, y) as in the tree and go to step 3.

If item (c) occurs, mark (x, y) as in the tree. If y is an exposed vertex, stop because an augmenting path of in-tree edges from v to y has been found. If y is a matched vertex, then mark the unique matching edge (y, z) incident to y in the tree. Label vertex y as an inner vertex and label vertex z as an outer vertex. Return to step 2.

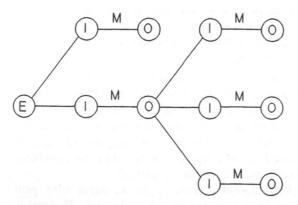

Fig. 7.12 Example of alternating tree. E = exposed, I = inner, O = outer, M = matching edge.

Step 3 (Odd Cycle). This step is reached only after an edge (x, y) with two outer endpoints has been marked as in the tree. This creates a cycle of in-tree edges. This cycle C must be an odd cycle since the vertices in C from x to y alternate between outer and inner vertices. Stop, because an odd cycle has been discovered.

Step 4 (Hungarian Tree). This step is reached only after no further marking is possible. The edges in the tree form an alternating tree called a *Hungarian tree*. Stop.

Note that the alternating tree subroutine stops with (a) an augmenting path, (b) an odd cycle, or (c) a Hungarian tree.

EXAMPLE 1

Let us generate an alternating tree rooted at exposed vertex 1 in Fig. 7.13.

Step 1. Vertex 1 is labeled outer.
Step 2.

Edge under Examination	Outer Vertices	Inner Vertices	In-Tree Edges
Initial	1	None	None
(1, 8)	1, 7	8	(1, 8), (8, 7)
(7, 4)	1, 7, 4	8, 4	(1, 8), (8, 7)
			(7, 4), (4, 5)
(1, 2)	1, 7, 5, 3	8, 4, 2	(1, 8), (8, 7)
			(7, 4), (4, 5)
			(1, 2), (2, 3)
(5, 6)	1, 7, 5, 3, 6	8, 4, 2	(1, 8), (8, 7)
			(7, 4), (4, 5)
			(1, 2), (2, 3)
			(5, 6)

Stop, an augmenting path (1, 8), (8, 7), (7, 4), (4, 5), (5, 6) has been found.

Note that often in step 2 there are several edges that can be examined. The choice is arbitrary. If edge (7, 6) had been examined instead of edge (7, 4), the augmenting path (1, 8), (8, 7), (7, 6) would have been discovered. Hence, the result of the subroutine may depend on the arbitrary choice of which edge to examine next. Lastly, note that if step 2 marks a pair of edges as in the tree, one of these edges is a matching edge; the other is not a matching edge.

The maximum-cardinality matching algorithm can be initialized with any matching. The algorithm selects an exposed vertex and grows an alternating tree rooted at this vertex using the alternating tree subroutine. If an augmenting path is discovered, the roles of the edges in this path are reversed. This generates a new matching with greater cardinality and matches the exposed root of the alternating tree.

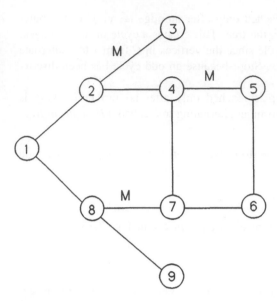

Fig. 7.13 M = matching edge

If the alternating tree turns out to be a Hungarian tree, then no augmenting path exists from this root. (See the proof of the algorithm to verify this.) If an odd cycle is discovered, this cycle is shrunk into a single vertex. The algorithm is continued on the new smaller graph resulting from shrinking this odd cycle.

After each exposed vertex has been examined by the alternating tree subroutine, the resulting matching is a maximum-cardinality matching for the terminal graph. Next, the algorithm judiciously expands out all shrunken cycles inducing a matching on the edges of each of these cycles. After all shrunken cycles have been expanded out, the process is repeated for another unexamined exposed vertex.

With this as motivation, we can now formally state the maximum-cardinality matching algorithm.

Maximum-Cardinality Matching Algorithm

Step 1 (Initialization). Denote the graph under consideration by G_0. Select any matching M_0 for graph G_0. All exposed vertices are called *unexamined*. Let $i = 0$.

Step 2 (Examination of an Exposed Vertex). If all vertices are incident to a matched edge, stop. If graph G_0 has only one unexamined exposed vertex, go to step 7. Otherwise, select any unexamined exposed vertex v. Use the alternating tree subroutine to grow an alternating tree rooted at vertex v.

If the alternating tree subroutine terminates with an augmenting path, go to step 3. If the alternating tree subroutine terminates with an odd cycle, go to step 4. If the alternating tree subroutine terminates with a Hungarian tree, go to step 5.

Step 3 (Augmenting Path). This step is reached only after the alternating tree subroutine discovers an augmenting path. Reverse the roles in matching M_i of the edges in the augmenting path. This increases by one the cardinality of matching M_i and matches vertex v. Go to step 6.

Step 4 (Odd Cycle). This step is reached only after the alternating tree subroutine discovers an odd cycle. Let $i = i + 1$. Denote the odd cycle by C_i. Shrink C_i into an artificial vertex a_i. Each edge incident to a single vertex in C_i now becomes incident to a_i. Call the new graph G_i. Let matching M_i consist of all edges in M_{i-1} that are in G_i. (Note that all but one of the vertices in C_i are matched by edges in C_i. Hence, after C_i is shrunk into vertex a_i, at most one matching edge is incident to a_i.) Return to step 2, selecting as the exposed vertex for examination the image of vertex v in graph G_i. Note that most of the labeling and marking from the previous iteration of the alternating tree subroutine can be reused in the coming iteration of the alternating tree subroutine in step 2.

Step 5 (Hungarian Tree). This step is reached only after the alternating tree subroutine discovers a Hungarian tree rooted at exposed vertex v. Vertex v is now called *examined*. Go to step 6.

Step 6 (Exploding Shrunken Odd Cycles). During the repetitions of step 4, a sequence $G_1, G_2, \ldots, G_t$ of graphs was generated, and a sequence $a_1, a_2, \ldots, a_t$ of artificial vertices was generated. Also, a sequence $M_1, M_2, \ldots, M_t$ of matching in $G_1, G_2, \ldots, G_t$, respectively, was generated.
Let $M_t^* = M_t$.
For $j = t, t - 1, \ldots, 1$, generate from matching M_j^* in graph G_j a matching M_{j-1}^* in graph G_{j-1} as follows:

(a) If vertex a_j is matched in matching M_j^*, then let M_{j-1}^* consist of all edges in M_j^* together with the unique set of edges in odd cycle C_i that match all the exposed vertices in C_i (see Fig. 7.14). Go to step 2.

(b) If vertex a_j is exposed in matching M_j^*, then let M_{j-1}^* consist of all edges in M_j^* together with any set of edges in C_i that match all but one of the vertices in C_i. The vertex in C_i that remains exposed is chosen arbitrarily (see Fig. 7.15). Go to step 2.

Step 7. Matching M_0^* is a maximum-cardinality matching for the original graph G_0. Stop.

Proof. If the examination of an exposed vertex results in the discovery of an augmenting path in step 3, then it is easily seen from the explosion of shrunken odd cycles in step 6 that the new matching has one more edge than the preceding matching and that the exposed vertex under examination is no longer exposed.

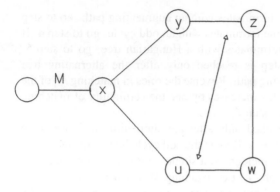

Fig. 7.14 Selecting matching edges in odd cycle C_i. Add the two edges (y, z) and (y, w) to M_{j-1}^*.

It remains to show that if the examination of an exposed vertex v results in a Hungarian tree in step 5, then no augmenting path exists from v in the final matching. Suppose no augmenting path exists from v to any exposed vertex after vertex v has been examined. Then the augmenting path that is discovered next by the algorithm cannot include any edge in the Hungarian tree resulting after the examination of vertex v; otherwise, this path could as well end at vertex v. Moreover, it can be seen that whatever happens in the matching to the edges outside the Hungarian tree rooted at v in no way allows the Hungarian tree rooted at v to be extended. Hence, it remains only to show that no augmenting path is rooted at v in the matching M_v resulting at the termination of the examination of vertex v.

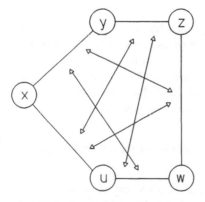

Fig. 7.15 Selecting matching edges in odd cycle C_i. Add any two nonadjacent edges to M_{j-1}^*; for example, if vertex y is to remain exposed, add the two edges (x, u) and (z, w) to the matching. All vertices in C_i are exposed to M_j^*.

Suppose that in matching M_v there is an augmenting path C joining two exposed vertices v and w in G_t. At least one of these two vertices, say vertex v, must have been examined by step 2 of the maximum-cardinality matching algorithm. This examination must have terminated at step 5 with a Hungarian tree rooted at vertex v. Traverse path C from v to w. Let (x, y) denote the first edge in C that is not in the Hungarian tree. Thus, vertex x must be an outer vertex of the tree. Vertex y can be (a) exposed, (b) unlabeled and matched, (c) labeled outer, or (d) labeled inner.

(a) Vertex y cannot be exposed; if so, step 2 would have terminated with an augmenting path from v to y rather than with a Hungarian tree rooted at v.

(b) Vertex y cannot be unlabeled and matched; if so, the alternating tree subroutine would have labeled vertex y as an inner vertex.

(c) Vertex y cannot be an outer vertex; if so, edge (x, y) would have been marked by the alternating tree subroutine, creating an odd cycle.

(d) If vertex y is an inner vertex, then the alternating path from root v to vertex y forms part of another augmenting path C' from v to w. Augmenting path C' consists of the alternating path from v to y in the Hungarian tree together with the portion of C from y to w.

Repeat the above analysis for augmenting path C'. This leads either to a contradiction as in cases (a), (b), and (c), or to another augmenting path C''. Each augmenting path generated by this process must have fewer edges not in the Hungarian tree than did the preceding augmenting path. Thus, we ultimately reach a contradiction or show that an augmenting path has been marked by the alternating tree subroutine instead of a Hungarian tree. Consequently, if an augmenting path exists, the examination of vertex v cannot have terminated with a Hungarian tree. Q.E.D.

EXAMPLE 2

Let us find a maximum-cardinality matching for the graph in Fig. 7.16a. Note that this graph possesses nine vertices, and consequently no matching can contain more than four edges.

Step 1. Start with the matching shown in Fig. 7.16a.

Step 2. Exposed vertex 1 is selected for examination. Alternating tree subroutine: Label vertex 1 outer. Mark edges (1, 4) and (4, 2). Label vertex 4 inner and label vertex 2 outer. Mark edge (1, 2) between outer vertices 1 and 2. An odd cycle (1, 4), (4, 2), (2, 1) has been found. Go to step 4.

Step 4. Shrink the odd cycle into an artificial vertex a_1. The new graph G_1 is shown in Fig. 7.16b. Return to step 2.

Step 2. Exposed vertex a_1 (the image of 1 in G_1) is selected for examination. Alternating tree subroutine: Label vertex a_1 outer. Mark edge $(a_1, 3)$. An augmenting path $(a_1, 3)$ has been found. Go to step 3.

Step 3. Add edge $(a_1, 3)$ to the matching. See Fig. 7.16c. Go to step 6 and explode a_1. Return to step 2.

Step 2. Exposed vertex 7 is selected for examination. Alternating tree subroutine: Label vertex 7 outer. Mark edges (7, 8) and (8, 6). Label vertex 8 inner and label vertex 6 outer. Mark edge (6, 9). An augmenting path (7, 8), (8, 6), (6, 9) has been found. Go to step 3.

Step 3. Remove edge (8, 6) from the matching. Add edges (7, 8) and (6, 9) to the matching. See Fig. 7.16d. Go to step 2.

Step 2. Only vertex 5 is exposed. Go to step 7.

Step 7. The final matching is shown in Fig. 7.16e.

Note that neither of the edges in the original matching appears in the terminal matching. Also, note that other maximum-cardinality matchings are possible. For example, {(1, 2), (3, 6), (9, 8), (4, 7)} is also a maximum-cardinality matching.

The terminal matching depends heavily on the choice of which exposed vertex is examined and on the arbitrary choices of edges colored in the alternating tree subroutine.

7.4 MAXIMUM-WEIGHT MATCHING IN A BIPARTITE GRAPH: THE ASSIGNMENT PROBLEM

The maximum-weight assignment problem is a weighted matching problem on a bipartite graph $G(S, T, E)$. It can be formulated as the following linear program:

$$\max \sum_i \sum_j c_{ij} x_{ij}$$

$$\sum_j x_{ij} \leq 1 \quad \text{for all } i \in S$$

$$\sum_i x_{ij} \leq 1 \quad \text{for all } j \in T$$

$$x_{ij} \geq 0 \quad \text{for all } i \in S, j \in T$$

This problem is similar in structure to the transportation problem and, in fact, can easily be shown to be a special case of the transportation problem. An optimal extreme point always exists that is integer. Therefore, $x_{ij} = 1$ if and only if the edge from $i \in S$ to $j \in T$ is included in the optimal matching. The constraints guarantee that at most one edge in any feasible solution is incident to any vertex.

An efficient algorithm for the assignment problem was developed by Kuhn (1955). The algorithm relies on a result by a Hungarian mathematician and hence is called the *Hungarian algorithm.* To motivate the algorithm, we first write down the dual to the linear programming formulation of the assignment problem. Associate dual variables u_i with vertices $i \in S$ and v_j with vertices $j \in T$. The dual is

$$\min \sum u_i + \sum v_j$$

$$u_i + v_j \geq c_{ij} \quad \text{for all } i, j$$

$$u_i, v_j \geq 0 \quad \text{for all } i, j$$

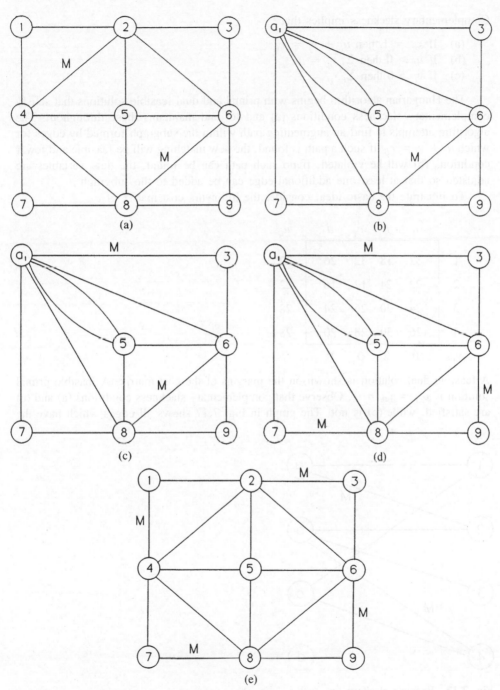

Fig. 7.16 Example of the alternating tree subroutine.

Complementary slackness implies that

(a) If $x_{ij} = 1$ then $u_i + v_j = c_{ij}$.
(b) If $u_i > 0$ then $\Sigma_j x_{ij} = 1$.
(c) If $v_j > 0$ then $\Sigma_i x_{ij} = 1$.

The Hungarian algorithm begins with primal and dual feasible solutions that satisfy complementary slackness conditions (a) and (c) and maintains these throughout. The algorithm attempts to find an augmenting path within the subgraph formed by edges for which $u_i + v_j = c_{ij}$. If such a path is found, the new matching will be feasible and fewer conditions (b) will be violated. If no such path can be found, the dual variables are adjusted so that at least one additional edge can be added to the subgraph.

To illustrate the basic idea, consider the following cost matrix of c_{ij}:

	a	b	c	d	u_i
1	32	18	32	26	32
2	22	24	12	16	22
3	24	30	26	24	28
4	26	30	28	20	28
v_j	0	2	0	0	

A feasible dual solution is shown on the margins of the cost matrix. A feasible primal solution is $x_{2a} = x_{4b} = 1$. Observe that complementary slackness conditions (a) and (c) are satisfied, while (b) is not. The graph in Fig. 7.17 shows all edges which have the

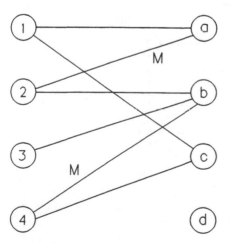

Fig. 7.17 Example of alternating path.

property that $u_i + v_j = c_{ij}$. We can see that an augmenting path exists along vertices 3-b-4-c. After augmenting, we have the matching shown in Fig. 7.18. Notice that the new primal solution $x_{2a} = x_{3b} = x_{4c} = 1$ is primal feasible and that the dual solution has not changed. Complementary slackness conditions (a) and (c) remain satisfied; in addition, condition (b) corresponding to vertex 3 now is satisfied. In the graph of Fig. 7.18, no augmenting path exists. At this point, the Hungarian algorithm will adjust the dual variables. Suppose that we subtract $\delta = 4$ from each u_i and add $\delta = 4$ to $v_j, j = 1, 2, 3$. The new dual solution is shown below.

	a	b	c	d	u_i
1	32	18	32	26	28
2	22	24	12	16	18
3	24	30	26	24	24
4	26	30	28	20	24
v_j	4	6	4	0	

Since we have subtracted δ from the dual variable associated with the vertices in S and added δ to the vertices in T which were endpoints of an edge in Fig. 7.18 (for which $u_i + v_j = c_{ij}$), this property is maintained for these edges. In addition, however, the dual-variable change caused edge $(3, d)$ to satisfy $u_3 + v_d = c_{3d}$; this edge is now a member of the subgraph (shown in Fig. 7.19). We have a new augmenting path d-3-b-2-a-1, which upon augmentation yields the solution in Fig. 7.20.

The tricky parts of the algorithm are the details necessary to find augmenting paths

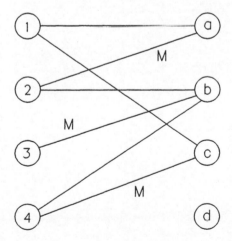

Fig. 7.18 No augmenting path exists.

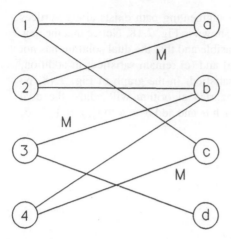

Fig. 7.19 Subgraph after dual-variable change.

and perform the proper changes in the dual variables. These are handled by labeling vertices in the graph. A formal statement of the algorithm is given below.

Hungarian Algorithm for Maximum-Weight Matching

   Initialize. Set $u_i = \max_j[c_{ij}]$, $i \in S$; $v_j = 0$, $\pi_j = \infty$, $j \in T$.
   Construct the subgraph consisting of all edges for which $u_i + v_j = c_{ij}$. For each $i \in S$, choose the first edge (i, j) such that j is not matched and place it in the initial matching M. All vertices are unscanned and unlabeled.

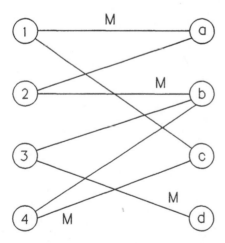

Fig. 7.20 Matching after augmentation.

Step 1. Label each exposed vertex $i \in S$ with $p(i) = 0$.

Step 2. Select any unscanned but labeled vertex $i \in S$, or $j \in T$ having $\pi_j = 0$. If none exist, then go to step 5.

Step 3. If the vertex chosen in step 2 is $i \in S$, then for each edge (i, j) not in M, label vertex $j \in T$ with $p(j) = i$ (replacing any existing label) if $u_i + v_j - c_{ij} < \pi_j$ and replace π_j with $u_i + v_j - c_{ij}$. If the vertex chosen in step 2 is $j \in T$, then determine if j is exposed. If vertex j is exposed, go to step 4. If not, then there is an edge (i, j) in M. Label vertex $i \in S$ with $p(i) = j$. In either case ($i \in S$ or $j \in T$) return to step 2.

Step 4. An augmenting path ending at vertex $i \in S$ or $j \in T$ has been found. Trace this path using the predecessor function $p(\ )$. Augment the matching by adding to M all edges not currently in M and deleting any edges that are currently in M. Set $\pi_j = \infty$ for all $j \in T$, erase all labels, and return to step 1.

Step 5. Compute $\delta_1 = \min\{u_i, i \in S\}$, $\delta_2 = \min\{\pi_j > 0, j \in T\}$, and $\delta = \min\{\delta_1, \delta_2\}$. Set $u_i = u_i - \delta$ for each labeled vertex $i \in S$. Set $v_j = v_j + \delta$ for all $j \in T$ with $\pi_j = 0$. Set $\pi_j = \pi_j - \delta$ for each labeled vertex $j \in T$ with $\pi_j > 0$. If $\delta = \delta_2$, then go to step 2. Otherwise the maximum-weight matching has been found.

We shall illustrate this algorithm using the example above. Denote the set of unscanned and labeled vertices by L. The initial set of dual variables and matching are shown below.

	a	b	c	d	u_i
1	32	18	32	26	32
2	22	24	12	16	24
3	24	30	26	24	30
4	26	30	28	20	30
v_j	0	0	0	0	
π_j	∞	∞	∞	∞	

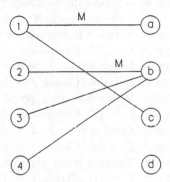

Step 1. Vertices 3 and 4 are exposed. Label $p(3) = 0$, $p(4) = 0$.

Step 2. $L = \{3, 4\}$. Choose $i = 3$.

Step 3. Label vertices $j \in T$: $p(a) = 3, p(b) = 3, p(c) = 3, p(d) = 3$. The new π vector is $(6, 0, 4, 6)$.

Step 2. $L = \{4, b\}$. Choose $i = 4$.

Step 3. Label $j \in T$: $p(a) = 4$, $p(c) = 4$. $\pi = (4, 0, 2, 6)$.

Step 2. $L = \{b\}$. Choose $j = b$.

Step 3. Vertex b is not exposed. Label vertex 2, $p(2) = b$.

Step 2. $L = \{2\}$. Choose $i = 2$.

Step 3. Label $j \in T$: $p(a) = 2$. $\pi = (2, 0, 2, 6)$.
Step 2. Go to step 5.
Step 5. $\delta_1 = \min\{32, 24, 30, 30\} = 24$; $\delta_2 = \min\{2, 2, 6\} = 2$; $\delta = 2$. Vertices 2, 3, and 4 are labeled. Subtract δ from the dual values of these vertices. Add δ to v_b. Subtract δ from π_a, π_c, and π_d. This results in the following:

	a	b	c	d	u_i
1	32	18	32	26	32
2	22	24	12	16	22
3	24	30	26	24	28
4	26	30	28	20	28
v_j	0	2	0	0	
π_j	2	0	0	4	

Step 2. $L = \{c\}$. Choose $j = c$.
Step 3. Vertex c is exposed. Go to step 4.
Step 4. Augment the matching by tracing back: $p(c) = 4$, $p(4) = 0$.
$\pi = (\infty, \infty, \infty, \infty)$. $M = \{(1, a), (2, b), (4, c)\}$.
Step 1. Vertex 3 is exposed; label $p(3) = 0$.
Step 2. $L = \{3\}$. Choose $i = 3$.
Step 3. Label $p(a) = 3$, $p(b) = 3$, $p(c) = 3$, $p(d) = 3$. $\pi = (4, 0, 2, 4)$.
Step 2. $L = \{b\}$. Choose $j = b$.
Step 3. Vertex b is not exposed. Label $p(2) = b$.
Step 2. $L = \{2\}$. Choose $i = 2$.
Step 3. Label $p(a) = 2$. $\pi = (0, 0, 2, 4)$.
Step 2. $L = \{a\}$. Choose $j = a$.
Step 3. Vertex a is not exposed. Label $p(1) = a$.
Step 2. $L = \{1\}$. Choose $i = 1$.
Step 3. Label $p(c) = 1$, $\pi = (0, 0, 0, 4)$.
Step 2. $L = \{c\}$. Choose $j = c$.
Step 3. Vertex c is not exposed. Label $p(4) = c$.
Step 2. $L = \{4\}$. Choose $i = 4$.
Step 3. No new labels.
Step 2. Go to step 5.
Step 5. $\delta_1 = \min\{32, 22, 28, 28\} = 22$; $\delta_2 = \min\{4\} = 4$; $\delta = 4$. Vertices 1, 2, 3, and 4 are labeled. Subtract δ from the dual values of these vertices. Add δ to v_a and v_2. Subtract δ from π_d. This results in the following:

	a	b	c	d	u_i
1	32	18	32	26	28
2	22	24	12	16	18
3	24	30	26	24	24
4	26	30	28	20	24
v_j	4	6	4	0	
π_j	0	0	0	0	

Step 2. $L = \{d\}$. Choose $j = d$.

Step 3. Vertex d is exposed. Go to step 4.

Step 4. Augment the matching. $M = \{(1, a), (2, b), (3, d), (4, c)\}$.

$\pi = (\infty, \infty, \infty, \infty)$.

Step 1. No vertices $i \in S$ are exposed.

Step 2. Go to step 5.

Step 5. $\delta = \delta_1$. Terminate.

The Hungarian algorithm has time complexity $O(n^3)$. Some computational improvements that remain within this time bound are reported by Jonker and Volgenant (1986).

7.5 MAXIMUM-WEIGHT MATCHING ALGORITHM

This section presents an algorithm due to Edmonds and Johnson (1970) for finding a maximum-weight matching of a graph $G = (X, E)$. Like the maximum-cardinality matching algorithm of Section 7.3, the basic operation of this algorithm is the generation of an alternating tree.

As we saw in the last two sections, the key issue in finding a maximum-cardinality matching was the existence of augmenting paths. A similar result holds for maximum-weight matchings:

A *weighted augmenting* path is an alternating path in which

1. The total weight of the nonmatching edges exceeds the total weight of the matching edges.
2. The first vertex in the path is exposed if the first edge in the path is not a matching edge.
3. The last vertex in the path is exposed if the last edge in the path is not a matching edge.

Observe that if the roles in the matching of the edges in a weighted augmenting path are reversed, the resulting matching has greater weight than the original matching. For

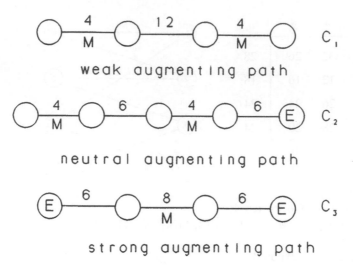

Fig. 7.21   Weighted augmenting paths.

example, in Fig. 7.21, paths C_1, C_2, and C_3 are alternating paths. In each path, the nonmatching edge(s) weigh 12 units and the matching edge(s) weigh 8 units. All are weighted augmenting paths. Path C_1 is called a *weak augmenting path* because it contains more matching edges than nonmatching edges. Path C_2 is called a *neutral augmenting path* because it contains an equal number of matching and nonmatching edges. Path C_3 is called a *strong augmenting path* because it contains more nonmatching edges than matching edges.

Analogous to Theorem 7.1, we now show:

Theorem 7.2.   A matching M is a maximum-weight matching if, and only if, M possesses no weighted augmenting paths.

Proof.   If matching M possesses a weighted augmenting path C, then M cannot be a maximum-weight matching since the matching M' generated by reversing the roles of the edges in C has greater weight than matching M.

To prove the rest of the theorem, let M^* be any matching with greater weight than M. As in Theorem 7.1, consider the set of all edges that are in exactly one of these two matchings M and M^*. We know from the proof of Theorem 7.1 that each connected component of these edges must be either an even cycle or a path. Since all cycles are paths, we can regard each connected component as a path. Since M^* weighs more than M, in one of these paths the edges in M^* must weigh more than the edges in M. This path is a weighted augmenting path for matching M.   Q.E.D.

Let $V = \{V_1, V_2, \ldots, V_z\}$ denote the set of all odd-cardinality vertex subsets. Let T_m denote the set of all edges with both endpoints in vertex set V_m. Let $T = \{T_1, T_2, \ldots, T_z\}$. Let the number of vertices in V_m be denoted by $2n_m + 1$.

No matching can contain more than n_m members of set T_m.

Let $a(i, j)$ denote the weight of edge (i, j). Let $x(i, j) = 1$ if edge (i, j) is in the matching; otherwise, let $x(i, j) = 0$.

To understand the maximum-weight matching algorithm, we must first examine the following linear programming formulation of the maximum-weight matching problem:

Maximize

$$\sum_{(i, j)} a(i, j)x(i, j) \tag{4}$$

$$\sum_{j} [x(i, j) + x(j, i)] \leq 1 \quad \text{(for all } i \in X) \tag{5}$$

$$\sum_{(i, j) \in T_m} x(i, j) \leq n_m \quad \text{(for } m = 1, 2, ..., z) \tag{6}$$

$$0 \leq x(i, j) \quad \text{[for all } (i, j)] \tag{7}$$

[Note that an edge joining vertices i and j is denoted by either (i, j) or (j, i).]

Constraint (5) requires that no more than one matching edge be incident to each vertex i. Constraint (6) requires that no more than n_m edges from T_m be present in the matching. The objective function (4) equals the total weight of the matching edges.

Every matching satisfies constraints (5)–(7). However, it is virtually impossible to enumerate all these constraints for graphs of even moderate size. Fortunately, the maximum-weight matching algorithm produces a matching that is an optimal solution for the linear programming problem (4)–(7). How do we know that this solution is an optimal solution to the linear programming problem (4)–(7)? We construct a feasible solution to the dual linear programming problem that, together with the solution for the primal, satisfies all complementary slackness conditions.

The dual linear programming problem for the primal linear programming problem (4)–(7) is

Minimize

$$\sum_{i \in X} y_i + \sum_{m=1}^{z} n_m z_m \tag{8}$$

$$y_i + y_j + \sum_{m:(i, j) \in T_m} z_m \geq a(i, j) \quad \text{[for all } (i, j)] \tag{9}$$

$$y_i \geq 0 \quad \text{(for all } i \in X) \tag{10}$$

$$z_m \geq 0 \quad \text{(for all } m = 1, 2, ..., z) \tag{11}$$

The dual variable associated with the primal constraint (5) for vertex i is denoted by y_i. The dual variable associated with the primal constraint (6) for T_m is denoted by z_m.

The complementary slackness conditions for this pair of primal-dual linear programming problems are

$$x(i, j) > 0 \Rightarrow y_i + y_j + \sum_{m:(i, j) \in T_m} z_m = a(i, j) \quad \text{[for all } (i, j)] \tag{12}$$

$$y_i > 0 \Rightarrow \sum_{\substack{j \in X}} [x(i, j) + x(j, i)] = 1 \quad \text{(for all } i \in X) \tag{13}$$

$$z_m > 0 \Rightarrow \sum_{(i, j) \in T_m} x(i, j) = n_m \quad \text{(for } m = 1, 2, \ldots, z) \tag{14}$$

How does the maximum weight matching algorithm work? The algorithm starts with a null matching [all $x(i, j) = 0$] and feasible values for the dual variables y_i, $i \in X$, and z_m, $m = 1, 2, \ldots, z$, that satisfy complementary slackness conditions (12) and (14). Only conditions (13) remain unsatisfied.

At each iteration of the algorithm, the matching and/or the values of the dual variables are changed so that all the conditions (5)–(7), (9)–(11), (12), and (14) remain satisfied and so that condition (13) is satisfied for at least one more dual variable y_i. Since there are only $|X|$ dual variables y_i, after not more than $|X|$ iterations all conditions (13) become satisfied. By complementary slackness, the resulting matching must be a maximum-weight matching.

Let us examine condition (13) more closely. Condition (13) states that if the dual variable y_i for vertex i is positive, then vertex i must be matched. Thus, only exposed vertices with positive dual variables violate condition (13).

The algorithm identifies an exposed vertex v with $y_v > 0$ and uses the alternating tree subroutine to grow an alternating tree rooted at vertex v. As we have seen in Section 7.3, the subroutine terminates with (a) an augmenting path, (b) an odd cycle, or (c) a Hungarian tree. If the subroutine finds an augmenting path, then this path is a strong augmenting path. The roles of the edges in this path are reversed. This increases the total weight of the matching and matches vertex v. Consequently, vertex v satisfies condition (13). If the subroutine finds an odd cycle, this cycle is shrunk into an artificial vertex and the algorithm is continued on the resulting graph. If the subroutine finds a Hungarian tree, the dual variables are changed so that all primal, dual, and complementary slackness conditions, except possibly condition (13), remain satisfied and so that another edge can be added to the alternating tree. Ultimately, either y_v is reduced to zero so that condition (13) becomes satisfied or vertex v is matched.

During the course of the algorithm, odd cycles are shrunk into artificial vertices. Eventually, all artificial vertices are expanded out into their original odd cycles. However, the vertices need not be expanded out in the same order in which they were generated. Due to these shrinkings and expansions, the algorithm will produce a sequence of graph $G_0, G_1, \ldots, G_t$.

With this as background, we are now prepared to state formally the *maximum-weight matching algorithm*.

Maximum-Weight Matching Algorithm

Step 1 (Initialization). Initially, let matching M_0 contain no edges and let all dual variables $z_m = 0$, $m = 1, 2, \ldots, z$. Choose any initial values for the dual variables y_i, $i \in X$, such that $y_i + y_j \geq a(i, j)$ for all edges (i, j). (For instance, you could let each y_i equal half the maximum edge weight.) Let $k = 0$. Denote the original graph by $G_k = (X_k, E_k)$.

Step 2 (Examination of an Exposed Vertex). Select any nonartificial, exposed vertex v in graph G_i with $y_v > 0$. If no such vertex exists, go to step 6. Otherwise, let E^* consist of all edges (i, j) in G_k such that

$$y_i + y_j + \sum_{(i, j) \in T_m} z_m = a(i, j) \tag{15}$$

Using the alternating tree subroutine, grow an alternating tree rooted at v using only edges in E^*. If the subroutine finds an augmenting path, go to step 3. If the subroutine finds an odd cycle, go to step 4. If the subroutine finds a Hungarian tree, go to step 5.

Step 3 (Augmenting Path). This step is reached only after the alternating tree subroutine finds an augmenting path. Reverse the roles in matching M_k of the edges in this path. Vertex v is no longer exposed. Return to step 2.

Step 4 (Odd Cycle). This step is reached only after the alternating tree subroutine finds an odd cycle. Let $k = k + 1$. Denote this odd cycle by C_k. Shrink the odd cycle C_k into an artificial vertex a_k. Denote the new graph by $G_k = (X_k, E_k)$. Let M_k be the matching in G_k consisting of all edges in M_{k-1} that are in G_k.

In all future labeling, let all vertices subsumed into artificial vertex a_k carry the same label as a_k.

Return to step 2 and continue to grow an alternating tree rooted at the image of vertex v in G_k even if this vertex is artificial. Note that the labeling and marking of the last iteration of the alternating tree subroutine can be salvaged for the next iteration.

Step 5 (Hungarian Tree). This step is reached only after the alternating tree subroutine finds a Hungarian tree.

Let

$$\delta_1 = \min\{y_i + y_j - a(i, j)\} \tag{16}$$

where the minimization is taken over all (i, j), where $i \in X_0$ is an outer vertex and $j \in X_0$ is unlabeled.

Let

$$\delta_2 = \frac{1}{2} \min\{y_i + y_j - a(i, j)\} \tag{17}$$

where the minimization is taken over all (i, j), where $i \in X_0$ is an outer vertex, $j \in X_0$ is an outer vertex, and i and j are not inside the same artificial vertex.

Let

$$\delta_3 = \frac{1}{2} \min\{z_m\} \tag{18}$$

where the minimization is taken over all odd-cardinality vertex sets V_m that are shrunk into an artificial vertex a_k that is labeled inner.

Let

$$\delta_4 = \min\{y_i\} \tag{19}$$

where the minimization is taken over all vertices $i \in X_0$ that are labeled outer. Lastly, let

$$\delta = \min\{\delta_1, \delta_2, \delta_3, \delta_4\} \tag{20}$$

Adjust the dual variables as follows:

(a) Outer vertex variables y_i are decreased by δ.
(b) Inner vertex variables y_i are increased by δ.
(c) For each outer artificial vertex in G_k, increase its dual variable z_m by 2δ.
(d) For each inner artificial vertex in G_k, decrease its dual variable z_m by 2δ.

If $\delta = \delta_1$, then the edge (i, j) that determined δ_1 enters E^*. This edge can now be marked by the alternating tree subroutine. Return to step 2 and continue to grow an alternating tree rooted at v.

If $\delta = \delta_2$, then the edge (i, j) that determined δ_2 enters E^*. This edge can now be marked by the alternating tree subroutine creating an odd cycle. Return to step 2 and continue to grow an alternating tree rooted at v.

If $\delta = \delta_3$, then some dual variable z_i becomes zero. Expand the artificial vertex corresponding to this dual variable back to its original odd cycle. Let $k = k + 1$. Call the resulting graph $G_k = (X_k, E_k)$. Let matching M_k consist of all edges in M_{k-1} together with the n_i edges of T_i that match the $2n_i$ exposed vertices of V_i. (The remaining vertex of V_i is matched in M_k since all inner artificial vertices in G_{k-1} are matched in M_{k-1}.)

Return to step 2 and continue to grow an alternating tree rooted at v.

If $\delta = \delta_4$, then the dual variable y_i of some outer vertex i becomes zero. The path in the alternating tree from root v to vertex i is a neutral augmenting path. Reverse the roles in matching M_k of the edges in this path. Vertex v becomes matched and vertex i becomes exposed, which is all right since $y_i = 0$. Return to step 2.

Step 6 (Expansion of Artificial Vertices). This step is reached only after all vertices violating condition (13) have been examined by step 2. Consider all artificial vertices remaining in the terminal graph. Expand out each artificial vertex in reverse order (the last to be generated is expanded first, etc.) and induce a maximum matching on the resulting odd cycle.

The terminal matching is a maximum-weight matching for the original graph G_0. Stop.

Proof of the Maximum-Weight Matching Algorithm.   The algorithm starts with a matching with no edges and maintains a matching throughout all iterations. We need only prove that the terminal matching is a maximum-weight matching. This is accomplished by showing that the terminal values for the dual variables y_i, $i \in X$, and z_m, $m = 1, 2, \ldots,$ z, satisfy dual feasibility (9)–(11) and complementary slackness (12)–(14).

Since equations (16)–(20) ensure that no dual variable is ever reduced to a negative value, and since step 1 selects initial dual-variable values that are nonnegative, conditions (10) and (11) are satisfied at all times by the algorithm.

To verify that condition (9) is satisfied at all times by the algorithm, note that

1. Initially, condition (9) is satisfied.
2. Edge (i, j) must be (a) in an artificial vertex, (b) not in an artificial vertex and in E^*, or (c) not in an artificial vertex and not in E^*.

If (a) occurs, then a dual-variable change does not change the left side of condition (9) since both y_i and y_j change by δ and the dual variable for the artificial vertex containing (i, j) changes by -2δ.

If (b) occurs, then a dual-variable change increases the inner vertex dual variable and decreases the outer vertex dual variable. Hence, the left side of (9) remains unchanged. If (c) occurs, then

$$y_i + y_j + \sum_{m:(i,j) \in T_m} z_m > a(i, j)$$

If both i and j are not labeled, or if i and j have different labels, then the dual-variable change preserves the above inequality. If one of the vertices i, j is labeled inner and the other is unlabeled, or if both vertices i and j are labeled inner, then the left side of (9) increases after a dual-variable change. If one of the vertices i and j is labeled outer and the other is unlabeled, then by equation (16) the dual variable will not reverse the above inequality. If both vertices i and j are outer vertices then by equation (17) the dual variable change will not reverse the preceding inequality.

Thus, under all dual-variable changes condition (9) remains satisfied.

To verify condition (12), we must consider two cases: (a) matching edge (i, j) is not in an artificial vertex at the beginning of step 6, or (b) matching edge (i, j) is contained in an artificial vertex at the beginning of step 6.

If (a) occurs, then edge $(i, j) \in E^*$ and condition (12) is satisfied. If (b) occurs, then edge (i, j) is contained in some artificial vertex after the last dual-variable change has been made, and condition (9) holds with equality. Hence, condition (12) is satisfied by the terminal dual-variable values.

Condition (13) is satisfied for all vertices at the end of the algorithm; otherwise, step 2 would have been repeated for any vertex violating condition (13).

Lastly, it remains to show that condition (14) is satisfied. The only way that a dual variable z_m can become positive is to shrink the vertices in V_m into an artificial vertex. In step 6, each artificial vertex is expanded and a maximum matching is induced on the corresponding odd cycle. Thus, condition (14) is satisfied. Q.E.D.

EXAMPLE 3

Let us find a maximum-weight matching for the graph in Fig. 7.22.

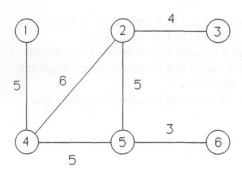

Fig. 7.22 Maximum-weight matching algorithm.

Step 1. Since the largest edge weight equals 6, let $y_i = 6/2 = 3$ for all vertices i. Let all $z_i = 0$. Initially, no edges are in the matching.

Step 2. Exposed vertex 4 is selected for examination. Alternating tree subroutine: $E^* = \{(4, 2)\}$. Vertex 4 is labeled outer. Vertex 2 is labeled, and an augmenting path (4, 2) has been discovered. Go to step 3.

Step 3. Add edge (4, 2) to the matching.

Step 2. Exposed vertex 5 is selected for examination. Alternating tree subroutine: $E^* = \{(4, 2)\}$. Vertex 5 is labeled outer. No further labeling is possible. The alternating tree consisting entirely of vertex 5 is Hungarian. Go to step 5.

Step 5. Perform a dual-variable change.

$$\delta_1 = \min\{y_4 + y_5 - a(4, 5), y_2 + y_5 - a(2, 5), y_6 + y_5 - a(6, 5)\}$$
$$= \min\{3 + 3 - 5, 3 + 3 - 5, 3 + 3 - 3\} = 1$$

$$\delta_2 = \infty, \qquad \delta_3 = \infty$$

$$\delta_4 = y_5 = 3$$

$$\delta = \min\{\delta_1, \delta_2, \delta_3, \delta_4\} = \min\{1, \infty, \infty, 3\} = 1$$

Thus, y_5 decreases by 1. (See Fig. 7.23.)

Step 2. Continued examination of exposed vertex 5. Alternating tree subroutine: $E^* = \{(4, 2), (4, 5), (5, 2)\}$. Vertex 5 is labeled outer. Vertex 2 is labeled inner, vertex 4 is labeled outer, and edges (5, 2) and (4, 2) are marked. Next, edge (4, 5) joining outer vertices 4 and 5 is marked. An odd cycle has been discovered. Go to step 4.

Step 4. Shrink the odd cycle (5, 2), (4, 2), (4, 5) into an artificial vertex a_1. Denote the dual variable for the odd vertex set $\{5, 2, 4\}$ by z_{a_1}. The new graph resulting from this shrinking is shown in Fig. 7.24.

Step 2. Examination of exposed artificial vertex a_1. Alternating tree subroutine: $E^* = \{(4, 2), (4, 5), (5, 2)\}$. Vertex a_1 is labeled outer. (Hence, vertices 4, 2, 5 are also labeled outer.) No further labeling is possible. The tree consisting entirely of vertex a_1 is Hungarian. Go to step 5.

Step 5. Perform a dual-variable change.

	y_1	y_2	y_3	y_4	y_5	y_6	z_7	Matching
Initialization	3	3	3	3	3	3	0	Empty
Examine 4	3	3	3	3	3	3	0	(4, 2)
Examine 5	3	3	3	3	2	3	0	(4, 2)
Examine a_1	3	2	3	2	1	3	2	(1, a_1)
Examine 3	3	2	2	2	1	3	2	(1, a_1)
	2	3	1	3	2	3	0	(1, 4), (5, 2)
Examine 6	2	3	1	3	2	1	0	(1, 4), (6, 5)
								(2, 3)

Fig. 7.23 Dual-variable changes, maximum-weight matching algorithm.

$$\delta_1 = \min\{y_2 + y_3 - a(2, 1), \, y_1 + y_4 - a(1, 4), \, y_5 + y_6 - a(5, 6)\}$$
$$= \min\{3 + 3 - 4, \, 3 + 3 - 5, \, 2 + 3 - 3\} = \min\{2, 1, 2\} = 1$$

$$\delta_2 = \infty$$

$$\delta_3 = \infty \quad \text{(there are no inner artificial vertices)}$$

$$\delta_4 = \min\{y_2, \, y_4, \, y_5\} = \min\{3, 3, 2\} = 2$$

$$\delta = \min\{1, \infty, \infty, 2\} = 1$$

Thus, $y_4 = 3 - 1 = 2$, $y_5 = 2 - 1 = 1$, $y_2 = 3 - 1 = 2$, $z_{a_1} = 2\delta = 2$. Note that for each edge in E^*, condition (9) is satisfied with equality; e.g., for edge (4, 5),

$$y_4 + y_5 + z_{a_1} = 2 + 1 + 2 = 5 = a(4, 5)$$

Step 2. Continued examination of exposed artificial vertex a_1. Alternating tree subroutine: $E^* = \{(4, 5), (5, 2), (2, 4), (1, 4)\}$. Vertex a_1 is labeled outer. Thus, vertices 2, 5, 4 are also labeled outer. Edge $(1, a_1)$ is marked, and an augmenting path has been found. Go to step 3.

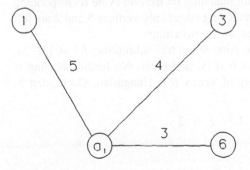

Fig. 7.24 Graph resulting from shrinking an odd cycle.

Step 3. Add edge $(1, a_1)$ to the matching.

Step 2. Examination of exposed vertex 3. Alternating tree subroutine: $E^* = \{(4, 5),$ $(5, 2), (2, 4), (1, 4)\}$. Vertex 3 is labeled outer. No further labeling. The tree consisting entirely of vertex 3 is Hungarian. Go to step 5.

Step 5. Perform a dual-variable change.

$$\delta_1 = y_3 + y_2 - a(3, 4) = 3 + 2 - 4 = 1$$

$$\delta_2 = \infty$$

$$\delta_3 = \infty$$

$$\delta_4 = y_3 = 3$$

$$\delta = \min\{1, \infty, \infty, 3\} = 1$$

Thus, $y_3 = 3 - 1 = 2$.

Step 2. Continued examination of exposed vertex 3. Alternating tree subroutine: $E^* = \{(4, 5), (5, 2), (2, 4), (1, 4), (3, 2)\}$. Vertex 3 is labeled outer. Vertex a_1 is labeled inner, and vertex 1 is labeled outer. Edges $(3, a_1)$ and $(a_1, 1)$ are marked. (Since a_1 is an inner vertex, vertices 2, 5, 4 are also labeled inner.) No further labeling is possible. The tree is Hungarian. Go to step 5.

Step 5. Perform a dual-variable change.

$$\delta_1 = \infty$$

$$\delta_2 = \infty$$

$$\delta_3 = \tfrac{1}{2} z_{a_1} = 1$$

$$\delta_4 = \min\{y_3, y_1\} = \min\{2, 3\} = 2$$

$$\delta = \min\{\infty, \infty, 1, 2\} = 1$$

Thus, $y_3 = 2 - 1 = 1, y_1 = 3 - 1 = 2, y_2 = 2 + 1 = 3, y_4 = 2 + 1 = 3, y_5 = 1 + 1 = 2$, and $z_{a_1} = 2 - 2(1) = 0$. Since z_{a_1} has returned to zero, artificial vertex a_1 must be expanded. Expanding vertex a_1 yields the original graph in Fig. 7.22.

Now we must induce a maximum matching on the odd cycle corresponding to vertex a_1. Since edge $(4, 1)$ is a matching edge, only vertices 5 and 2 are left exposed. Thus, edge $(5, 2)$ is added to the matching.

Step 2. Examination of exposed vertex 6. Alternating tree subroutine: $E^* = \{(4, 5),$ $(5, 2), (2, 4), (1, 4), (2, 3)\}$. Vertex 6 is labeled outer. No further labeling is possible. The tree consisting entirely of vertex 6 is Hungarian. Go to step 5.

Step 5. Perform a dual-variable change.

$$\delta_1 = y_6 + y_5 - a(6, 5) = 3 + 2 - 3 = 2$$

$$\delta_2 = \infty$$

$$\delta_3 = \infty$$

$$\delta_4 = y_6 = 3$$
$$\delta = \min\{2, \infty, \infty, 3\} = 2.$$

Thus, $y_6 = 3 - 2 = 1$. All other dual variables remain unchanged.

Step 2. Continued examination of exposed vertex 6. Alternating tree subroutine: E^* consists of all edges in graph G. Vertex 6 is labeled outer. Vertex 5 is labeled inner, and vertex 2 is labeled outer. Edges $(6, 5)$ and $(5, 2)$ are marked. Edge $(2, 3)$ is marked next. An augmenting path $(6, 5), (5, 2), (2, 3)$ has been found. Go to step 3.

Step 3. Reverse the roles of the edges in path $(6, 5), (5, 2), (2, 3)$. Thus edge $(5, 2)$ leaves the matching, and edges $(6, 5)$ and $(2, 3)$ enter the matching. The matching now consists of edges $(1, 4), (6, 5)$, and $(2, 3)$.

Step 2. There are no more unexamined, exposed vertices. Go to step 6.

Step 6. No artificial vertices remain in the final graph. Stop.

The maximum-weight matching is $(1, 4), (6, 5), (2, 3)$ with a total weight $5 + 3 + 4 = 12$.

Notice that at termination all dual variables $z_m = 0$, and $\Sigma \, y_i = 2 + 3 + 1 + 3 + 2 + 1 = 12$. Thus, the value of the primal objective function (4) is 12, and the value of the dual objective function (8) is also 12. Consequently, both primal and dual solution must be optimal.

APPENDIX: USING NETSOLVE TO FIND OPTIMAL MATCHINGS AND ASSIGNMENTS

The NETSOLVE weighted matching algorithm determines an optimal matching in a network without loops. (If loops are present, they are ignored.) The network can be directed or undirected, but edge direction is ignored. Edge weights are specified in the COST data field. A matching may contain isolated nodes (that is, nodes which are not incident to any matched edge), incurring a certain "self-loop weight" defined below. An optimal matching is a matching that has minimum or maximum total weight, where the weight is the sum of all edge weights in the matching and self-loop weights of isolated nodes. Because the matching algorithm requires integer data, all weights are internally rounded by NETSOLVE to the precision set by the current DECIMALS value and then scaled to integer values. The output will list the actual weights.

There are two ways in which self-loop weights can be defined. In the first, the self-loop weight of a node is given by the associated SUPPLY data field for that node. In the second, signified by the presence of the keyword SLCOST, a common self-loop weight value is incurred for matching any node to itself. In calculating a minimum-weight matching (signified by the keyword MIN), the user should normally define the common self-loop weight as a large positive value; for a maximum-weight matching (signified by the keyword MAX), the user should employ a large negative value for the common self-loop weight.

For example, to determine a minimum-weight matching for a network, the command line

MATCH MIN SLCOST = 1000

can be used. Here the weight associated with matching any node to itself is 1000. If the command line

MATCH MIN

were used instead, the values found in the SUPPLY data field would be used as weights for matching nodes to themselves. In a similar way, the keyword MAX when used in place of MIN in the above examples will generate a maximum-weight matching for the network.

The first line of output displays the total weight of the optimal matching. If the terminal is ON, then the edges and nodes comprising this matching are displayed, together with the associated edge/self-loop weights.

To illustrate the use of the weighted matching algorithm, we will present an example of the basketball scheduling problem discussed in Section 7.1. This application was studied by Campbell and Chen (1976) and Ball and Webster (1977), who investigated scheduling in the Southeastern Conference (SEC). The SEC had formerly consisted of ten teams: Alabama (AL), Auburn (AU), Florida (F), Georgia (G), Kentucky (K), Louisiana State University (L), Mississippi State (M), Ole Miss (O), Tennessee (T), and Vanderbilt (V). A matrix of distances between home sites is given in Fig. 7.25. These values represent the edge weights of the complete graph corresponding to these ten teams.

Figure 7.26 shows a listing of the edge data for the NETSOLVE network. Applying the matching algorithm using the command line

>MATCH MIN SLCOST = 1000

we obtain the solution shown in Fig. 7.27. Thus, the optimal pairings are Alabama and Mississippi State, Auburn and Florida, Georgia and Tennessee, Kentucky and Vanderbilt, and Louisiana State and Ole Miss. Notice that it is not optimal to pair Mississippi State and Ole Miss even though they are only 110 miles apart.

The NETSOLVE assignment algorithm determines a minimum-cost or maximum-value assignment of supply nodes to demand nodes in a directed network. Supply nodes have a value of $+1$ in the SUPPLY data field; demand nodes have a value of -1. All nodes must be either supply or demand nodes, and each edge must be directed from a supply node to a demand node. The total cost of an assignment is given by the sum of the COST data fields over the edges used to assign a distinct supply node to each demand node. (Note that there must be at least as many supply nodes as demand nodes.)

The algorithm is executed on the current network by issuing the command line

	AL	AU	F	G	K	L	M	O	T	V
AL	—	160	444	272	455	322	88	172	312	238
AU		—	297	200	521	419	186	279	362	350
F			—	335	714	604	523	608	526	584
G				—	397	594	351	396	208	295
K					—	761	501	477	189	217
L						—	286	296	634	544
M							—	110	384	266
O								—	404	248
T									—	178

Fig. 7.25 Distance matrix for SEC teams.

ASSIGN

or

ASSIGN MIN

If the keyword MAX is used in the second form above, then a maximum-value assignment (based on the COST field) will be produced. In either case, output consists of the total cost (or value) of the optimal assignment and (if the terminal is ON) a list of the edges used to assign supplies to demands. Assignments are indicated by a FLOW value of 1 in the edge list display.

In addition, the user will be asked whether a sensitivity analysis should be conducted on the problem. This analysis produces dual variables for each node and reduced costs for each edge. Also, tolerance intervals are produced for each relevant edge cost, such that varying the individual edge cost within that range will not change the optimality of the current solution.

Particular assignments can be forced or prohibited by appropriate specification of the LOWER and UPPER data fields on the individual edges. Note that for any edge we must have UPPER at least as large as LOWER. The LOWER value can be 0 (no restriction) or 1 (forcing an assignment). The UPPER value can be 0 (prohibiting an assignment) or greater than or equal to 1 (no restriction). The normal default values assigned to LOWER and UPPER are such that no restrictions are placed on the solution. Thus, the user who wishes to solve an assignment problem having neither required nor prohibited assignments need only specify values for the SUPPLY and COST data fields.

FROM	TO	COST
AL	AU	160.
AL	F	444.
AL	G	272.
AL	K	455.
AL	L	322.
AL	M	88.
AL	O	172.
AL	T	312.
AL	V	238.
AU	F	297.
AU	G	200.
AU	K	521.
AU	L	419.
AU	M	186.
AU	O	279.
AU	T	362.
AU	V	350.
F	G	335.
F	K	714.
F	L	604.
F	M	523.
F	O	608.
F	T	526.
F	V	584.
G	K	397.
G	L	594.
G	M	351.
G	O	396.
G	T	208.
G	V	295.
K	L	761.
K	M	501.
K	O	477.
K	T	189.
K	V	217.
L	M	286.
L	O	296.
L	T	634.
L	V	544.
M	O	110.
M	T	384.
M	V	266.
O	T	404.
O	V	248.
T	V	178.

Fig. 7.26 Edge data for matching algorithm.

```
TOTAL COST OF MINIMUM MATCHING:          1106.

  EDGES (UNDIRECTED) IN THE MINIMUM MATCHING
     FROM        TO                  COST
     ----        --               --------
     AL          M                     88.
     AU          F                    297.
     G           T                    208.
     K           V                    217.
     L           O                    296.
```

Fig. 7.27 Optimal solution to SEC matching problem.

	SEA	CAL	MIL	DET	CLE	NY	BOS
OAK	678	337	1845	2079	2161	2586	2704
TEX	1670	1246	843	982	1015	1383	1555
KC	1489	1363	483	630	694	1113	1254
MIN	1399	1536	297	528	622	1028	1124
DET	1932	1979	237	0	95	509	632
TOR	2124	2175	434	206	193	366	463
BOS	2469	2611	860	632	563	187	0

Fig. 7.28 Distance matrix between home cities.

```
NAME              SUPPLY
----            --------
BOSCLE              1.00
BOSKC              -1.00
CALOAK             -1.00
CLETEX             -1.00
DETCWS              1.00
DETTOR             -1.00
KCNYY               1.00
MILBAL             -1.00
MINCAL              1.00
NYYCWS             -1.00
OAKSEA              1.00
SEAMIN             -1.00
TEXBAL              1.00
TORMIL              1.00
```

Fig. 7.29 Node data for umpire scheduling problem.

FROM	TO	COST	LOWER	UPPER
BOSCLE	BOSKC	0.00	0.00	999999.00
BOSCLE	CLETEX	563.00	0.00	999999.00
DETCWS	BOSKC	632.00	0.00	999999.00
DETCWS	CALOAK	1979.00	0.00	999999.00
DETCWS	CLETEX	95.00	0.00	999999.00
DETCWS	DETTOR	0.00	0.00	999999.00
DETCWS	MILBAL	237.00	0.00	999999.00
DETCWS	NYYCWS	509.00	0.00	999999.00
DETCWS	SEAMIN	1932.00	0.00	999999.00
KCNYY	BOSKC	1254.00	0.00	999999.00
KCNYY	CALOAK	1363.00	0.00	999999.00
KCNYY	CLETEX	694.00	0.00	999999.00
KCNYY	DETTOR	630.00	0.00	999999.00
KCNYY	MILBAL	483.00	0.00	999999.00
KCNYY	NYYCWS	1113.00	0.00	999999.00
KCNYY	SEAMIN	1489.00	0.00	999999.00
MINCAL	BOSKC	1124.00	0.00	999999.00
MINCAL	CALOAK	1536.00	0.00	999999.00
MINCAL	CLETEX	622.00	0.00	999999.00
MINCAL	DETTOR	528.00	0.00	999999.00
MINCAL	MILBAL	297.00	0.00	999999.00
MINCAL	NYYCWS	1028.00	0.00	999999.00
MINCAL	SEAMIN	1399.00	0.00	999999.00
OAKSEA	CALOAK	337.00	0.00	999999.00
OAKSEA	SEAMIN	678.00	0.00	999999.00
TEXBAL	BOSKC	1555.00	0.00	999999.00
TEXBAL	CALOAK	1246.00	0.00	999999.00
TEXBAL	CLETEX	1015.00	0.00	999999.00
TEXBAL	DETTOR	982.00	0.00	999999.00
TEXBAL	MILBAL	843.00	0.00	999999.00
TEXBAL	NYYCWS	1383.00	0.00	999999.00
TEXBAL	SEAMIN	1670.00	0.00	999999.00
TORMIL	BOSKC	463.00	0.00	999999.00
TORMIL	CALOAK	2175.00	0.00	999999.00
TORMIL	CLETEX	193.00	0.00	999999.00
TORMIL	DETTOR	206.00	0.00	999999.00
TORMIL	MILBAL	434.00	0.00	999999.00
TORMIL	NYYCWS	366.00	0.00	999999.00
TORMIL	SEAMIN	2124.00	0.00	999999.00

Fig. 7.30 Edge data for umpire scheduling problem.

We will illustrate the use of the assignment algorithm to solve the umpire scheduling problem discussed in Section 7.1.

A more complete discussion of the umpire scheduling problem can be found in Evans (1988). Figure 7.28 shows the distance matrix between the home cities from series 2 to series 3 in Fig. 7.6. These distances represent the arc costs in the assignment problem.

```
ASSIGNMENT PROBLEM:    MINIMUM COST IS        3281.00

FROM        TO            LOWER       FLOW         UPPER        COST
----        --            --------    --------     --------     --------
BOSCLE      BOSKC          0.00        1.00      999999.00         0.00
TEXBAL      CALOAK         0.00        1.00      999999.00      1246.00
KCNYY       CLETEX         0.00        1.00      999999.00       694.00
DETCWS      DETTOR         0.00        1.00      999999.00         0.00
M1NCAL      MILBAL         0.00        1.00      999999.00       297.00
TCRMIL      NYYCWS         0.00        1.00      999999.00       366.00
OAKSEA      SEAMIN         0.00        1.00      999999.00       678.00
```

Fig. 7.31 NETSOLVE solution to umpire scheduling problem.

Figures 7.29 and 7.30 provide a listing of the NETSOLVE node data and edge data for the assignment problem illustrated in Fig. 7.7. The output from the assignment algorithm is shown in Fig. 7.31. The total distance travelled by all umpire crews is 3281. The solution shows that the crews in Boston and Detroit in series 2 remain there; the crew in Texas goes to California; the crew in Kansas city travels to Cleveland; the crew in Minnesota goes to Milwaukee; the crew in Toronto travels to New York; and the crew in Oakland is assigned to Seattle.

EXERCISES

1. Construct a graph in which a maximum-cardinality matching is not a maximum-weight matching.

2. Consider the graph in Fig. 7.32 (ignore the edge weights). Apply a maximum-flow algorithm to find the maximum-cardinality matching.

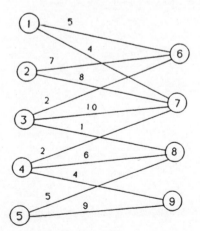

Fig. 7.32

3. Find a maximum-cardinality matching for the crew scheduling example shown in Fig. 7.5.

4. Consider the graph in Fig. 7.33 (ignore the edge weights). Find a maximum-cardinality matching.

5. Use the Hungarian algorithm to find the *minimum* cost assignment for the following cost matrix.

   0 7 9 5 6
   2 0 0 0 0
   0 7 7 9 8
   0 6 4 3 5
   0 7 8 1 4

6. Modify the Hungarian algorithm so that it will find a minimum cost assignment without having to transform the problem into a maximization. Apply your answer to Exercise 5.

7. A machine shop possesses six different drilling machines. On a certain day, five jobs that need drilling arrive. The number of person-hours required to perform each job on each of the machines is given in the table at the top of p. 271. Find the best way to assign each job to a different machine.

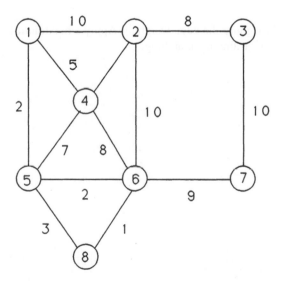

Fig. 7.33

	Job				
	A	B	C	D	E
Machine 1	5	7	6	4	9
2	8	10	3	4	7
3	6	11	5	4	7
4	5	8	7	3	9
5	3	6	4	2	7
6	3	7	5	3	7

8. The United Nations sponsors a sister-cities program in which cities are paired off for cultural and educational exchange programs. This year, 10 new cities have applied. What is the best way to pair them off so that the total distance between all sister cities is minimized? The intercity distances are as follows.

	1	2	3	4	5	6	7	8	9	10
From city 1	0	80	70	70	60	45	90	110	85	155
2		0	75	95	90	80	90	160	70	45
3			0	65	70	60	100	80	80	55
4				0	80	80	70	170	200	250
5					0	110	170	190	270	300
6						0	100	150	110	200
7							0	75	95	100
8								0	90	100
9									0	50

9. Find a maximum-weight matching in the graph in Fig. 7.32 using both the Hungarian algorithm and the weighted matching algorithm.

10. Find a maximum-weight matching in the graph in Fig. 7.33.

11. Find a minimum-weight matching in the graph in Fig. 7.34.

12. How does the maximum-cardinality matching algorithm remove an edge that has incorrectly been placed into solution? How does the maximum-weight matching algorithm accomplish this? How does the minimum-weight covering algorithm accomplish this?

13. Consider the following greedy heuristic algorithm for the maximum-weighted matching problem:

    Step 1. Select the edge with the largest weight in the current graph and add it to the matching.

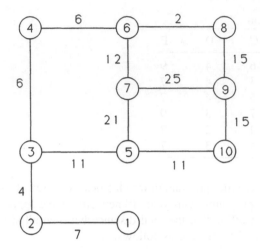

Fig. 7.34 Graph for Exercise 11.

Step 2. Delete the endpoints of the edge selected in step 1 and all edges incident to these vertices. If the graph contains at least two vertices, return to step 1; otherwise, stop.

Investigate implementations of this algorithm using different data structures and their time complexities. Construct examples where this algorithm finds and does not find optimal solutions.

REFERENCES

Balinksi, M., 1969. Labelling to Obtain a Maximum Matching. In *Combinatorial Mathematics and Its Applications* (eds. R. C. Bose and T. A. Dowling), University of North Carolina Press, Chapel Hill, pp. 585–602.

Ball, B. C., and D. B. Webster, 1977. Optimal Scheduling for Even-Numbered Team Athletic Conferences, *AIIE Trans.*, *9*, pp. 161–169.

Berge, C., 1957. Two Theorems in Graph Theory, *Proc. Natl. Acad. Sci. U.S.A.*, *43*, pp. 842–844.

Brown, J.R. (undated). Maximum Cardinality Matching, Kent State University, unpublished manuscript.

Campbell, R. T., and D.S. Chen, 1976. A Minimum Distance Basketball Scheduling Problem. In *Management Science in Sports* (eds. R. Machol et al.), North-Holland, Amsterdam, pp. 15–25.

Edmonds, J., 1965. Paths, Trees, and Flowers, *Can. J. Math.*, *17*, pp. 449–467.

Edmonds, J., 1965. Maximum Matching and Polyhedra with 0-1 Vertices, *J. Res. Natl. Bur. Stand.* *69B*(1,2), pp. 125–130.

Edmonds, J., and E. Johnson, 1970. Matching: A Well Solved Class of Integer Linear Programs, *Combinatorial Structures and Their Applications*, Gordon and Breach, New York, pp. 89–92.

Evans, James R. "A Microcomputer-Based Decision Support System for Scheduling Umpires in the American Baseball League," *Interfaces*, Vol. 18, No. 6, November–December 1988, pp. 42–51.

Jonker, R., and T. Volgenant, 1986. Improving the Hungarian Assignment Algorithm, *Oper. Res. Lett.*, *5*, pp. 171–175.

Kuhn, H. W., 1955. The Hungarian Method for the Assignment Problem, *Naval Research Logistics Quarterly*, *1*, pp. 83–97.

White, L. J., 1967. A Parametric Study of Matchings and Coverings in Weighted Graphs, Ph.D. Thesis, University of Michigan.

8

THE POSTMAN AND
RELATED ARC ROUTING
PROBLEMS

8.1 INTRODUCTION AND EXAMPLES

Many practical routing problems involve finding paths or cycles that traverse a set of arcs in a graph. In general, we call such problems *arc routing problems*. Arc routing problems have many practical applications. The following examples describe the major classes of arc routing problems.

The Chinese Postman Problem

Before starting his or her route, a postal carrier must pick up the mail at the post office, then deliver the mail along each block on the route, and finally return to the post office. To make the job easier and more productive, every postal carrier would like to cover the route with as little walking as possible. Thus, the problem is to determine how to cover all the streets assigned and return to the starting point in the shortest total distance. If we construct a graph $G = (X, E)$ in which each edge represents a street on which mail must be delivered and each vertex represents an intersection, this problem is equivalent to finding a cycle in G which traverses each edge *at least* once in minimum total distance.

This problem was first discovered by a Chinese mathematician, Kwan Mei-Ko, and is popularly known as the *Chinese postman problem (CPP)*. Many problems can be modeled as a CPP. For example, routing street sweepers, snowplows, interstate lawn mowers, police patrol cars, electric line inspectors, or automated guided vehicles in a factory involves determining minimum-cost paths that cover edges of the graph that models the situation. The classic Chinese postman problem assumes that all the edges of the graph are undirected. Some problems involve graphs in which all edges are directed (the *directed postman problem*) or in which only some of the edges are directed (*the mixed postman problem*). We shall consider each of these situations in this chapter.

The Capacitated Chinese Postman Problem

The capacitated Chinese postman problem (CCPP) arises when each arc has associated with it a positive demand and the vehicles to be routed have a finite capacity. For instance, in applications involving road salting, trucks can carry only a maximum amount of salt, say 10 tons. The salt spreading rate usually is fixed, say at 600 lb per lane-mile. Therefore, the demand on each road segment can be computed as the distance of the road segment times the salt spreading rate. One truck may not be able to service all the roads in a county or district due to its limited capacity. The CCPP is to find a set of routes from a single depot that service all edges in the graph at minimal cost and subject to the constraint that the total demand on each route does not exceed the capacity of the vehicle.

Capacitated Arc Routing Problems

Most practical generalizations of the CCPP are called capacitated arc routing problems (CARP). In a CARP, some of the demands on edges could be zero. This is a common situation in road salting applications at the county and state levels of government. A county, for example, is responsible only for county roads but can use state highways or other municipal roads for traveling. Therefore, while the graph defining the problem may include a large number of edges, only a subset of them must actually be serviced. The other edges can be used to "deadhead" between edges requiring service, that is, travel on the roads without performing any service.

We shall see that the Chinese postman problem can be solved rather easily in polynomial time. The CCPP and CARP problems are NP-hard; thus we normally resort to heuristic procedures to obtain a solution. Before we present specific algorithms for solving these problems, we need to study some basic results about cycles in graphs.

8.2 EULER TOURS

In Chapter 1 we discussed the Königsberg bridge problem, which was solved by the mathematician Leonhard Euler. Recall that this problem was to determine whether one could traverse each edge of a graph (i.e., each bridge) exactly once and return to the starting point. For the Königsberg bridge graph, Euler showed that this was impossible. We call any cycle in a graph that crosses each edge exactly once an *Euler tour*. Any graph that possesses an Euler tour is called an *Euler graph*. If an Euler tour exists in a postman problem, observe that the number of times that a postman arrives at a vertex must equal the number of times that the postman leaves that vertex. If the postman does not repeat any edges incident to a vertex, then this vertex must have an even number of edges incident to it, or even degree. With this reasoning, Euler proved the following result:

Theorem 8.1.   An undirected graph is Euler if and only if all vertices have even degree.

If you examine the graph of the Königsberg bridge problem, you will see that the graph is not Euler, and consequently no Euler tour exists.

Figure 8.1 shows an example of an Euler graph. There are several different Euler tours in the graph starting from vertex s. For example, each of the following four routes is an Euler tour:

Route 1: $(s, 1)$, $(1, 2)$, $(2, 3)$, $(3, 4)$, $(4, 2)$, $(2, s)$
Route 2: $(s, 1)$, $(1, 2)$, $(2, 4)$, $(4, 3)$, $(3, 2)$, $(2, s)$
Route 3: $(s, 2)$, $(2, 3)$, $(3, 4)$, $(4, 2)$, $(2, 1)$, $(1, s)$
Route 4: $(s, 2)$, $(2, 4)$, $(4, 3)$, $(3, 2)$, $(2, 1)$, $(1, s)$

Each of these four routes traverses each edge exactly once; thus, the total length of each route is $3 + 2 + 1 + 3 + 3 + 7 + 6 = 22$. The postman cannot do any better than this.

If a graph possesses an Euler tour, then the Chinese postman problem is solved; we need only find some Euler tour. If a graph has no Euler tour, then at least one edge must be crossed more than once. In a vehicle routing context, we call this *deadheading*, since the vehicle is not performing any productive work. In Fig. 8.2, for example, there is no way for the postman to traverse edge $(2, 3)$ only once. No Euler tour exists. A shortest tour that traverses each edge at least once is $(s, 1)$, $(1, 2)$, $(2, 3)$, $(3, 4)$, $(4, 5)$, $(5, 3)$, $(3, 2)$, $(2, s)$. Can you find another optimal solution? The total length of the tour is the length of all the edges $(3 + 2 + 5 + 1 + 3 + 7 + 6)$ plus the length of deadheading (5), or 32. It is easy to see that since the sum of all edge lengths is constant, the Chinese postman problem can be interpreted as minimizing the amount of deadheading necessary in the graph.

Constructing Euler Tours

Let us suppose that we have an Euler graph $G_1 = (X, E_1)$. How can we construct an Euler tour in G_1? A simple algorithm for finding an Euler tour proceeds as follows:

Step 1. Begin at any vertex s and construct a cycle C. This can be done by traversing

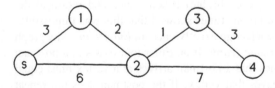

Fig. 8.1   Graph with an Euler tour.

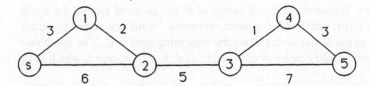

Fig. 8.2 Graph with no Euler tour.

any edge (s, x) incident to vertex s and marking this edge "used." Next traverse any unused edge incident to vertex x. Repeat this process of traversing unused edges until you return to vertex s. (This process must return to vertex s since every vertex has even degree and every visit to a vertex leaves an even number of unused edges incident to that vertex. Hence, every time a vertex is entered, there is an unused edge for departing from that vertex.)

Step 2. If C contains all the edges of G, stop. If not, then the subgraph G' in which all edges of C are removed must be Euler since each vertex of C must have an even number of incident edges. Since G is connected, there must be at least one vertex v in common with C.

Step 3. Starting at v, construct a cycle in G', say C'.

Step 4. Splice together the cycles C and C', calling the combined cycle C. Return to step 2.

We illustrate this method using the graph in Fig. 8.3. If we begin at vertex 1, we might construct the cycle C consisting of edges a, f, h, and i. Let v be vertex 2. The remaining graph has a unique cycle C' consisting of edges b, c, d, g, and e. We splice the

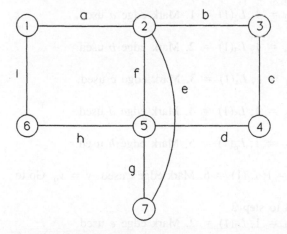

Fig. 8.3 Euler tour example.

cycles together as follows. Begin at the initial vertex of C and proceed around the cycle until the common vertex v is reached. At this point, traverse C' until you return to vertex v. Then continue from v to the initial vertex along the remaining path in C. The combined cycle in the example above would consist of edges a, b, c, d, g, e, f, h, and i, which is an Euler tour in G.

The basic ideas in this algorithm were formalized by Edmonds and Johnson (1973). Their implementation maintains a list of vertices $L_x(1)$, $L_x(2)$, . . . , $L_x(k)$, . . . to visit next when vertex x is reached the kth time.

Step 1. Select any vertex v_0. Let $v = v_0$ and $k_x = 0$ for all vertices x. All edges are labeled "unused."

Step 2. Randomly select an unused edge incident to vertex v. Mark this edge "used." Let y be the vertex at the other end of the edge. Set $k_y = k_y + 1$ and $L_y(k_y) = v$. If vertex y has any incident, unused edges, go to step 3. Otherwise vertex y must be v_0. In this case, go to step 4.

Step 3. Set $v = y$ and return to step 2.

Step 4. Let v_0 be any vertex which has at least one used edge and one unused edge incident to it. Set $v = v_0$ and return to step 2. If no such vertex exists, go to step 5.

Step 5. To construct the the tour, start at the original vertex v_0. The first time vertex x is reached, leave it by going to vertex $L_x(k_x)$. Set $k_x = k_x - 1$ and continue, each time going from vertex x to vertex $L_x(k_x)$.

EXAMPLE 1

We will illustrate this algorithm using Fig. 8.3.

Step 1. $v_0 = 1 = v$. $k_x = 0$ for $x = 1, \ldots, 7$.

Step 2. Select edge a; $y = 2$; $k_2 = 1$; $L_2(1) = 1$. Mark edge a used.

Step 3. $v = 2$.

Step 2. Select edge b; $y = 3$; $k_3 = 1$; $L_3(1) = 2$. Mark edge b used.

Step 3. $v = 3$.

Step 2. Select edge c; $y = 4$; $k_4 = 1$; $L_4(1) = 3$. Mark edge c used.

Step 3. $v = 4$.

Step 2. Select edge d; $y = 5$; $k_5 = 1$; $L_5(1) = 4$. Mark edge d used.

Step 3. $v = 5$.

Step 2. Select edge h; $y = 6$; $k_6 = 1$; $L_6(1) = 5$. Mark edge h used.

Step 3. $v = 1$.

Step 2. Select edge i; $y = 1$; $k_1 = 1$; $L_1(1) = 6$. Mark edge i used. $y = v_0$. Go to step 4.

Step 4. Set $v_0 = 2 = v$. Return to step 2.

Step 2. Select edge e; $y = 7$; $k_7 = 1$; $L_7(1) = 2$. Mark edge e used.

Step 3. $v = 7$.

Step 2. Select edge g; $y = 5$; $k_5 = 2$; $L_5(2) = 7$. Mark edge g used.

Step 3. $v = 5$.

Step 2. Select edge f; $y = 2$; $k_2 = 2$; $L_2(2) = 5$. Mark edge f used. Go to step 4 and determine that no new vertex exists. Go to step 5.

Step 5. Begin at $v_0 = 1$. Go to $L_2(2) = 5$; set $k_2 = 1$. Next, go to $L_5(2) = 7$; set $k_5 = 1$. The remaining vertices to be visited in order are 2, 1, 6, 5, 4, 3, 2.

8.3 THE POSTMAN PROBLEM FOR UNDIRECTED GRAPHS

This section describes how to solve the postman problem for any undirected graph $G = (X, E)$. We have seen that if G is an Euler graph, any Euler tour solves the postman problem and no deadheading is necessary. If G is not an Euler graph, we seek to minimize the amount of deadheading that is required. Let $a(i, j)$ be the length of edge (i, j) in G.

In any postman route, the number of times the postman enters a vertex equals the number of times the postman leaves that vertex. Consequently, if vertex x does not have even degree, then at least one edge incident to vertex x must be repeated by the postman.

Let $f(i, j)$ denote the number of times that edge (i, j) is repeated by the postman. Edge (i, j) is traversed $f(i, j) + 1$ times by the postman. Of course, $f(i, j)$ must be a nonnegative integer. Note that $f(i, j)$ contains no information about the direction of travel across edge (i, j).

Construct a new graph $G^* = (X, E^*)$ that contains $f(i, j) + 1$ copies of each edge (i, j) in graph G. Clearly, an Euler tour of graph G^* corresponds to a postman route in graph G.

The postman wishes to select values for the $f(i, j)$ variables so that

(a) Graph G^* is an even graph.

(b) $\Sigma\ a(i, j)f(i, j)$, the total length of repeated edges, is minimized.

If vertex x is an odd-degree vertex in graph G, an odd number of edges incident to vertex x must be repeated by the postman, so that in graph G^* vertex x has even degree. Similarly, if vertex x is an even-degree vertex in graph G, an even number of edges (zero is an even number) incident to vertex x must be repeated by the postman, so that in graph G^* vertex x has even degree. Recall from Chapter 1, Exercise 5, that graph G contains an even number of vertices with odd degree.

If we trace out as far as possible a path of repeated edges starting from an odd-degree vertex, this path must necessarily end at another odd-degree vertex. Thus, the repeated edges form paths whose initial and terminal vertices are odd-degree vertices. Of course, any such path may contain an even-degree vertex as one of its intermediate vertices. Consequently, the postman must decide (a) which odd-degree vertices will be joined together by a path of repeated edges and (b) the precise composition of each such path.

One method of solving the problem is to arbitrarily join the odd-degree vertices by paths of repeated edges and use the following theorem that was proved by Mei-Ko:

Theorem 8.2. A feasible solution to the postman problem is optimal if and only if

(i) no more than one duplicate edge is added to any original edge and
(ii) the length of the added edges in any cycle does not exceed one-half the length of the cycle.

Proof. Necessity is proved in a constructive fashion. Suppose that a feasible solution violates condition (i). Then we may eliminate two redundant added edges and obtain a new feasible solution with a shorter total length. Next, suppose that a feasible solution violates condition (ii). Then the length of the cycle without the added edges is less than one-half the length of the cycle. Eliminate an added edge from each place in the cycle where there was one, and add one to each place in the cycle where there was not one. Clearly, the total length of the added edges in the cycle has been reduced and all vertex degrees remain even. Sufficiency follows from the following lemmas, which we state without proof.

Lemma 8.1. If two feasible solutions satisfy (i) and (ii), then the lengths of their added edges are equal.

Lemma 8.2. Optimal solutions always exist.

The constructive proof of Theorem 8.2 provides a method of successively improving feasible solutions until an optimal solution is found. The only problem is that the number of cycles that must be checked in condition (ii) grows exponentially in the size of the graph. Therefore, this algorithm cannot be performed in polynomial time. However, a polynomial algorithm that employs other algorithms that we have seen in other chapters provides an elegant solution to the problem.

By performing either the Floyd or Dantzig shortest-path algorithm of Section 4.4, the postman can determine a shortest path between each pair of odd-degree vertices in graph G.

The postman can determine which pairs of odd-degree vertices are to be joined by a path of repeated edges as follows: Construct a graph $G' = (X', E')$ whose vertex set consists of all odd-degree vertices in G and whose edge set contains an edge joining each pair of vertices. Let the weight of each edge equal a very large number minus the length of a shortest path between the corresponding two vertices in graph G as found by the Floyd or Dantzig algorithm.

Next, the postman should find a maximum-weight matching for graph G' using the maximum-weight matching algorithm in Section 7.4. Since graph G' has an even number of vertices and each pair of vertices in G' is joined by an edge, the maximum-weight matching will cover each vertex exactly once. This matching matches together odd-degree vertices in graph G. The edges in a shortest path joining a matched pair of odd-degree vertices should be repeated by the postman. Since this matching has maximum total weight, the resulting postman route must have minimum total length.

Thus, we can solve the postman problem for an undirected graph by using the Floyd

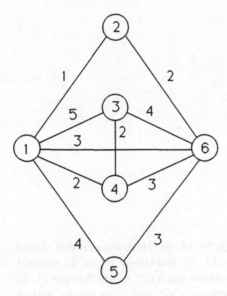

Fig. 8.4

	1	2	3	4	5	6
1	0	1	4	2	4	3
2	1	0	5	3	5	2
3	4	5	0	2	7	4
4	2	3	2	0	6	3
5	4	5	7	6	0	3
6	3	2	4	3	3	0

Fig. 8.5 Shortest path length matrix.

or Dantzig algorithm and the maximum-weight matching algorithm. No new algorithm is needed.

EXAMPLE 2

Let us find an optimal postman route for the undirected graph in Fig. 8.4. Notice that vertices 1, 3, 4, and 6 have odd degree. The length of a shortest path between all pairs of odd vertices is shown in Fig. 8.5. The reader should verify these values using either the Floyd or Dantzig algorithm.

Form the graph G' shown in Fig. 8.6. The vertices of G' are the odd-degree vertices 1, 3, 4, and 6 of graph G. All possible edges are present in G'. Happily, since G' does not have many vertices, we can find a minimum-weight matching of all the edges in G' by enumeration rather than by using the maximum-weight matching algorithm. Three matchings are possible:

Matching	Weight
(1, 3), (4, 6)	4 + 3 = 7
(1, 4), (3, 6)	2 + 4 = 6
(1, 6), (3, 4)	3 + 2 = 5

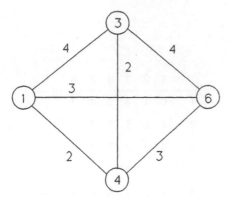

Fig. 8.6 Graph G'.

Consequently, the minimum-weight matching is (1, 6), (3, 4). Thus, the postman should repeat the shortest path from 1 to 6, which is edge (1, 6), and should repeat the shortest path from 3 to 4, which is edge (3, 4). Figure 8.7 shows graph G^*, in which edges (1, 6) and (3, 4) have each been duplicated once. All vertices in G^* have even degree, and an optimal postman route for the original graph in Fig. 8.4 corresponds to an Euler tour of graph G^* in Fig. 8.7. The technique described in case A for even graphs can be applied to graph G^*. An optimal route is (1, 2), (2, 6), (6, 5), (5, 1), (1, 3), (3, 6), (6, 4), (4, 3),

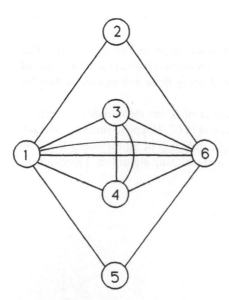

Fig. 8.7 Graph G^*.

(3, 4), (4, 1), (1, 6), (6, 1), which traverses each edge in G^* exactly once and traverses each edge in G at least once. Only edge (1, 6) and (3, 4) are repeated in graph G. The total length of this route is 34 units, which is 5 units more than the sum of the edge lengths.

Note that if the maximum-weight matching algorithm had been used, each edge weight in graph G' would have been set equal to a large number, say M, minus the length of a shortest path between the corresponding two endpoints in G. Thus, the matching (1, 3), (4, 6) would have a total weight equal to $(M - 4) + (M - 3) = 2M - 7$. Matching (1, 4), (3, 6) would have total weight equal to $(M - 2) + (M - 4) = 2M - 6$. Matching (1, 6), (3, 4) would have total weight equal to $(M - 3) + (M - 2) = 2M - 5$ and would be selected as the maximum-weight matching.

The large M values can be viewed simply as a device for converting the problem of finding a minimum-weight matching that covers all vertices into a maximum-weight matching problem.

8.4 THE POSTMAN PROBLEM FOR DIRECTED GRAPHS

In this section we shall study the postman problem for a directed graph $G = (X, A)$. A directed graph corresponds to a physical situation in which all streets are one-way streets. The direction of an arc specifies the direction in which the corresponding street must be traversed.

In Chapter 1 we defined the number of arcs directed into vertex x in a directed graph as the *in-degree* of vertex x and the number of arcs directed away from vertex x as the *out-degree* of vertex x. The in-degree and out-degree of vertex x are denoted as $d^-(x)$ and $d^+(x)$, respectively. If the in-degree equals the out-degree for all vertices x in graph G, then G is called *symmetric*.

Unlike the postman problem for undirected graphs, the postman problem may have no solution for a directed graph. For example, consider the graph in Fig. 8.8. Once the postman arrives at either vertex 1 or vertex 2, he cannot return to vertex s because no arcs

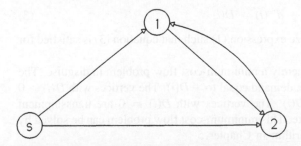

Fig. 8.8 No postman route exists.

leave set $\{1, 2\}$ for a vertex not in this set. In general, no solution exists for the postman problem whenever there is a set S of vertices with the property that no arcs go from a vertex in S to a vertex not in S. If no such set S exists, it is always possible for the postman to complete his route, no matter how long it takes, since he cannot be trapped anywhere.

As before, the number of times a postman enters a vertex must equal the number of times the postman leaves that vertex. Consequently, if vertex x has more arcs entering it than leaving it [that is, $d^-(x) > d^+(x)$], the postman must repeat some of the arcs leaving x. Similarly, if vertex x has more arcs leaving it than entering it [that is, $d^+(x) > d^-(x)$], the postman must repeat some arcs entering x. Thus, if for some vertex x, $d^+(x) \neq d^-(x)$, no Euler tour is possible.

Two cases must be considered separately:

Case A: Graph G is symmetric [that is, $d^+(x) = d^-(x)$ for all x].
Case B: Graph G is not symmetric.

Case A. If graph G is symmetric, it is possible for the postman to perform his route without repeating any arcs; i.e., the optimal solution to the postman problem is an Euler tour.

An Euler tour in a directed graph $G = (X, A)$ can be found by using techniques similar to the ones used for finding an Euler tour in an even undirected graph in Section 8.2. The only difference is that the arcs selected when leaving a vertex must be directed out of that vertex.

Case B. As before, let $f(i, j)$ denote the number of times that the postman repeats arc (i, j). The postman wants to select nonnegative integer values for the $f(i, j)$ variables so as to minimize

$$\sum_{i} a(i, j) f(i, j) \tag{1}$$

the total length of repeated arcs such that he enters and leaves each vertex x the same number of times, that is,

$$d^-(i) + \sum_{j} f(j, i) = d^+(i) + \sum_{j} f(i, j) \tag{2}$$

Rewriting equation (2) yields

$$\sum_{j} [f(i, j) - f(j, i)] = d^-(i) - d^+(i) \equiv D(i) \tag{3}$$

Thus, the postman wishes to minimize expression (1) such that equation (3) is satisfied for all vertices i in graph G.

This minimization problem is merely a minimum-cost flow problem in disguise. The vertices with $D(i) < 0$ are sinks with demand equal to $-D(i)$. The vertices with $D(i) > 0$ are sources with supply equal to $D(i)$. The vertices with $D(i) = 0$ are transshipment vertices. All arc capacities are infinite. This minimum-cost flow problem can be solved by using any minimum-cost flow algorithm in Chapter 5.

Since all right-side values in equation (3) are integers, we know that the minimum-

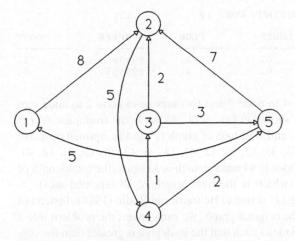

Fig. 8.9 Postman problem in a directed graph.

cost flow algorithm will produce optimal values for the $f(i, j)$ that are nonnegative integers.

After finding the optimal integer values for the $f(i, j)$ variables, create a graph G^* with $f(i, j) + 1$ copies of arc (i, j) for all $(i, j) \in A$. By equation (3), graph G^* is symmetric. The technique described for case A can now be applied to find an Euler tour of graph G^*. An Euler tour of graph G^* corresponds to a postman route in graph G that traverses each arc (i, j) $f(i, j) + 1$ times. Since the optimal values of $f(i, j)$ minimize expression (1), this Euler tour in graph G^* must correspond to an optimal postman route in graph G.

EXAMPLE 3

Let us find an optimal postman route for the directed graph in Fig. 8.9. Examining the inner and outer degrees, we find that

$d^-(1) = 1 = d^+(1)$; vertex 1 is a transshipment vertex.
$d^-(2) = 3 > d^+(2) = 1$; vertex 2 is a source with $3 - 1 = 2$ units supply.
$d^-(3) = 1 < d^+(3) = 2$; vertex 3 is a sink with $2 - 1 = 1$ unit demand.
$d^-(4) = 1 < d^+(4) = 2$; vertex 4 is a sink with $2 - 1 = 1$ unit demand.
$d^-(5) = 2 = d^+(5)$; vertex 5 is a transshipment vertex.

Figure 8.10 shows the minimum-cost flow problem that results. Using NETSOLVE, we obtain the optimal solution shown below.

MINIMUM COST FLOW PROBLEM: MINIMUM COST IS   11.

FROM	TO	LOWER	FLOW	UPPER	COST
4	3	0.	1.	999999.	1.
2	4	0.	2.	999999.	5.

Thus, we add one arc from node 4 to node 3 and two arcs from node 2 to node 4 to create a symmetric graph G^*. This is shown in Fig. 8.11. As we know from case A, G^* possesses an Euler tour. For example, an Euler tour of graph G^* and an optimal postman route of graph G is (1, 2), (2, 4), (4, 3), (3, 2), (2, 4), (4, 3), (3, 5), (5, 2), (2, 4), (4, 5), (5, 1). The total length of this tour is 44 units, which is 33 units (the total length of all arcs in this graph) plus 11 units (which is the total length of all repeated arcs).

An alternative way of determining G^* is due to Beltrami and Bodin (1974). Instead of creating a network flow problem on the original graph, we can convert the problem into a transportation problem. First, locate nodes i such that the in-degree is greater than the out-degree. The difference, $s(i)$, represents the number of additional arcs that must be directed out of node i. Also, locate all nodes j for which the out-degree is greater than the in-degree. Call this difference $d(j)$. This represents the number of additional arcs that must be directed into node j. Now find the shortest distance from each node i to each node j, and call this distance $c(i, j)$. Solve the following transportation problem:

$$\min \sum_i \sum_j c(i, j)g(i, j) \tag{4}$$

$$\sum_j g(i, j) = s(i) \quad \text{for all } i \tag{5}$$

$$\sum_i g(i, j) = d(j) \quad \text{for all } j \tag{6}$$

$$g(i, j) \geq 0 \quad \text{for all } i, j \tag{7}$$

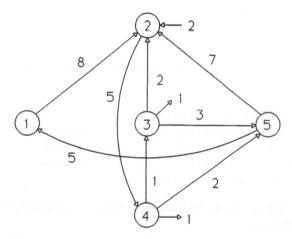

Fig. 8.10 Network flow problem for directed CPP.

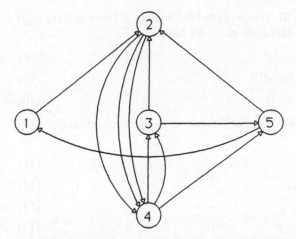

Fig. 8.11 Graph G^*.

Here, $g(i, j)$ represents the number of times that the shortest path between nodes i and j must be added to the graph.

In Example 3, we see that node 2 has an in-degree greater than its out-degree, with $s(2) = 2$. Nodes 3 and 4 have out-degree greater than the in-degree; $d(3) = d(4) = 1$. The shortest path between node 2 and node 3 consists of arcs $(2, 4)$ and $(4, 3)$ and has distance $c(2, 3) = 6$; the shortest path between node 2 and node 4 consists of the arc $(2, 4)$ and has distance $c(2, 4) = 5$. The transportation network is shown in Fig. 8.12 and has a trivial solution: $g(2, 3) = g(2, 4) = 1$. Thus, we add the arcs $(2, 4)$ twice and $(4, 3)$ once to create G^*.

An algorithm that works directly on the network without requiring the solution of a transformed minimum-cost flow problem or transportation problem was proposed by Lin and Zhao (1988). Such an algorithm is advantageous if it is embedded within a larger algorithm or decision support system, for instance. This approach is based on a slightly different linear programming formulation than the one discussed previously.

Define $h(i, j)$ to be the number of times that arc (i, j) is traversed in the Euler tour. [In

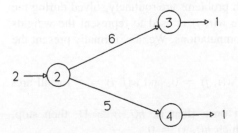

Fig. 8.12 Transportation network for Example 3.

the previous formulation, the variable $f(i, j)$ represented the number of times that arc (i, j) is *repeated* in the Euler tour.] With this definition, we have

$$\min \Sigma \, a(i, j)h(i, j) \tag{8}$$

$$-\Sigma \, h(i, j) + \Sigma \, h(j, i) = 0 \quad \text{for all } i \tag{9}$$

$$h(i, j) \geq 1 \text{ and integer} \quad \text{for all } (i, j) \tag{10}$$

Associate dual variables $p(i)$ with each constraint (9) and $r(i, j)$ with each constraint (10). The dual linear program is

$$\max \Sigma \, r(i, j) \tag{11}$$

$$-p(i) + p(j) + r(i, j) \leq a(i, j) \tag{12}$$

$$r(i, j) \geq 0 \tag{13}$$

$$p(i) \text{ unrestricted} \tag{14}$$

Now consider the complementary slackness conditions

$$[-p(i) + p(j) + r(i, j) - a(i, j)]h(i, j) = 0 \tag{15}$$

$$[h(i, j) - 1]r(i, j) = 0 \tag{16}$$

Since any feasible solution to the primal problem has $h(i, j) \geq 1$, from (15) we must have

$$r(i, j) = a(i, j) + p(i) - p(j) \tag{17}$$

[Note that (12) is automatically satisfied.] Also,

$$\text{if } h(i, j) = 1, \quad \text{then } r(i, j) \geq 0$$

and $\tag{18}$

$$\text{if } h(i, j) \geq 2, \quad \text{then } r(i, j) = 0$$

Hence, primal feasibility, dual feasibility, and complementary slackness optimality conditions are summarized by (9), (10), (17), and (18).

The algorithm begins with an initial solution that satisfies all (9), (17), and (18) but not (10) and proceeds to adjust the primal and dual solutions until (10) is satisfied while not violating any other condition. Shortest-path problems are routinely solved during the course of the algorithm. An auxiliary variable $w(i, j)$ is used to represent the weights associated with the arcs for the shortest-path computations. We now formally present the algorithm.

> *Step 1.* Set $h(i, j) = 0$, $r(i, j) = a(i, j)$, $w(i, j) = 0$, and $w(j, i) = \infty$ for all arcs (i, j). Set $p(i) = 0$ for all nodes i.
>
> *Step 2.* If each arc is traversed at least once [that is, $h(i, j) \geq 1$], then stop. Otherwise, select an arc (t, s) for which $h(t, s) = 0$.

Step 3. Using the weights $w(i, j)$, find a shortest path from s to each node i in the network. Call this path $P(s, i)$ and denote the shortest distance from s to i as $d(i)$. Consider the cycle formed by the union of $P(s, t)$ and the arc (t, s). Traverse the cycle in the direction of arc (t, s). For each arc in the cycle that is in the same direction as (t, s), set $h(i, j) = h(i, j) + 1$; for each arc in the opposite direction to (t, s), set $h(i, j) = h(i, j) - 1$.

Step 4. For all arcs (i, j), set $r(i, j) = \min\{d(i), d(t)\} - \min\{d(j), d(t)\} + r(i, j)$. If $h(i, j) = 0$, set $w(i, j) = 0$ and $w(j, i) = \infty$. If $h(i, j) = 1$, set $w(i, j) = r(i, j)$ and $w(j, i) = \infty$. Finally, if $h(i, j) \geq 2$, set $w(i, j) = w(j, i) = 0$. Return to step 2.

EXAMPLE 4

Let us consider the problem in Fig. 8.9.

Step 1. All variables are initialized. Observe that all complementary slackness conditions except (10) hold. The table below provides a summary.

Arc	$r(i, j)$	$h(i, j)$	$w(i, j)$	$w(j, i)$
1, 2	8	0	0	∞
2, 4	5	0	0	∞
3, 2	2	0	0	∞
3, 5	3	0	0	∞
4, 3	1	0	0	∞
4, 5	2	0	0	∞
5, 1	5	0	0	∞
5, 2	7	0	0	∞

Step 2. Choose $(t, s) = (1, 2)$. Using the weights $w(i, j)$, find shortest paths from node 2 to all other nodes. The shortest-path arborescence and distance labels are shown below.

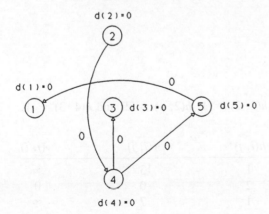

Traversing the cycle 1-2-4-5-1 in the direction of arc (1, 2), we set $h(1, 2) = 1$, $h(2, 4) = 1$, $h(4, 5) = 1$, and $h(5, 1) = 1$.

Step 4. Update $r(i, j)$ and $w(i, j)$ values. These are summarized below.

Arc	$r(i, j)$	$h(i, j)$	$w(i, j)$	$w(j, i)$
1, 2	8	1	8	∞
2, 4	5	1	5	∞
3, 2	2	0	0	∞
3, 5	3	0	0	∞
4, 3	1	0	0	∞
4, 5	2	1	2	∞
5, 1	5	1	5	∞
5, 2	7	0	0	∞

Step 2. Choose arc $(3, 2) = (t, s)$.

Step 3. Find shortest paths from node 2 to all others. The shortest-path arborescence and distance labels are shown below.

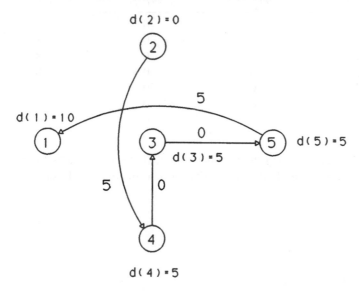

Traversing the cycle 3-2-4-3, we set $h(3, 2) = 1$, $h(2, 4) = 2$, and $h(4, 3) = 1$.

Step 4. Update $r(i, j)$ and $w(i, j)$.

Arc	$r(i, j)$	$h(i, j)$	$w(i, j)$	$w(j, i)$
1, 2	13	1	13	∞
2, 4	0	2	0	0
3, 2	7	1	7	∞

Arc	$r(i, j)$	$h(i, j)$	$w(i, j)$	$w(j, i)$
3, 5	3	0	0	∞
4, 3	1	1	1	∞
4, 5	2	1	2	∞
5, 1	5	1	5	∞
5, 2	12	0	0	∞

Step 2. Choose arc $(3, 5) = (t, s)$.

Step 3. Find the shortest paths from node 5 to all others. The shortest-path arborescence and distance labels are shown below.

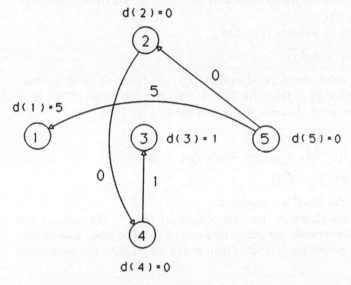

$d(2) = 0$

$d(1) = 5$

$d(3) = 1$

$d(5) = 0$

$d(4) = 0$

Traversing the cycle 3-5-2-4-3, we set $h(3, 5) = 1$, $h(5, 2) = 1$, $h(2, 4) = 3$, and $h(4, 3) = 2$. At this point, all $h(i, j) \geq 1$ so we may stop. We see that the solution corresponds to G^* in Fig. 8.11.

If we compute the final values of $r(i, j)$ in step 4, we obtain the following:

Arc	$r(i, j)$
1, 2	14
2, 4	0
3, 2	8
3, 5	4
4, 3	0
4, 5	2
5, 1	4
5, 2	12

Since the dual objective function is $\Sigma\ r(i, j)$, we see that at termination, $\Sigma\ r(i, j) = 44$, which is the value of the optimal primal solution.

Proof of the Algorithm: To show that this algorithm correctly solves the directed CPP, we must show that $r(i, j)$, $h(i, j)$, and $p(i)$ satisfy the optimality conditions (9), (10), (17), and (18). It is sufficient to prove that $r(i, j)$ satisfies the dual constraints for all (i, j) during the course of the algorithm. First, we show that $r(i, j) \geq 0$. We show this by induction. Let k be the iteration counter. Clearly, when $k = 0$, $r^0(i, j) = a(i, j) \geq 0$. Suppose that $r^{k-1}(i, j) \geq 0$. In step 4 of the algorithm, we have $d^k(j) \leq d^k(i) + w^k(i, j)$ and $w^k(i, j) \leq r^k(i, j)$. Therefore, it follows that $d^k(i) + r^{k-1}(i, j) - d^k(j) \geq 0$. Combining these results, we have $r^k(i, j) \geq 0$.

Next, we show that $r(i, j)$ satisfies (12). Let

$$p(i) = p^k(i) = p^{k-1}(i) + \delta p^k(i)$$

where $\delta p^k(i) = \min\{d^k(i), d^k(t)\}$. From the algorithm, $p^0(i) = d^0(i) = 0$ for all i. Thus, $p^1(i) = 0$ and $\delta p^1(i) = 0$ for all i. Thus, for $k = 1$, $r^k(i, j) = \delta p^k(i) + r^{k-1}(i, j) - \delta p^k(j) = p^k(i) + a(i, j) - p^k(j)$. Assume this is true for $k - 1$. Then

$$
\begin{aligned}
r^k(i, j) \quad &= \delta p^k(i) + r^{k-1}(i, j) - \delta p^k(j) \\
&= \delta p^k(i) + [p^{k-1}(i) + a(i, j) - p^{k-1}(j)] - \delta p^k(j) \\
&= p^k(i) + a(i, j) - p^k(j)
\end{aligned}
$$

Thus (12) is satisfied and the proof is completed.

It can be shown that this algorithm has time complexity $O(kn^2)$. The constant k is related to the structure of the network; for sparse networks, k could be much smaller than m and n. In these cases, its performance is better than that of the previous two approaches we have discussed.

8.5   THE POSTMAN PROBLEM FOR MIXED GRAPHS

In this section, we shall consider the postman problem in a graph G in which some arcs are directed and some arcs are not directed (a mixed graph). If an arc is directed, the postman must traverse this arc only along its direction (a one-way street). If an arc is not directed, the postman may traverse this arc in either (or, if necessary, both) directions (a two-way street). An undirected arc is not considered when calculating the inner and outer degrees of the vertices.

Can the postman always find a route in a mixed graph? Not always. It might happen that the graph contains a set S of vertices with the property that all arcs joining a vertex in S to a vertex not in S are directed toward the vertex in S. In this case, once the postman reaches a vertex in S he can never reach a vertex outside S. He is trapped, and no solution exists for the postman problem. If no set S with this property exists, the postman can keep on traveling, no matter how long it takes, until he has traversed all arcs and returned to his starting vertex.

Three conditions must be met for a mixed graph to contain an Euler tour. First, G must be connected. Second, every node of G must have even degree (ignoring the directions). Third, for every subset S of nodes, the difference between the number of directed arcs from S to its complement and the number of directed arcs from the complement of S to S is less than or equal to the number of undirected arcs joining S and its complement. Because there are of order $O(2^n)$ subsets S, verifying this third condition is difficult. Notice that for undirected graphs, the second condition implies the third; for directed graphs, the third condition is equivalent to the graph being symmetric.

For a mixed graph $G = (X, A)$, three cases must be treated separately:

Case A: Graph G is even and symmetric.
Case B: Graph G is even but not symmetric.
Case C: Graph G is neither even nor symmetric.

Case A. This is the easiest case to analyze since the solution technique for this case is a composite of the solution techniques for even directed and even undirected graphs presented in Sections 8.4 and 8.3, respectively.

Starting with any directed arc in graph G, generate a cycle of directed arcs as done in Section 8.4 for even directed graphs. (Since G is symmetric, this is possible.) Repeat this procedure until all directed arcs have been used. (This is possible since the unused arcs always form an even, symmetric graph.)

Next, repeat this procedure using only the undirected arcs in graph G. (Again, this is always possible, since the unused arcs form an even graph.) After all the arcs in graph G have been used, splice together all the cycles generated above into one cycle C. Cycle C forms an Euler tour of graph G and is an optimal solution to the postman problem for case A.

Case B. Graph G is even but not symmetric. In this case, it is not easy to know in advance if the postman must repeat any arcs. For example, an Euler tour is an optimal solution for the even, nonsymmetric graph in Fig. 8.13 with undirected arc (1, 2)

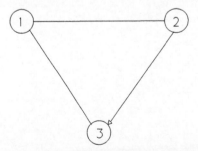

Fig. 8.13 Mixed graph with an Euler tour.

traversed from 1 to 2 and undirected arc (3, 1) traversed from 3 to 1. On the other hand, no Euler tour can be optimal for the even, nonsymmetric graph in Fig. 8.14 since arc (6, 1) must be repeated twice so that the postman can exit vertex 1 along arcs (1, 2) and (1, 4), and also (1, 5).

The mixed postman algorithm arbitrarily selects a direction for each undirected arc in graph G. This transforms graph G into an even directed graph G_D, and the solution technique for even, directed graphs given in Section 8.4 can be applied to graph G_D. However, due to the arbitrary choice of arc directions, some modifications are needed to correct some arc directions.

Mixed Postman Algorithm

Let $G = (X, A)$ be any even, mixed graph. An optimal postman route (if one exists) can be found as follows.

Let U denote the set of all undirected arcs in graph G; let V denote the set of all directed arcs in graph G. Tentatively, select a direction for each arc in U. Call the resulting directed graph G_D. For each vertex i in G_D, calculate

$$D(i) = d^-(i) - d^+(i) \tag{19}$$

If $D(i) < 0$, vertex i is a sink with demand equal to $-D(i)$. If $D(i) > 0$, vertex i is a source with supply equal to $D(i)$. If $D(i) = 0$, vertex i is a transshipment vertex.

If all vertices in G_D are transshipment vertices, graph G_D is an even, symmetric directed graph and the solution technique of Section 8.4 can be applied to graph G_D. This technique yields an Euler tour of G_D which corresponds to an optimal postman route of graph G.

Otherwise, construct a graph $G' = (X, A')$ as follows:

(a) For each arc $(i, j) \in V$, place an arc (i, j) in A' with infinite capacity and cost equal to the length of (i, j).

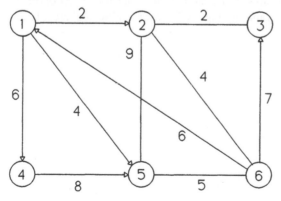

Fig. 8.14 Mixed graph with no Euler tour.

(b)   For each arc $(i, j) \in U$, create two directed arcs (i, j) and (j, i) in A'. Let each of these arcs have infinite capacity and cost equal to the length of (i, j).

(c)   For each arc $(i, j) \in U$, create a directed arc $(j, i)_1$ in A' whose direction is the reverse of the direction assigned this arc in G_D. These arcs are called *artificial* arcs. Assign each artificial arc a zero cost and a capacity equal to two.

Using the source supplies and sink demands defined above for graph G_D, apply a minimum-cost flow algorithm to find a minimum-cost flow in graph G' that satisfies all sink demands.

If no such flow exists, then no postman route exists. Otherwise, let $f(i, j)$ denote the number of flow units sent through arc (i, j) in G' in the minimum-cost flow produced by the minimum-cost flow algorithm. Recall that each optimal flow value is a nonnegative integer. In the proof of this algorithm, it will be shown that each artificial arc carries either zero or two flow units.

Create a graph G^* as follows:

(a)   For each nonartificial arc (i, j) in G' place $f(i, j) + 1$ copies of arc (i, j) in graph G^*.

(b)   If the flow in an artificial arc is two units, place one copy of this arc in graph G^*.

(c)   If the flow in an artificial arc is zero, reverse the direction of this arc and place one copy of this arc in graph G^*. (Thus, if no units traverse an artificial arc, the tentative direction assigned to this arc in G_D is retained; if two flow units traverse an artificial arc, the tentative direction assigned to this arc in G_D is reversed.)

Graph G^* is an even, symmetric, directed graph. The solution technique for even, symmetric, directed graphs presented in Section 8.4 can now be applied to find an Euler tour of graph G^*. This Euler tour of graph G^* corresponds to an optimal postman route of the original graph G.

Proof.   Since graph G is not necessarily symmetric, some vertices may have a surplus of incoming arcs; other vertices may have a surplus of outgoing arcs. Ideally, we would like to assign a direction to all the undirected arcs in G so that the resulting directed graph is symmetric. Then the solution technique for even, symmetric, directed graphs presented in Section 8.4 could be applied to find an Euler tour of graph G.

However, it can happen that there is no way to direct the undirected graphs so that the resulting graph is symmetric. In this case, some of the arcs (directed or undirected) must be repeated by the postman. Of course, the postman wants to select the repeated arcs so that their total length is as small as possible.

The algorithm selects a tentative direction for each undirected arc in G. The resulting directed graph is called G_D. The solution technique of Section 8.4 for even directed graphs could be applied to graph G_D. However, the solution generated by this technique depends on the tentative directions, and it is always possible that the tentative directions in graph G_D will lead to a nonoptimal solution to the postman problem.

The algorithm generates a graph G' and, using the same source supplies and sink demands as in graph G_D, finds a minimum-cost flow that satisfies all these sink demands.

There are three kinds of arcs in graph G':

(a) Arcs corresponding to directed arcs in G (these arcs have nonzero cost and infinite capacity).
(b) Nonartificial arcs corresponding to undirected arcs in G (these arcs also have nonzero cost and infinite capacity).
(c) Artificial arcs corresponding to undirected arcs in G (these arcs have zero cost and a capacity equal to two).

The number $f(i, j)$ carried by an arc of type (a) or (b) in the minimum-cost flow equals the number of times the corresponding directed arc is repeated by the postman. Thus, the postman will traverse each arc $(i, j) \in V f(i, j) + 1$ times, and the postman will traverse each arc $(i, j) \in U$ a total of $f(i, j) + f(j, i) + 1$ times.

As shown later, each artificial arc [type (c)] carries either zero or two flow units. If an artificial arc $(j, i)_1$ carries two flow units in the minimum-cost flow, this arc decreases the supply at j by two units and increases the supply at i by two units. The same effect could have been achieved by selecting the reverse tentative direction for this arc in graph G_D. Hence, the algorithm reverses the tentative direction of this arc.

If an artificial arc $(j, i)_1$ carries no flow units in the minimum-cost flow, this arc has no effect on the supplies at vertices i and j. This is equivalent to retaining the tentative direction given this arc in graph G_D. Hence, the algorithm retains the tentative direction given this arc.

From the minimum-cost flow values $f(i, j)$ for the arcs in graph G', the algorithm generates a directed graph G^*. It remains to show that

(a) Graph G^* is even and symmetric.
(b) An Euler tour of graph G^* corresponds to an optimal postman route of graph G.
(c) If the minimum-cost flow algorithm cannot find any flow that satisfies all sink demands in graph G', no postman route exists for graph G.

Proof of (a). Since graph G_D is an even graph, $d^+(i) + d^-(i)$ is an even number for all vertices i in G_D. Thus, $d^+(i)$ and $d^-(i)$ are both odd or are both even. In either case, $D(i)$ must be even. Consequently, all supplies and demands in graph G' are even numbers (zero is an even number). Also, all arc capacities are even numbers. Hence, a minimum-cost flow algorithm will produce an optimal flow in which all flow values are even numbers. Thus, flow units in a minimum-cost flow will travel in pairs.

The value of $d(i)$ in G^* equals the value of $d(i)$ in G plus the number of flow units that enter or leave vertex i along nonartificial arcs. Since all flow units travel in pairs, it follows that $d(i)$ is even in graph G^*. Thus, graph G^* is even.

A pair of flow units arriving at vertex i via a nonartificial arc increase $d^-(i)$ in G^* by two units. A pair of flow units leaving vertex i along a nonartificial arc increase $d^+(i)$ by two units. A pair of flow units arriving at vertex i via an artificial arc cause the tentative direction of this arc to be reversed, which has the effect of increasing $d^-(i)$ by two units. A pair of flow units leaving vertex i along an artificial arc cause the tentative direction of this arc to be reversed, which has the effect of increasing $d^+(i)$ by two units. Thus, each flow unit arriving at vertex i increases $d^-(i)$ by one unit, and each flow unit leaving vertex i increases $d^+(i)$ by one unit. Since the minimum-cost flow satisfies equation (3) in Section 8.4 for all vertices in graph G, it follows that graph G^* is symmetric.

Proof of (b). Suppose that the postman route produced by the mixed postman algorithm is not optimal. Then there exists another postman route whose duplicated arcs have an even smaller total length. This route must correspond to a flow in graph G' with even lower cost than the flow generated by a minimum-cost flow algorithm, which is a contradiction.

Proof of (c). Since $\Sigma_i\, d^-(i) = \Sigma_i\, d^+(i)$ in any graph G_D, the total of the source supplies in graph G_D must equal the total of the sink demands in graph G_D. Thus, all source supplies must be shipped out in order to satisfy all sink demands.

If graph G' contains an arc from i to j, then graph G' must contain an arc with infinite capacity from i to j. Suppose that the minimum-cost flow algorithm presented in Section 6.6 is used and terminates without satisfying all sink demands. Let S denote the set of all vertices that were labeled after the last iteration of the minimum-cost flow algorithm. From the flow-augmenting algorithm, which is a subroutine of the minimum-cost flow algorithm, we know that all arcs from a vertex in S to a vertex not in S must carry a full capacity flow. This is impossible since some of these arc capacities are infinite. Thus, no such arcs can exist, and all arcs with one labeled endpoint and one unlabeled endpoint are directed into set S. Consequently, once the postman reaches set S he cannot leave set S and no postman route is possible.   Q.E.D.

EXAMPLE 4

Let us find an optimal postman route for the even, nonsymmetric, mixed graph G shown in Fig. 8.14. First, directions are arbitrarily selected for each undirected arc in graph G. The resulting graph G_D is shown in Fig. 8.15. In graph G_D

$d^+(1) = 3,\ d^-(1) = 1,\ D(1) = -2;$ vertex 1 is a sink with demand 2

$d^+(2) = 2,\ d^-(2) = 2,\ D(2) = 0;$ vertex 2 is a transshipment vertex

$d^+(3) = 0,\ d^-(3) = 2,\ D(3) = 2;$ vertex 3 is a source with supply 2

$d^+(4) = 1,\ d^-(4) = 1,\ D(4) = 0;$ vertex 4 is a transshipment vertex

$d^+(5) = 2,\ d^-(5) = 2,\ D(5) = 0;$ vertex 5 is a transshipment vertex

$d^+(6) = 2,\ d^-(6) = 2,\ D(6) = 0;$ vertex 6 is a transshipment vertex

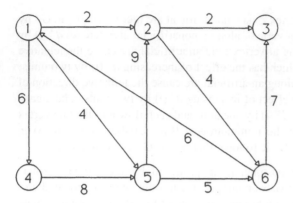

Fig. 8.15 Graph G_D.

Graph G' is shown in Fig. 8.16. The first number next to each arc denotes its cost. If the arc capacity is finite, it is denoted by the second number next to the arc.

A minimum-cost flow algorithm is now required to find a minimum-cost way of sending 2 flow units from source 3 to sink 1. (Notice that all supplies and demands are even numbers and that the total supply at the sources equals the total demand at the sinks.) By inspection, we can see that the minimum-cost flow consists of sending both flow units from 3 to 1 along the path $(3, 2)_1$, $(2, 6)$, $(6, 1)$. The total cost is $0 + 4 + 6 = 10$ units for each flow unit, or 20 units. Thus, the optimal flow values are $f(3, 2)_1 = f(2, 6) = f(6, 1) = 2$, and all other flow values equal zero. Since $f(3, 2)_1 = 2$, we must reverse the tentative direction of this arc so that it is directed from 3 to 2. All other arbitrary directions

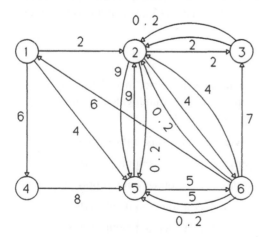

Fig. 8.16 Graph G'.

are retained. Moreover, arcs (2, 6) and (6, 1) must be repeated twice since $f(2, 6) = f(6, 1) = 2$.

Graph G^* is shown in Fig. 8.17. Note that graph G^* is an even, symmetric, directed graph. The solution technique for even, symmetric directed graphs given in Section 8.4 can now be applied to find an Euler tour of graph G^*. This tour corresponds to an optimal postman route in graph G. Since the minimum-cost flow in G' costs 20 units, the optimal postman route in graph G will repeat arcs with a total length of 20 units.

Case C. Graph G is neither even nor symmetric.

As far as the authors know, no efficient optimal solution technique is currently available for this case. This problem could be approached as a two-stage problem: First make the graph even in an optimal way; then make the even graph symmetric using the optimal procedure given in Section 8.4, case B. However, there is no guarantee that optimally solving each stage will lead to an optimum solution for the overall problem.

8.6 EXTENSIONS TO THE POSTMAN PROBLEM

In this section we shall consider two major extensions to the Chinese postman problem that we described in the opening section of this chapter, namely the capacitated Chinese postman problem and the capacitated arc routing problem.

The Capacitated Chinese Postman Problem

The CCPP was introduced by Christofides (1973). Given a graph $G = (X, E)$ in which each edge (x, y) has associated a cost $a(x, y)$ and demand $q(x, y)$, and a set of vehicles having fixed capacity W, find a minimal-cost set of cycles each of which passes through a

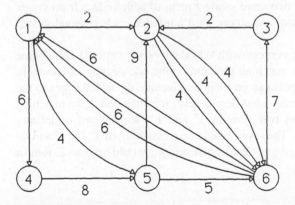

Fig. 8.17 Graph G^*.

fixed vertex (the depot) and such that the total demand on each cycle does not exceed the vehicle capacity W.

Golden and Wong (1981) show that the problem of finding a solution whose cost is guaranteed to be less than 1.5 times the optimal value is NP-hard. Therefore, we must resort to heuristics for a practical solution.

Christofides' algorithm sequentially creates feasible cycles, removes them from the graph, and solves (if necessary) a matching problem so that all vertices in the remaining graph have even degree. The algorithm is described as follows. We assume that vertex 1 is the depot.

> *Step 1.* If the current graph contains no edges, stop. Otherwise, starting at vertex 1, attempt to construct a feasible cycle (i.e., one in which the sum of demands does not exceed W). The cycle should be such that when its edges are removed from the graph, the remaining graph should be connected (except possibly for isolated vertices). This can be done by sequentially constructing a path from vertex 1 to vertex x and adding edge (x, y), which when removed does not disconnect the graph (except for isolated vertices) and such that the total demand on the path from vertex 1 to vertex x plus the demand on the shortest path (using demands as lengths) from vertex y to vertex 1 does not exceed W. This step is repeated until no more feasible cycles can be found; then proceed to step 2. (If none are found at all, go to step 3.)
>
> *Step 2.* Remove the cycles formed in step 1 from the graph. If the graph contains any added edges (from step 3 of the algorithm), remove these also and go to step 3. Note that after the first iteration of step 1, no added edges will exist.
>
> *Step 3.* One of two cases will hold.

(a) If the remaining graph has all even or zero vertex degrees (except for vertex 1, whose degree must be even but not zero), then add the two shortest paths of added edges from vertex 1 to the vertex nearest 1. Return to step 1. If there are still no feasible cycles, add two more shortest paths of added edges from vertex 1 to the second nearest vertex, and so on, until a feasible cycle is found in step 1.

(b) If the remaining graph has vertices with odd degrees (or vertex 1 has degree zero), solve a minimal-cost matching problem on the odd-degree vertices as in the uncapacitated Chinese postman problem and include the added edges in the graph. In the case where vertex 1 has degree zero, it is included in the matching problem by splitting it into two vertices $1'$ and $1''$, with a cost of infinity between these two vertices. The result would be a matching of $1'$ and $1''$ to two other vertices so that the degree of the original vertex 1 would become 2. Return to step 1.

EXAMPLE 5

We shall illustrate this heuristic with the example used by Christofides in Fig. 8.18. Costs are shown on the edges, all $q(x, y) = 1$, and $W = 5$.

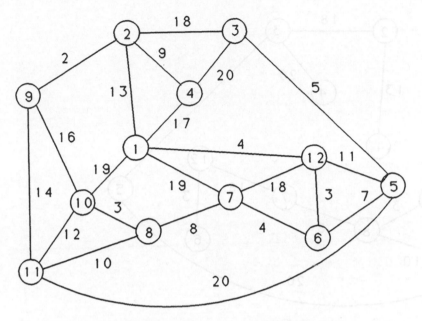

Fig. 8.18 Capacitated Chinese postman problem example.

Step 1. Two feasible cycles that leave the remaining graph connected are 1-10-9-2-4-1 and 1-12-5-6-7-1 with costs of 63 and 45, respectively.

Step 2. The cycles formed in step 1 are removed from the graph. The remaining graph is shown in Fig. 8.19.

Step 3. Since the remaining graph has odd-degree vertices, case (b) applies. Applying the matching algorithm, we add the paths 4-2-9, 8-7-6, and 1-12-6-5-3. These are shown in Fig. 8.20.

Step 1. A feasible cycle in the new graph is 1-2-9-11-10-8-7-6-12-1 having cost 63.

Step 2. Removing the cycle constructed in step 1 yields the graph shown in Fig. 8.21.

Step 3. The graph contains odd-degree vertices and again case (b) applies. Vertex 1 is split into two vertices, 1' and 1". The minimum-cost matching consists of paths 1'-12, 1"-12-6-5-3, and 2-4. The new graph is shown in Fig. 8.22.

Step 1. We construct the cycle 1-12-7-8-11-5-3-5-6-12-1 having cost 84.

Step 2. Removing the cycle, we are left with the graph in Fig. 8.23.

Step 3. From case (b), we find the matching 1'-2 and 1'-4.

Step 1. We find the final cycle 1-2-3-4-1 with cost 68 and the algorithm stops. The total cost of the solution is 323.

There is considerable latitude in constructing cycles in step 1. Christofides does not suggest any explicit procedure for finding them. Of course, how these cycles are determined would affect the performance of the heuristic. In computational experiments by Golden et al. (1982) it was suggested that computer-based algorithms may not be as

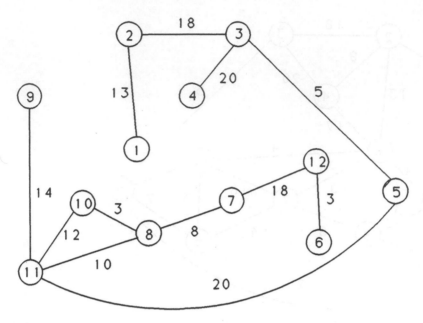

Fig. 8.19 Graph after step 1, first iteration.

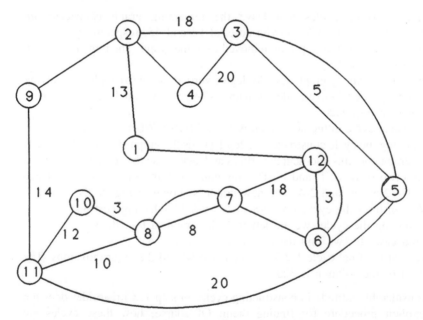

Fig. 8.20 Optimal matching of odd nodes.

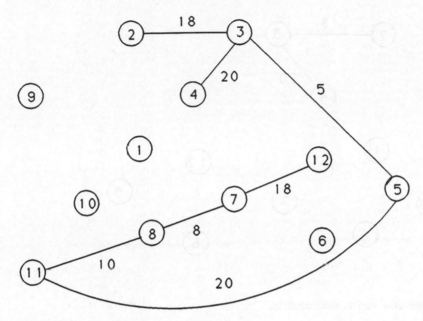

Fig. 8.21 Graph after step 1, second iteration.

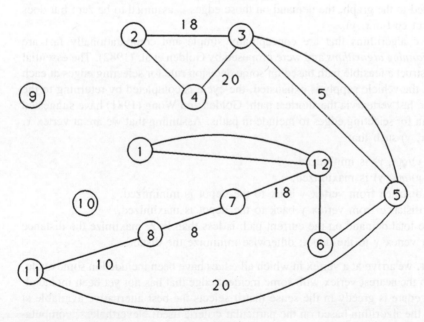

Fig. 8.22 Optimal matching of odd nodes.

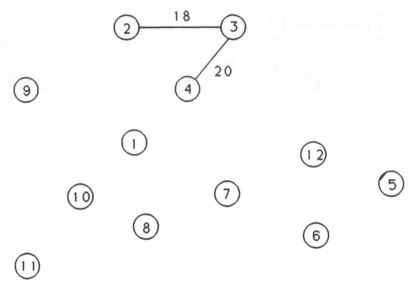

Fig. 8.23 Graph after step 1, third iteration.

effective as human visualization in finding "good" cycles. Also, note that whenever edges are added to the graph, the demand on those edges is assumed to be zero but does incur the travel cost $a(x, y)$.

Alternative algorithms that are conceptually simple and computationally fast are called *path-scanning algorithms* and were proposed by Golden et al. (1982). The essential idea is to construct a feasible path based on some decision rule for selecting edges at each vertex. When the vehicle supply is exhausted, the cycle is completed by returning to the depot from the last vertex via the shortest path. Golden and Wong (1981) have suggested various criteria for selecting edges to include in paths. Assuming that we are at vertex x, select edge (x, y) such that

1. $a(x, y)/q(x, y)$ is minimized.
2. $a(x, y)/q(x, y)$ is maximized.
3. The distance from vertex y back to the depot is minimized.
4. The distance from vertex y back to the depot is maximized.
5. If the total demand on the current path is less than $W/2$, maximize the distance from vertex y to the depot; otherwise minimize the distance.

If at any point, we arrive at a vertex in which all edges have been included in some cycle, we proceed to the nearest vertex with some incident edge that has not yet been included.

This procedure is greedy in the sense that it selects the best alternative available at each stage of the algorithm based on the particular criteria used. Nevertheless, computational experiments have shown that this procedure is quite effective. We shall illustrate this heuristic using the first selection rule for the graph in Fig. 8.18. Solutions using the remaining rules are left as exercises.

EXAMPLE 6

Since $q(x, y) = 1$, selection rule 1 is equivalent to choosing the shortest edge incident to the current vertex on the path. The steps of the procedure are summarized below.

Cycle 1

Current Vertex	Edge Selected	Remaining Capacity
1	(1, 12)	4
12	(12, 6)	3
6	(6, 7)	2
7	(7, 8)	1
8	(8, 10)	0

Return to 1 via the shortest path 10-1. The total length of the cycle is 41.

Cycle 2

Current Vertex	Edge Selected	Remaining Capacity
1	(1, 2)	4
2	(2, 9)	3
9	(9, 11)	2
11	(11, 8)	1

At this point, there are no edges incident to vertex 8 from which to select. We proceed to the closest vertex having an edge that has not been selected. This is along the path 8-10. We select edge (10, 11) and then return to vertex 1 along the path 11-9-2-1. The total length of this cycle is 83.

Cycle 3

Current Vertex	Edge Selected	Remaining Capacity
1	(1, 4)	4
4	(4, 2)	3
2	(2, 3)	2
3	(3, 5)	1
5	(5, 6)	0

Return along the path 6-12-4. The total cost of the cycle is 63.

Cycle 4

Current Vertex	Edge Selected	Remaining Capacity
1	(1, 7)	4
7	(7, 12)	3
12	(12, 5)	2
5	(5, 11)	1

Proceed to vertex 10 along the path 11-10. Select edge (10, 9) and return to vertex 1 along the shortest path 9-2-1. The length of this cycle is 111.

Cycle 5

The last edge to be selected is (1, 10). We return via this edge for a total length of 38.
    The total cost of all five cycles is $41 + 83 + 63 + 111 + 38 = 336$.

Capacitated Arc Routing Problem

The CARP differs from the CCPP only in that some demands $q(x, y)$ may be zero. The path-scanning heuristic for the CCPP may be applied with some simple modification. Obviously, in selecting edges to include in the path, only edges with positive demands need be considered. Second, edges with zero demands should be considered when solving the shortest-path problems. Another different type of algorithm was proposed by Golden and Wong (1981) but will not be described here.

EXERCISES

1. Construct an Euler tour for the undirected graph shown in Fig. 8.24 using
    (a) the technique of generating cycles and splicing them together,
    (b) Edmonds and Johnson's algorithm.

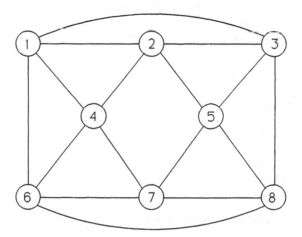

Fig. 8.24 Graph for Exercise 1.

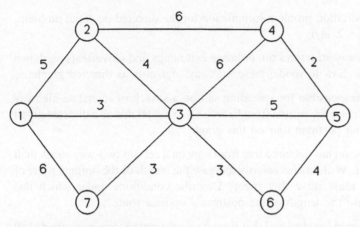

Fig. 8.25 Graph for Exercise 3.

2. Find an optimal postman route for the directed graph in Fig. 8.17.

3. Find an optimal postman route for the graph shown in Fig. 8.25.

4. Solve the (uncapacitated) Chinese postman problem on the graph in Fig. 8.18.

5. When solving the postman problem for an undirected graph, show that at most k iterations of the Floyd or Dantzig algorithm are ever needed, where k is the number of even-degree vertices in the graph. (Hint: an odd-degree vertex will never be an intermediate vertex in any path of repeated edges.)

6. Find the solution to the directed Chinese postman problem for the graph in Fig. 8.26 using
    (a) Minimum-cost flow problem formulation
    (b) Transportation problem formulation
    (c) The algorithm of Lin and Zhao

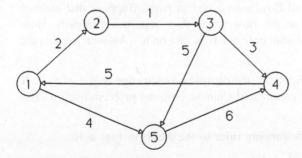

Fig. 8.26 Graph for Exercise 6.

7. In using the transportation problem formulation for the directed postman problem, prove that $\Sigma\, s(i) = \Sigma\, d(j)$.

8. Two common highway structures are ordinary exit ramps and cloverleafs as shown in Fig. 8.27. Show how to model these highway segments as directed graphs.

9. A state district is responsible for spreading salt on a stretch of interstate highway shown in Fig. 8.28. Use the results from Exercise 8 to model this as a directed graph and find the optimal postman tour on this graph.

10. The city administration has declared that from now on a certain two-way street shall be a one-way street. Will this necessarily increase the length of the optimal route of the postman who must serve this street? Describe conditions under which this change will increase the length of the postman's optimal route.

11. The city administration has declared that from now on a certain one-way street shall be a two-way street. Will this necessarily decrease the length of the optimal route of the postman who must serve this street? Describe conditions under which this change will decrease the length of the postman's optimal route.

12. Prove that the postman need never repeat an arc in an undirected graph more than once. Is this also true for directed graphs?

13. Suppose that the postman wishes to start his route at one vertex and finish at a different vertex. How can this be incorporated in the postman problem?

14. Suppose that the postman must traverse street A before traversing street B. How can this be incorporated into the postman problem for the following?
    (a) An undirected graph
    (b) A directed graph
    (c) A mixed graph

15. Solve the postman problem for the mixed graph shown in Fig. 8.14 using tentative directions different from those selected for the example that used this graph.

16. In the minimum-cost flow produced by the minimum-cost flow algorithm to solve the mixed postman problem, all flow units travel in pairs. Suppose that another minimum-cost flow exists and in this flow not all flow units travel in pairs. How should you treat an artificial arc with only one flow unit on it? (Answer: Let this arc be undirected in graph G^*.)

17. Suppose that for the graph in Fig. 8.25, the demand on each edge is $q(x, y) = 1$. If $W = 3$, find a solution to the capacitated Chinese postman problem using Christofides' algorithm.

18. Apply the remaining four path-scanning rules to the graph in Fig. 8.18.

19. Consider the graph in Fig. 8.29. Solid edges represent roads that require service

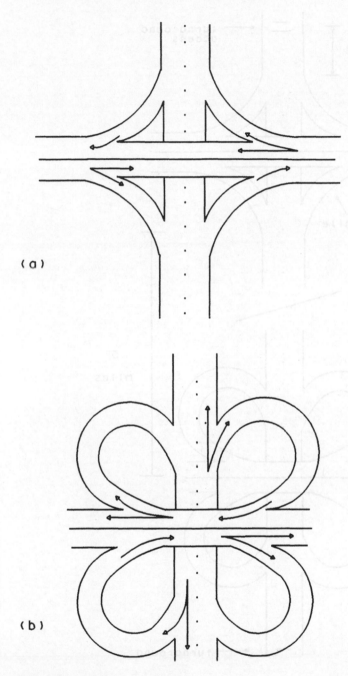

(a)

(b)

Fig. 8.27 Graph for Exercise 8.

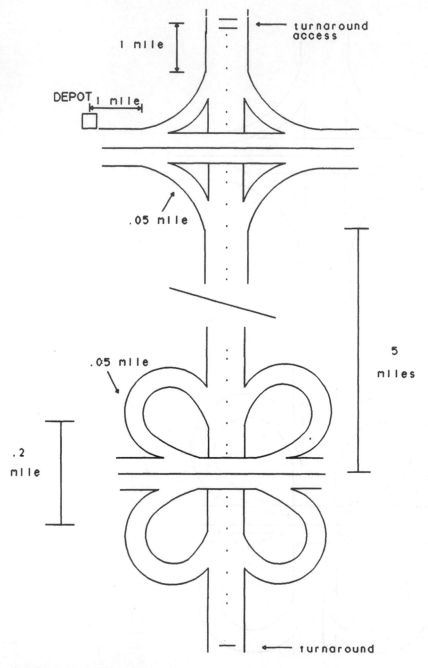

Fig. 8.28 Graph for Exercise 9.

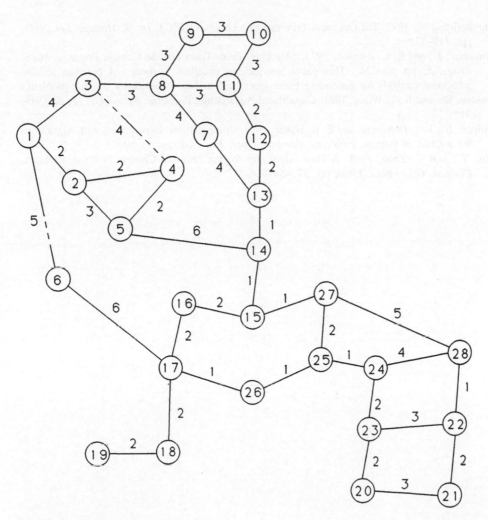

Fig. 8.29 Graph for Exercise 19.

(salt spreading, for instance), and dashed lines represent roads that do not require service but can be traveled on. Assume that $q(x, y) = 5*a(x, y)$ for the edges that require service and that vehicles with capacities of 125 units are available. Use each of the path-scanning rules to find routes for this capacitated arc routing problem.

REFERENCES

Beltrami, E., and L. Bodin, 1974. Networks and Vehicle Routing for Municipal Waste Collection, *Networks*, 4(1), pp. 65–94.

Christofides, N., 1973. The Optimum Traversal of a Graph, *OMEGA: Int. J. Manage. Sci.*, *1*(6), pp. 719–732.

Edmonds, J., and E. L. Johnson, 1973. Matching, Euler Tours and the Chinese Postman, *Math. Prog.*, *5*, pp. 88–124. (This paper provides an excellent treatment of postman results, additional methods for generating Euler tours, and a long bibliography for this problem.)

Golden, B., and R. T. Wong, 1981. Capacitated Arc Routing Problems, *Networks*, *11*, pp. 305–315.

Golden, B., J. S. DeArmon, and E. K. Baker, 1983. Computational Experiments with Algorithms for a Class of Routing Problems, *Comput. Oper. Res.*, *10*, pp. 47–69.

Lin, Y., and Y. Zhao, 1988. A New Algorithm for the Directed Chinese Postman Problem, *Comput. Oper. Res.*, *15*(6), pp. 577–584.

9

THE TRAVELING SALESMAN AND RELATED VERTEX ROUTING PROBLEMS

9.1 INTRODUCTION AND EXAMPLES

In the previous chapter we studied problems of routing along the edges of a graph. In this chapter we will study problems of routing on the vertices of a graph. These represent important and practical classes of problems in network and graph optimization. Probably the most famous routing problem is the *traveling salesman problem*, which can be described as follows. A traveling salesman is required to call at each town in his district before returning home. The salesman would like to schedule his visits so as to travel as little as possible. Thus, the salesman encounters the problem of finding a route that minimizes the total distance (or time or cost) needed to visit all the towns in the district.

The traveling salesman problem can be rephrased in terms of a graph. Construct a graph $G = (X, A)$ whose vertices correspond to the towns in the salesman's district and whose arcs correspond to the roads joining two towns. Let the length $d(x, y) \geq 0$ of each arc (x, y) equal the distance (or time or cost) of the corresponding journey along arc (x, y). A cycle that includes each vertex in G *at least once* is called a *salesman cycle*. A cycle that includes each vertex in G *exactly* once is called a *Hamiltonian cycle* after the Irish mathematician Sir William Rowan Hamilton, who first studied these problems in 1859. Hamiltonian cycles often are called *traveling salesman tours* in the graph. The *general salesman problem* is the problem of finding a salesman cycle with the smallest possible total length. The *(traveling) salesman problem*, or TSP, is the problem of finding a Hamiltonian cycle with the smallest possible total length.

Salesman problems can be formulated on either undirected or directed graphs. On an undirected graph we seek a cycle through the vertices; on a directed graph we seek a directed cycle. Problems on undirected graphs commonly are called *symmetric* traveling salesman problems. Those on directed graphs are called *asymmetric* traveling salesman problems. Of course, we could replace an undirected edge by two oppositely directed arcs each having the same length as the undirected edge they replace. Thus, we need only be concerned with algorithms that solve the directed case. However, the undirected case lends itself to some important algorithmic simplifications. Thus, we shall consider both types of problems in this chapter.

Many practical problems can be formulated as salesman problems. We now present some useful applications.

Collection and Delivery Problems

Many examples involving periodic collection from or delivery to a set of discrete locations have a natural traveling salesman analogy. For example, companies such as Federal Express, United Parcel Service, and the U.S. Postal Service must deliver overnight letters and packages by 10:30 a.m. on each business day. Each truck must visit each customer exactly once. Minimizing the total time traveled along each route maximizes the productivity of the drivers. Another example is a soft drink vendor who must replenish the machines and collect the money periodically. You can easily think of many more similar situations.

Unless the road network contains many one-way streets, the time to travel from vertex x to vertex y will generally be equal to the time to travel from vertex y to vertex x. Therefore, the graph will be undirected, and we have a *symmetric* traveling salesman problem.

Computer Wiring (Lenstra and Rinnooy Kan, 1975)

This example arose at the Institute for Nuclear Physical Research in Amsterdam. A computer interface consists of several modules. On each module are located several pins. The position of each module is determined in advance. A given subset of pins has to be interconnected by wires. In view of possible future changes or corrections and of the small size of the pin, at most two wires can be attached to any pin. In order to avoid signal crosstalk and to improve the ease and neatness of wirability, the total wire length has to be minimized.

This problem is easily seen to be a salesman problem by letting each pin correspond to a vertex of a graph and $d(x, y)$ be the distance between two pins. Observe that if we removed the restriction that at most two wires can be attached to any pin, the problem could be solved by finding a minimum spanning tree.

Circuit Board Drilling (Magirou, 1986)

Metelco S.A. is a Greek manufacturer of printed circuit boards (PCBs). Holes must be drilled in PCBs at locations that correspond to pins where electronic components will later be soldered on the board. A typical PCB might have 500 pin placements. Most of the drilling is performed by programmable drilling machines. Finding the optimum sequence of drilling is easily represented as a traveling salesman problem. By solving salesman problems for PCBs on a personal computer, Metelco improved the time required to drill boards by 30%.

Order Picking in a Warehouse (Ratliff and Rosenthal, 1983)

An order consists of a subset of items stored in a warehouse. When receiving an order, the warehouse dispatches a vehicle to pick the items in the order and transport them back to the packing and shipping area. Figure 9.1 shows a common aisle configuration in a warehouse. We can easily see that the problem of picking the order in the least amount of time is a salesman problem by defining a graph in which each vertex corresponds to the location of an item or to the ends of each aisle as shown in Fig. 9.2. An order-picking tour is a cycle in the graph that includes each of the vertices corresponding to item locations at least once.

Sequencing in Job Shops

In a manufacturing facility, many different jobs must be processed on a single machine. The setup time for job x depends on which job is processed immediately before it. This

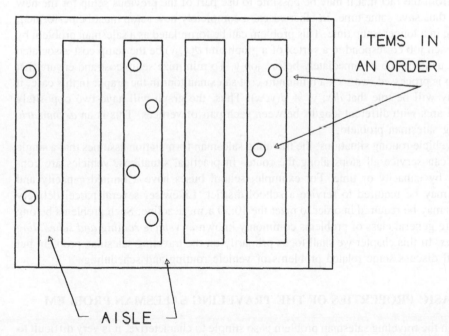

Fig. 9.1 Warehouse aisle configuration.

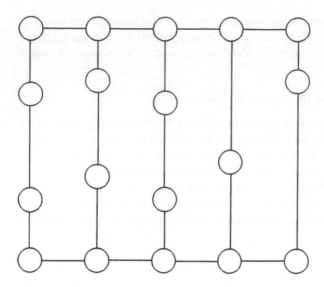

Fig. 9.2 Graph representation of order picking.

results from the fact that it may be possible to use part of the previous setup for the new job and thus save some time. In other cases, a completely new setup must be performed, resulting in a longer setup time. This problem can be formulated as a salesman problem by letting each job correspond to a vertex of a graph and $d(x, y)$ be the setup cost associated with sequencing job x immediately before job y. To minimize setup cost and ensure that each job is processed, we seek a minimum-cost salesman tour in the graph. In this case, it generally will be true that $d(x, y) \neq d(y, x)$. Thus, the graph will have two oppositely directed arcs with different lengths between each pair of vertices. This is an *asymmetric traveling salesman problem*.

In vehicle routing situations, the traveling salesman formulation assumes that a single vehicle can service all stops along the route. In practical situations, vehicles are constrained by capacity or time. For example, school buses have a limited capacity and several may be required to service a school district. Likewise, several parcel delivery vehicles may be required in order to meet the 10:30 a.m. deadline. Such problems belong to a more general class of problems commonly known as *vehicle routing and scheduling problems*. In this chapter we shall focus primarily on the traveling salesman problem but also will discuss some related problems of vehicle routing and scheduling.

9.2 BASIC PROPERTIES OF THE TRAVELING SALESMAN PROBLEM

Although the traveling salesman problem is so simple to characterize, it is very difficult to solve. The TSP belongs to the class of NP-hard problems that we discussed in Chapter 2. Thus it is unlikely that any efficient algorithm will be developed to solve it. Because of its simplicity, however, the TSP has been one of the most studied problems in this class. We first discuss some fundamental issues regarding salesman problems.

A salesman cycle with least total length is called an *optimum salesman cycle* and is an optimum solution for the general salesman problem. A Hamiltonian cycle with least total length is called an *optimum Hamiltonian cycle* and is an optimum solution to the salesman problem. Of course, in a directed graph, all the arcs in a salesman or Hamiltonian cycle must be oriented in the same direction. That is, we must have a *directed cycle*.

An optimum salesman cycle need not be an optimum Hamiltonian cycle. For example, consider the graph shown in Fig. 9.3. The only Hamiltonian cycle in this graph is (1, 2), (2, 3), (3, 1), which has a total length equal to $1 + 20 + 1 = 22$ units. The (optimum) salesman cycle (1, 2), (2, 1), (1, 3), (3, 1) that passes through vertex 1 twice has total length equal to $1 + 1 + 1 + 1 = 4$ units. Thus, an optimum salesman cycle need not be an optimum Hamiltonian cycle.

When is the solution to the general salesman problem a Hamiltonian cycle? This is answered by the following theorem.

Theorem 9.1.   If for each pair x, y of vertices in graph G,

$$d(x, y) \leq d(x, z) + d(z, y) \quad \text{(for all } z \neq x, z \neq y) \tag{1}$$

then a Hamiltonian cycle is an optimum solution (if a solution exists) to the general salesman problem for graph G.

Condition (1) merely says that the direct distance from x to y is never more than the distance via any other vertex z. This is called the *triangle inequality*.

Proof.   Suppose that no optimum solution to the general salesman problem is a Hamiltonian cycle. Let C be any optimum salesman cycle. Since C is not a Hamiltonian cycle, some vertex, say vertex z, appears at least twice in cycle C. Suppose that the first time the salesman arrives at vertex z he arrives from vertex x and departs to vertex y. Alter cycle C

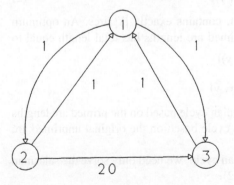

Fig. 9.3   Optimum salesman route.

so that the salesman travels from x directly to y, bypassing z. The resulting route C' is also a cycle since it visits every vertex at least once. Moreover, by (1), the total length of C' does not exceed the length of C. Replacing C by C' and repeating this argument, we generate another salesman cycle C'', etc. Eventually, this process leads us to an optimum cycle that is Hamiltonian since each successive cycle has one less arc than its predecessor.   Q.E.D.

From Theorem 9.1, it follows that if graph G satisfies the triangle inequality, then the optimum solutions for the salesman problem for graph G are also optimum solutions for the general salesman problem in graph G.

There is a simple way to spare ourselves the needless trouble of developing two solution techniques, one for the general salesman problem and one for the salesman problem. If graph G does not satisfy the triangle inequality, replace each arc length $d(x, y)$ that fails the triangle inequality with the length of a shortest path from x to y. Record that the arc from x to y no longer represents a direct journey from x to y but now represents a journey along a shortest path from x to y. Now, $d(x, y)$ all satisfy the triangle inequality.

If an optimum solution to the salesman problem for graph G contains an arc (x, y) whose length was shortened as specified above, replace arc (x, y) by a shortest path from x to y in the optimum solution. *Thus, we need solution techniques for only the salesman problem.* Any problem whose distances satisfy the triangle inequality can be called a TSP.

For example, in Fig. 9.3, $d(2, 3) = 20 > d(2, 1) + d(1, 3) = 1 + 1$. Thus, (1) fails for arc (2, 3). If the length of arc (2, 3) is reduced to 2, the length of a shortest path from 2 to 3, the only Hamiltonian cycle in the resulting graph is (1, 2), (2, 3), (3, 1), whose length is $1 + 2 + 1 = 4$. Replacing (2, 3) by (2, 1), (1, 3) yields the cycle (1, 2), (2, 1), (1, 3), (3, 1), which is an optimum salesman cycle for the original graph.

Suppose that instead of wishing to find a Hamiltonian cycle with the smallest total length, we wanted to find a Hamiltonian cycle with the *largest* total length. For example, the bank courier might want to maximize rather than minimize the total number of additional letters that he carries. Can this problem also be solved as a salesman problem?

Let M denote the largest arc length in graph G. Let

$$d'(x, y) = M - d(x, y) \geq 0 \quad \text{[for all } (x, y)] \tag{2}$$

Every Hamiltonian cycle in graph $G = (X, A)$, contains exactly $|X|$ arcs. An optimum (shortest) Hamiltonian cycle C based on the primed arc lengths has total length equal to

$$\sum_{(x, y) \in C} d'(x, y) = \sum_{(x, y) \in C} [M - d(x, y)]$$
$$= |X|M - \sum_{(x, y) \in C} d(x, y)$$

Thus, it follows that a minimum-length Hamiltonian cycle based on the primed arc lengths corresponds to a maximum-length Hamiltonian cycle based on the original unprimed arc lengths.

Thus, to find a maximum-length Hamiltonian cycle, we need only solve the salesman problem using arc lengths transformed as in (2).

Existence of a Hamiltonian Cycle

As we have seen, the salesman problem is solved by finding an optimum Hamiltonian cycle. Unfortunately, not all graphs contain a Hamiltonian cycle. Consequently, before proceeding to look for an optimum Hamiltonian cycle, we should at least try to establish whether the graph possesses any Hamiltonian cycles. This section describes several conditions under which a graph possesses a Hamiltonian cycle. A very extensive treatment of existence conditions for Hamiltonian cycle can be found in Berge (1973).

A graph is called *strongly connected* if for any two vertices x and y in the graph, there is a directed path from x to y. A subset X_i of vertices is called a *strongly connected vertex subset* if for any two vertices $x \in X_i$ and $y \in X_i$, there is a directed path from x to y in the graph and X_i is contained in no other set with the same property. The subgraph generated by a strongly connected vertex subset is called a *strongly connected component* of the original graph.

For example, the graph in Fig. 9.3 is strongly connected since there is a directed path from every vertex to every other vertex. The reader should verify this. Next, consider the graph in Fig. 9.4. This graph is not strongly connected because there is no directed path from vertex 4 to vertex 2. The vertices {1, 2, 3} form a strongly connected vertex subset since there is a directed path from each of these vertices to every other vertex in this set. Moreover, no other vertex can be added to this set without losing this property. For example, vertex 4 cannot be added to the set since there is no directed path from 4 to 1.

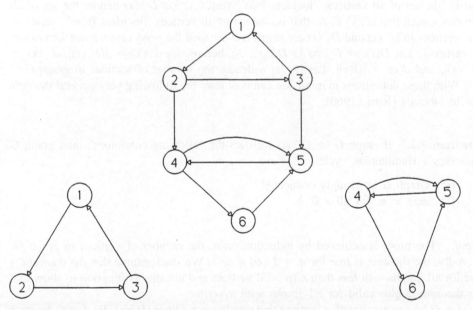

Fig. 9.4 Graph and its strongly connected components.

The subgraph generated by $\{1, 2, 3\}$ is shown in Fig. 9.4. This subgraph is a strongly connected component of the original graph.

There is a directed path from 4 to 5 and a path from 5 to 4. However, $\{4, 5\}$ is not a strongly connected vertex subset because vertex 6 can be added to this set without losing the strongly connected property. No other vertices can be added without losing this property. Hence $\{4, 5, 6\}$ is a strongly connected vertex subset. The strongly connected component generated by $\{4, 5, 6\}$ is also shown in Fig. 9.4.

If graph G is not strongly connected, then graph G does not possess a Hamiltonian cycle. This follows because a Hamiltonian cycle contains a directed path between each pair of vertices in the graph. Thus, a necessary condition for the existence of a Hamiltonian cycle is that graph G be strongly connected.

Recall that a loop is any arc whose head and tail are the same vertex, i.e., an arc of the form (x, x). No Hamiltonian cycle can contain a loop, and consequently the existence of a Hamiltonian cycle is not affected by adding or deleting loops from graph G.

If two vertices, say x and y, are joined by more than one arc $(x, y)_1, (x, y)_2, \ldots$, with the same direction, then the deletion of all of these arcs, except the shortest one, does not affect the existence of a Hamiltonian cycle in the graph or the length of an optimum Hamiltonian cycle (if one exists).

For these reasons, we shall henceforth assume that graph G contains no loops and not more than one arc from x to y for all x and y.

Let $D^-(x)$ denote the set of all vertices y in graph $G = (X, A)$ such that $(y, x) \in A$, that is, the set of all vertices "incident into" vertex x. Let $D^+(x)$ denote the set of all vertices y such that $(x, y) \in A$, that is, the set of all vertices "incident from" vertex x. The vertices in $D^-(x)$ and $D^+(x)$ are respectively called the *predecessors* and *successors* of vertex x. Let $D(x) = D^-(x) \cup D^+(x)$. As before, let $d^-(x) = |D^-(x)|$, $d^+(x) = |D^+(x)|$, and $d(x) = |D(x)|$. Lastly, let n denote the number of vertices in graph G.

With these definitions in mind, we can now state the following very general theorem due to Ghouila-Houri (1960):

Theorem 9.2.   If graph $G = (X, A)$ satisfies the following conditions, then graph G possesses a Hamiltonian cycle:

(I)   Graph G is strongly connected
(II) $d(x) \geq n$, for all $x \in X$.

Proof.   The proof is achieved by induction on n, the number of vertices in graph G. Trivially, the theorem is true for $n = 2$ and $n = 3$. We shall assume that the theorem is true for all graphs with less than n ($n > 3$) vertices and use this assumption to show that the theorem is also valid for all graphs with n vertices.

Let G be any graph with n vertices that satisfies conditions (I) and (II). Let C denote a simple cycle in G with the largest possible number of arcs. Let $x_1, x_2, \ldots, x_m$ denote the sequence in which C visits the vertices of G. If $m = n$, the theorem is true. Otherwise, $m < n$ and C is not a Hamiltonian cycle. Let $X_0 = \{x_1, x_2, \ldots, x_m\}$. Let $X_1, X_2, \ldots, X_p$ denote the strongly connected components of the subgraph generated by $X - X_0$.

Claim (a). Each strongly connected component $X_1, X_2, \ldots, X_p$ contains a Hamiltonian cycle.

To prove Claim (a), we need only show that the degree in X_i of each vertex $x \in X_i$ is at least $|X_i|$ for all $i = 1, 2, \ldots, p$. Then component X_i will satisfy both conditions (I) and (II), and by the induction hypothesis contain a Hamiltonian cycle.

We know that $d(x) \ge n$. Consider any vertex $y \in X_j, j \ne i$. Both arcs (x, y) and (y, x) cannot be present in A; otherwise, X_i and X_j would form one strongly connected component. For $k = 1, 2; \ldots, m$, both (x_k, x) and (x, x_k) cannot exist, otherwise cycle C could be extended to include vertex x. Consequently, the number of arcs joining x to a vertex not in X_i cannot exceed $|X| - |X_i|$. Thus, the degree of x in X_i is not less than X_i, and Claim (a) is true.

Claim (b). For $i = 1, 2, \ldots, p$,

$$|X_i| \le |X_0|$$

Otherwise, there would exist a Hamiltonian cycle in X_i that contains more arcs than cycle C.

Claim (c). At least $2 + |X_0| - |X_i| \ge 2$ arcs join each vertex $x \in X_i$ to vertices in X_0.

As mentioned in Claim (a), not more than $|X_j|$ arcs join vertex x and vertices in X_j for all $j \ne i, j \ne 0$. Also, not more than $2(|X_i| - 1)$ arcs can join vertex x to vertices in X_i. Since $d(x) \ge n$, it follows that at least $2 + |X_0| - |X_i|$ arcs join vertex x to vertices in X_0 and (c) follows.

Claim (d). There exists a component X_i such that there is an arc from a vertex in X_i to a vertex in X_0 and an arc from a vertex in X_0 to a vertex in X_i.

For $p = 1$, Claim (d) must be true since G is strongly connected.

For $p > 1$, suppose that all arcs joining X_1 and X_0 are directed from X_1 to X_0. Consider the subgraph generated by $X_1 \cup X_0$. Since a vertex in X_1 is joined at most once to each vertex not in $X_1 \cup X_0$; the vertices in X_1 satisfy condition (II) in this subgraph. It follows from Claim (a) that the vertices in X_0 also satisfy condition (II) in this subgraph.

Since G is strongly connected, there is a directed path P from vertex $x \in X_0$ to a vertex $y \in X_1$ such that the intermediate vertices of this path are not in $X_1 \cup X_0$. Path P must contain at least two arcs.

If arc (x, y) were added to the subgraph generated by $X_1 \cup X_0$, then this subgraph would be strongly connected. Consequently, this subgraph together with arc (x, y) would satisfy the induction hypothesis, and there would exist a Hamiltonian cycle of this subgraph that contains arc (x, y). Replace arc (x, y) with path P. The resulting cycle contains more arcs than cycle C, which is impossible. Hence, each component $X_1, X_2, \ldots, X_p$ must be joined to X_0 by arcs in both directions. This proves Claim (d).

Claim (e). Let X_1 be a component satisfying Claim (d). For each vertex $y \in X_1$, there is an arc from y to a vertex in X_0 and there is an arc from a vertex in X_0 to vertex y.

Let C_1 be a Hamiltonian cycle of X_1. Let $y_1, y_2, \ldots, y_q$ denote the order in which C_1 visits the vertices of X_1. [Cycle C_1 exists from Claim (a).] Suppose that there is no arc from any vertex in X_0 to vertex y_j. Follow cycle C_1 through the vertices $y_{j+1}, y_{j+2}, \ldots,$ until vertex y_{j+t} is encountered such that there is an arc (x_k, y_{j+t}) originating in X_0. Consider vertex y_{j+t-1}. No arc can be directed from y_{j+t-1} to $x_{k+1}, x_{k+2}, \ldots, x_{k+q}$. Otherwise, cycle C would be merged with the $q - 1$ arcs from y_{j+t} to y_{j+t-1} to create a cycle with more than m arcs, which is impossible. Consequently, no arcs are directed from X_0 into y_{j+t-1} and at most $m - q$ arcs are directed from y_{j+t-1} to vertices in X_0. Since $m = |X_0|$ and $q = |X_1|$, this contradicts Claim (c), and Claim (e) must be valid.

Claim (f) (Conclusion). As before, let $x_1, x_2, \ldots, x_m$ denote the order of the vertices visited by C, and let $y_1, y_2, \ldots, y_q$ denote the order of the vertices visited by C_1. Let

$$a(i, j) \quad = \quad \begin{cases} 1 & \text{if } (x_i, y_j) \in A \\ \\ 0 & \text{otherwise} \end{cases}$$

$$b(i, j) \quad = \quad \begin{cases} 1 & \text{if } (y_{j-1}, x_{i+1}) \in A \\ \\ 0 & \text{otherwise} \end{cases}$$

Since there are at least $q(m - q + 2)$ arcs joining X_0 and X_1 from Claim (c), it follows that

$$\sum_{i,j} a(i, j) + b(i, j) \geq q(m - q + 2)$$

For some j_0, it follows that

$$\sum_i a(i, j_0) + b(i, j_0) \geq m - q + 2 \tag{3}$$

From Claim (e), there is at least one arc (x_{i_0}, y_{j_0}) directed from X_0 into vertex y_{j_0}. Consequently, since no cycles with more than m arcs exist, it follows that $b(i_0, j_0) = 0$, $b(i_0 + 1, j_0) = 0, \ldots, b(i_0 + q - 1, j_0) = 0$. Thus, there are at most $m - q + 1$ arcs from y_{j_0-1} to vertices in X_0. If $a(i_0 + 1, j_0) = 1$, then $b(i_0 + q, j_0) = 0$. If $a(i_0 + 2, j_0) = 1$, then $b(i_0 + q + 1, j_0) = 0$, etc. Thus, the existence of each additional arc directed from X_0 to y_{j_0} prohibits the existence of at least one more arc from y_{j_0} to X_0.

Consequently,

$$\sum_i a(i, j_0) + b(i, j_0) \leq m - q + 1$$

which contradicts inequality (3). Thus, cycle C cannot contain only $m < n$ arcs, which is a contradiction.   Q.E.D.

It is easy to verify that a graph satisfies conditions (I) and (II) required by Theorem 9.2. Condition (I) is verified by applying either the Floyd or Dantzig shortest-path algorithm (see Chapter 4) to ascertain whether there is a path with finite length joining every pair of vertices in the graph. Condition (II) is verified simply by counting the number of arcs incident to each vertex in the graph.

If graph G is an undirected graph, then no direction is specified on each arc and the salesman is allowed to traverse an arc in either direction. The following result due to Chvátal (1972) describes a sufficient condition for an undirected graph to possess a Hamiltonian cycle.

Again, let n denote the number of vertices in the graph. Name the vertices x_1, x_2, . . . , x_n so that $d(x_1) \leq d(x_2) \leq \cdots \leq d(x_n)$

Theorem 9.3. If $n \geq 3$ and if

$$d(x_k) \leq k < \tfrac{1}{2}n \Rightarrow d(x_{n-k}) \geq n - k \tag{4}$$

then graph $G = (X, E)$ contains a Hamiltonian cycle.

Proof. Suppose that the theorem is false; i.e., graph G satisfies condition (4) but contains no Hamiltonian cycle. Consider any edge (x_i, x_j) not present in G. If the addition of this edge does not create a Hamiltonian cycle, add this edge to graph G. Repeat this process until no more edges can be added to graph G. Note that condition (4) remains satisfied after the addition of each new edge to graph G since the additional edge does not lower the degree of any vertex. Call the final graph $G^* = (X, E^*)$. We shall now use graph G^* to obtain a contradiction.

Let u and v be any two nonadjacent vertices such that $d(u) + d(v)$ is as large as possible. Without loss of generality, assume that $d(u) \leq d(v)$. Let C denote the longest simple path from u to v. Since no additional edges can be added to G^* without creating a Hamiltonian cycle, it follows that C contains all the vertices in X. Let $u = u_1, u_2, \ldots, u_{n-1}, u_n = v$ denote the order in which C visits the vertices of X.

Let S denote the set of all vertices u_i such that u_{i+1} is adjacent to u. Let T denote the set of all vertices adjacent to v. No vertex u_j can be a member of both S and T since if it were, the cycle $u_j, u_{j-1}, \ldots, u_1, u_{j+1}, u_{j+2}, \ldots, u_n, u_j$ would be a Hamiltonian cycle. Also, $S \cup T = \{u_1, u_2, \ldots, u_{n-1}\}$. Therefore, $d(u) + d(v) = |S| + |T| \leq n$. Consequently, $d(u) \leq \tfrac{1}{2} n$.

Since no vertex u_j is a member of both S and T, if $u_j \in S$, then u_j is not adjacent to v. By the maximality of $d(u) + d(v)$ it follows that $d(u_j) \leq d(u)$. Thus, there are at least $|S|$ vertices whose degree does not exceed the degree of u. Set $k = d(u)$. Thus, $d(x_k) \leq k$. By condition (4), it follows that $d(x_{n-k}) \geq n - k$. Thus, there must be at least $k + 1$ vertices whose degree is at least $n - k$. Since $d(u) = k$, vertex u is not adjacent to all of these $k + 1$ vertices. Thus, there is a vertex w that is not adjacent to u and $d(w) \geq n - k$. Consequently, $d(u) + d(w) > d(u) + d(v)$, which is a contradiction. Consequently, graph G^* must contain a Hamiltonian cycle.   Q.E.D.

Condition (4) is easy to verify. Merely order the vertices according to ascending degree and check if condition (4) is satisfied for the first $\tfrac{1}{2} n$ vertices.

Mathematical Programming Formulations

We may develop mathematical programming formulations for the TSP. Although we would never want to solve even a moderately sized problem in this fashion, such formulations often provide important insights for computing lower bounds and applying other solution techniques.

The most popular formulation for the asymmetric case is the following. Define $x_{ij} = 1$ if edge (i, j) is in the optimal tour and 0 otherwise, and $d_{ij} = d(i, j)$. Then we have

P1: $\min \sum \sum d_{ij} x_{ij}$ (5)

   $\sum x_{ij} = 1$ for all j (6)

   $\sum x_{ij} = 1$ for all i (7)

   a set of constraints which ensure that $\{x_{ij}\}$ is a tour   (8)

   $x_{ij} = 0, 1$ (9)

The number of constraints (8) generally is very large; this is what makes such an integer programming model difficult to solve. One such set of constraints (8) is

$$\sum_{i \in S} \sum_{j \in X - S} x_{ij} \geq 1 \quad \text{for every } S \subseteq X$$

Observe that the number of these constraints grows exponentially in the size of the problem.

For the symmetric TSP, we can pose an alternative formulation. Let each edge be denoted by a single index t and let d_t be the distance of edge t. Let $x_t = 1$ if edge t is part of the optimal tour. We have

P2: $\min \sum d_t x_t$ (10)

   $\sum x_t = n$ (11)

   $\sum_{t \in (S, T)} x_t \geq 1$ for every $(S, T), S \subseteq X$ (12)

   $\sum_{t \in A_i} x_t = 2$ for $i = 1, 2, \ldots, n$ (13)

where A_i is the set of edges incident to vertex i (the star of i).

   $x_i = 0, 1$ (14)

9.3   LOWER BOUNDS

This section presents methods for calculating lower bounds on the length of an optimum Hamiltonian cycle in either a directed or undirected graph G. If a lower bound is subtracted from the length of any Hamiltonian cycle, the difference equals the maximum amount by which this tour can exceed optimality. This is useful when evaluating nonoptimum tours.

Any feasible tour, or Hamiltonian cycle, is a subgraph of G. It must satisfy two properties:

1. For asymmetric problems, each vertex must have one arc directed into it and one arc directed out of it. For symmetric problems, we need only have two edges incident to each vertex.
2. The tour must be connected.

Any set of disjoint cycles that contain all vertices of G satisfies property 1. Any connected set of arcs satisfies property 2. Only Hamiltonian cycles satisfy both properties. If you reexamine the mathematical programming formulations presented in the previous section, you can see that the set of constraints ensures that these two properties are attained by any feasible solution.

A Lower Bound Based on the Assignment Problem

We will develop a lower bound for the traveling salesman problem based on solving an assignment problem. We will assume that the triangle inequality holds and will first consider the general asymmetric case.

Consider the family F of all subgraphs of G that satisfy property 1. (We shall assume that F is not empty.) Clearly, every Hamiltonian cycle is a member of F. Thus, the length of the shortest member of F is a lower bound on the length of an optimum tour. We denote this lower bound by L_1. Suppose that we remove, or *relax*, constraints (8) in the mathematical programming formulation P1. All feasible solutions to (6), (7), and (9) satisfy property 1. The resulting problem is

$$\min \sum \sum d_{ij} x_{ij} \tag{5}$$

$$\sum x_{ij} = 1 \quad \text{for all } j \tag{6}$$

$$\sum x_{ij} = 1 \quad \text{for all } i \tag{7}$$

$$x_{ij} = 0, 1 \tag{9}$$

We have seen this problem before in Chapter 7; it is precisely the assignment problem with a minimization objective. We may apply the Hungarian algorithm (in a minimization sense) to the distance matrix D. Essentially, we create a bipartite graph $G' = (S, T, A)$ as follows. For each vertex $x \in X$ in G, create two vertices, $x_s \in S$ and $x_t \in T$. If an arc (x, y) exists in G, direct an arc from x_s to y_t and give it a cost d_{xy}. A minimum-cost assignment in G' will have the property that exactly one arc is directed into and out of each vertex. The cost of this assignment is the lower bound L_1. If the arcs in the optimum assignment also satisfy property 2, then this solution must also solve the traveling salesman problem.

EXAMPLE 1

To illustrate the lower bound L_1, consider the graph G and distance matrix D shown in Fig. 9.5. The associated bipartite graph G' is shown in Fig. 9.6. For this small problem,

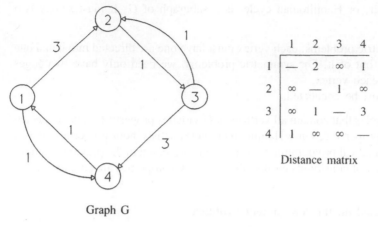

Distance matrix

Graph G

Fig. 9.5 Graph for Example 1.

we may find the optimal solution by inspection. The optimum assignment consists of arcs $(1_s, 4_t)$, $(2_s, 3_t)$, $(3_s, 2_t)$, and $(4_s, 1_t)$ having a cost of 4. Thus $L_1 = 4$. Observe that this solution forms two directed cycles in G and satisfies property 1 but not property 2. The unique Hamiltonian cycle in G is (1, 2), (2, 3), (3, 4), (4, 1) with a total length of 8.

   If the graph G contains undirected arcs, each undirected arc can be replaced by two oppositely directed arcs joining the same vertices. Let each of these directed arcs have length equal to the length of the original undirected arc. The resulting graph will contain only directed arcs, and the technique described above can be used to find the lower bound

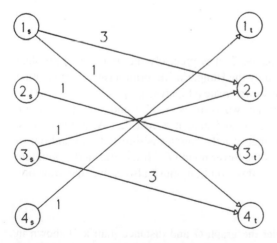

Fig. 9.6 Assignment graph for Example 1.

```
SHORTEST PATH TREE FROM 1

    NODE          DISTANCE      PREDECESSOR
    ----          --------      -----------
     1              0.00         1
     2              3.00         1
     3              4.00         1
     4              1.00         1
     5              2.00         1

SHORTEST PATH TREE FROM 2

    NODE          DISTANCE      PREDECESSOR
    ----          --------      -----------
     1              3.00         2
     2              0.00         2
     3              2.00         2
     4              4.00         1
     5              3.00         2

SHORTEST PATH TREE FROM 3

    NODE          DISTANCE      PREDECESSOR
    ----          --------      -----------
     1              4.00         3
     2              2.00         3
     3              0.00         3
     4              5.00         1
     5              2.00         3

SHORTEST PATH TREE FROM 4

    NODE          DISTANCE      PREDECESSOR
    ----          --------      -----------
     1              1.00         4
     2              4.00         1
     3              5.00         1
     4              0.00         4
     5              3.00         1

SHORTEST PATH TREE FROM 5

    NODE          DISTANCE      PREDECESSOR
    ----          --------      -----------
     1              2.00         5
     2              3.00         5
     3              2.00         5
     4              3.00         1
     5              0.00         5
```

Fig. 9.7 NETSOLVE solution of shortest paths for Example 2.

by solving the associated assignment problem. Since symmetric problems contain only undirected arcs, these comments apply.

EXAMPLE 2

Consider the matrix of distances between all pairs of five cities shown below.

	1	2	3	4	5
1	—	3	4	1	2
2		—	2	5	3
3			—	6	2
4				—	7
5					—

This matrix describes a complete graph on five vertices. We first check to see if the triangle inequality holds by finding the shortest paths between each pair of vertices. Using NETSOLVE and the SPATH algorithm, we find the shortest distances between all pairs of nodes as shown in Fig. 9.7. By replacing the distances that do not satisfy the triangle inequality by the shortest distances between those vertices, we have the distance matrix shown below.

	1	2	3	4	5
1	—	3	4	1	2
2		—	2	4	3
3			—	5	2
4				—	3
5					—

The assignment problem graph is shown in Fig. 9.8, and the optimum solution using NETSOLVE is given below.

```
       ASSIGNMENT PROBLEM:    MINIMUM COST IS          9.00

       FROM      TO             LOWER       FLOW        UPPER        COST
       ----      --           --------    --------    --------    --------
       4S        1T             0.00        1.00     999999.00       1.00
       5S        2T             0.00        1.00     999999.00       3.00
       2S        3T             0.00        1.00     999999.00       2.00
       1S        4T             0.00        1.00     999999.00       1.00
       3S        5T             0.00        1.00     999999.00       2.00
```

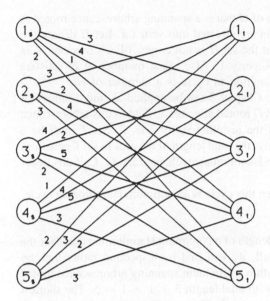

Fig. 9.8 Assignment problem for Example 2.

This solution consists of two subtours, 1-4-1 and 2-3-5-2, with a total distance of 9. Since this solution is not feasible for the traveling salesman problem, it does provide a lower bound on the optimum tour.

A Lower Bound from Minimum Spanning Trees and Arborescences

Now, let us compute another lower bound, L_2, on any feasible tour by relaxing property 1 and not specifying how many arcs must be incident to each vertex. We will first consider the asymmetric case.

In problem P1, suppose that we relax constraints (7). The resulting mathematical program is given below.

$$\min \sum \sum d_{ij} x_{ij} \tag{5}$$

$$\sum x_{ij} = 1 \quad \text{for all } j \tag{6}$$

$$\sum_{i \in S} \sum_{j \in X - S} x_{ij} \geq 1 \quad \text{for every } S \subseteq X \tag{8}$$

$$x_{ij} = 0, 1 \tag{9}$$

These constraints say that exactly one arc must be directed into each vertex (6) and that the subgraph must be connected (8).

Recall from Chapter 3 that an *arborescence* is a tree in which at most one arc is directed into any vertex. The unique vertex in an arborescence that has no arc directed into it is called the root. Consider any subgraph of G that is a spanning arborescence rooted at some vertex, say vertex x, together with an arc directed into vertex x. Let H denote the family of all such subgraphs. (Assume that the set H is not empty.) Each member of H satisfies property 2 but not necessarily property 1. Also, each member of H satisfies constraints (6) and (8). Every directed Hamiltonian cycle is a member of H.

The maximum branching algorithm of Chapter 3 can be applied to find a minimum total length spanning arborescence of graph G rooted at vertex x. Add to this arborescence the shortest arc directed into root x. Call the resulting set of arcs T. Set T must be a minimum-length member of H; consequently, the total length of the arcs in set T generates a lower bound on the length of an optimum Hamiltonian cycle in graph G. Call this lower bound L_2.

If set T forms a Hamiltonian cycle, then this cycle is an optimum Hamiltonian cycle.

EXAMPLE 3

Let us generate the lower bound L_2 for the length of an optimum Hamiltonian cycle for the graph in Fig. 9.5. Since the graph is small, we can find by inspection rather than by resorting to the maximum branching algorithm a minimum spanning arborescence rooted at vertex 1. It is (1, 2), (2, 3), and (1, 4) with total length $3 + 1 + 1 = 5$. The shortest arc incident into vertex 1 is arc (4, 1) with length 1. Thus, set T consists of the four arcs (1, 2), (2, 3), (1, 4), (4, 1), whose total length is $3 + 1 + 1 + 1 = 6$. Thus, $L_2 = 6$. Note that T does not form a Hamiltonian circuit.

Recall from Example 1 that $L_1 = 4 \neq L_2$. Thus, L_1 and L_2 are not always equal.

For the symmetric case, we use the mathematical programming formulation (10)–(14) to specify the relaxation. We relax constraints (13), which specify that exactly two edges must be incident to each vertex. The remaining constraints state that the subgraph must have n edges and be connected.

$$\min \sum d_t x_t \qquad (10)$$

$$\sum x_t = n \qquad (11)$$

$$\sum x_t \geq 1 \quad \text{for every } (S, T), S \subseteq X \qquad (12)$$

$$x_t = 0, 1 \qquad (14)$$

Now suppose that we remove any vertex x from G. Any spanning tree on the remaining vertices must consist of $n - 2$ edges and is necessarily connected. In particular, the minimum spanning tree will have the least cost among any connected set of $n - 2$ edges in this graph. If, in addition, we select the two shortest edges incident to vertex x and add these to the spanning tree, we have created a set U that satisfies constraints (11) and (12). This set U is called a *1-tree*.

EXAMPLE 4

Let us compute a lower bound L_2 based on the 1-tree relaxation of the traveling salesman problem. We will use the distance matrix from Example 2:

	1	2	3	4	5
1	—	3	4	1	2
2		—	2	4	3
3			—	5	2
4				—	3
5					—

If we remove vertex 1 from the graph, we are left with the following graph:

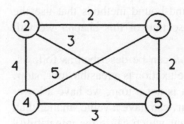

The two shortest edges connecting vertex 1 to the rest of the graph are (1, 4) and (1, 5). Adding these to the minimum spanning tree on this graph gives us the 1-tree shown below.

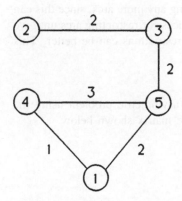

The total cost of this 1-tree is 10, and since it is not a feasible traveling salesman tour, it represents only a lower bound on the optimum solution.

The use of 1-trees as a bounding mechanism for the TSP was devised by Held and Karp (1970, 1971). They show that tighter bounds (that is, bounds that are closer to the optimal value) can be found using nonlinear programming techniques and transformations of the edge lengths. Any further discussion would take us too far afield. However, in the next section, we will see how these bounds are used to find optimum solutions for the traveling salesman problem.

9.4   OPTIMAL SOLUTION TECHNIQUES

Many solution techniques are available for the traveling salesman problem (e.g., Bellmore and Nemhauser, 1968; Garfinkel and Nemhauser, 1972; Held and Karp, 1970, 1971; Steckhan, 1970). A large number of these solution techniques rely heavily on advanced results in integer linear programming, nonlinear programming, and dynamic programming. A description of these techniques is beyond the scope of this book. In this section we shall consider only straightforward branch-and-bound methods that use the bounds described in the previous section. In the next section of this chapter we will describe various heuristic techniques.

The general idea of the *branch-and-bound algorithm* can be described as follows. First, solve a relaxation of the TSP. If the solution to the relaxation is a feasible tour, stop. It must be the optimum. If the solution to the relaxation is not a tour, we have a lower bound. No feasible tour, and in particular the optimal tour, can have a value smaller than this lower bound. Also, at least one of the arcs in this tour which cause the infeasibility cannot be included in the optimal solution. Consider solving a problem in which one of these arcs is forced out of the tour. The value of the optimal solution to the relaxation with this restriction cannot be any better than the initial solution. That is, the value of the lower bound cannot decrease. If the value of the new lower bound is at least as large as the value of any known feasible tour, we need not consider restricting any more arcs, since this can only increase the value of the solution. We repeat this process of restricting arcs until we find a better feasible tour, or we prove that no further restrictions can be better. This process is best seen through an example.

EXAMPLE 5

We will illustrate the branch-and-bound process for an asymmetric problem using the assignment-based lower bound L_1. Consider the distance matrix shown below.

	1	2	3	4	5
1	—	3	7	6	2
2	4	—	11	9	6
3	3	6	—	1	5
4	5	5	3	—	4
5	6	2	7	6	—

We will use NETSOLVE to find optimum assignments. Solving the assignment problem using this distance matrix yields the following solution.

ASSIGNMENT PROBLEM: MINIMUM COST IS   12.00

FROM	TO	LOWER	FLOW	UPPER	COST
2S	1T	0.00	1.00	999999.00	4.00
5S	2T	0.00	1.00	999999.00	2.00
4S	3T	0.00	1.00	999999.00	3.00
3S	4T	0.00	1.00	999999.00	1.00
1S	5T	0.00	1.00	999999.00	2.00

The lower bound $L_1 = 12$ and the solution consists of two subtours 1-5-2-1 and 3-4-3. At least one of these arcs must not be included in the optimal tours. We "branch" or partition the set of solutions into subsets, each of which excludes one of these arcs. This is shown in the *branch-and-bound tree* in Fig. 9.9. We can exclude these arcs from the solution by giving them a large cost in the assignment problem. With NETSOLVE, we simply use the CHANGE command to modify the appropriate arc cost. For example, setting the cost of arc (1, 5) to 99999, we have the following assignment problem solution:

ASSIGNMENT PROBLEM: MINIMUM COST IS   18.00

FROM	TO	LOWER	FLOW	UPPER	COST
2S	1T	0.00	1.00	999999.00	4.00
5S	2T	0.00	1.00	999999.00	2.00
1S	3T	0.00	1.00	999999.00	7.00
3S	4T	0.00	1.00	999999.00	1.00
4S	5T	0.00	1.00	999999.00	4.00

This solution has a value of 18 and corresponds to a feasible tour 2-3-4-5-2-1. We know that the optimal tour must have a total distance of no more than 18. This is an *upper bound*

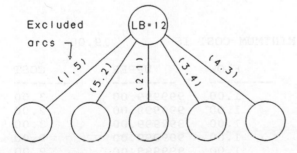

Fig. 9.9 Branch-and-bound tree, Example 5.

on the optimal solution. Next, consider excluding arc (5, 2). The assignment problem solution is shown below.

```
ASSIGNMENT PROBLEM:    MINIMUM COST IS         19.00

FROM      TO      LOWER        FLOW        UPPER          COST
----      --      --------     --------    --------       --------
5S        1T      0.00         1.00        999999.00      6.00
1S        2T      0.00         1.00        999999.00      3.00
4S        3T      0.00         1.00        999999.00      3.00
3S        4T      0.00         1.00        999999.00      1.00
2S        5T      0.00         1.00        999999.00      6.00
```

The optimal value of this solution is 19 and corresponds to the subtours 1-2-5-1 and 4-3-4. While this solution is not a feasible tour, the fact that the lower bound obtained by restricting arc (5, 2) is *greater* than the best known upper bound means that it will be fruitless to attempt to exclude any arcs in this solution and continue solving assignment problems. We can never find a tour with a distance less than 19.

If we solve the assignment problems by excluding the other arcs in the initial solution, we have the following.

Exclude arc (2, 1):

```
ASSIGNMENT PROBLEM:    MINIMUM COST IS         19.00

FROM      TO      LOWER        FLOW        UPPER          COST
----      --      --------     --------    --------       --------
5S        1T      0.00         1.00        999999.00      6.00
1S        2T      0.00         1.00        999999.00      3.00
4S        3T      0.00         1.00        999999.00      3.00
3S        4T      0.00         1.00        999999.00      1.00
2S        5T      0.00         1.00        999999.00      6.00
```

Exclude arc (3, 4):

```
ASSIGNMENT PROBLEM:    MINIMUM COST IS         19.00

FROM      TO      LOWER        FLOW        UPPER          COST
----      --      --------     --------    --------       --------
3S        1T      0.00         1.00        999999.00      3.00
5S        2T      0.00         1.00        999999.00      2.00
4S        3T      0.00         1.00        999999.00      3.00
2S        4T      0.00         1.00        999999.00      9.00
1S        5T      0.00         1.00        999999.00      2.00
```

Exclude arc (4, 3):

```
ASSIGNMENT PROBLEM:   MINIMUM COST IS        18.00
```

FROM	TO	LOWER	FLOW	UPPER	COST
2S	1T	0.00	1.00	999999.00	4.00
5S	2T	0.00	1.00	999999.00	2.00
1S	3T	0.00	1.00	999999.00	7.00
3S	4T	0.00	1.00	999999.00	1.00
4S	5T	0.00	1.00	999999.00	4.00

These lower bounds are summarized in Fig. 9.10 (p. 336). Since no lower bound is better than the best known upper bound, we have proved that the solution 1-3-4-5-2-1 must be optimal. If we had found a node in the branch-and-bound tree with a lower bound smaller than the best known upper bound, we would have had to continue branching by excluding more arcs until we either found a better tour or proved that we cannot.

EXAMPLE 6

We will use branch and bound to solve the symmetric problem in Example 2 using the 1-tree relaxation and lower bound L_2 as illustrated in Example 4. The initial 1-tree is shown again below.

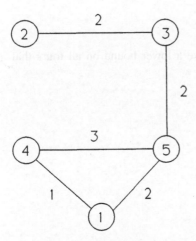

Note that vertex 5 has degree 3. Therefore, at least one of these incident edges must be excluded from the solution. The initial branch and bound tree is shown in Fig. 9.11. By excluding each of these edges and finding new 1-trees, we have the following.

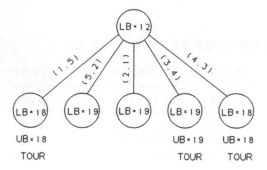

Fig. 9.10   Completed branch-and-bound tree, Example 6.

Exclude edge (5, 3): The 1-tree we obtain is

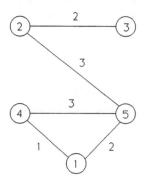

and has a cost of 11. Since it is not feasible, we have a lower bound on all tours that exclude edge (5, 3).

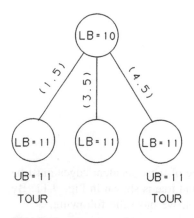

Fig. 9.11   Branch-and-bound tree, Example 6.

Exclude edge (5, 4): We find the 1-tree

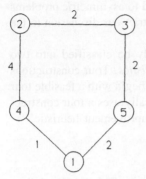

which also has a cost of 11, but represents the feasible tour 1-4-2-3-5-1.

Exclude edge (1, 5): We obtain the 1-tree

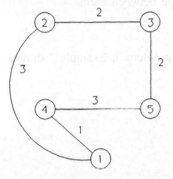

which is also a feasible tour, 1-2-3-5-4-1.

Since no lower bound in the branch-and-bound tree is less than 11, we need not continue further and have found alternative optimal solutions.

Interestingly, branch and bound was first developed in conjunction with the traveling salesman problem by Little et al. (1963) and has found widespread use in solving many other classes of problems. Although it is quite effective, branch and bound quickly can become computationally impractical for large problems. Thus, we must often resort to heuristics.

9.5 HEURISTIC ALGORITHMS FOR THE TSP

Since the traveling salesman problem is very difficult to solve optimally, many heuristics have been developed. These heuristic procedures generally are fast and provide solutions that usually are within a few percent of the optimum. Thus for problems of realistic sizes, heuristics represent a practical solution approach. We shall describe several different heuristics that have been proposed. Most of these can be applied to asymmetric problems with simple modifications (see the exercises); thus we shall restrict our discussion to the symmetric case.

Heuristics for the traveling salesman problem can broadly be classified into two groups: *tour construction heuristics* and *tour improvement heuristics*. Tour constructions heuristics generate a feasible tour. Tour improvement heuristics begin with a feasible tour and attempt to make improvements on it. In practice, one typically uses a tour construction heuristic to find an initial tour and then applies a tour improvement heuristic.

Tour Construction Heuristics

A simple heuristic to build a tour is the *nearest-neighbor heuristic*. This is essentially a greedy algorithm. Begin with any vertex x and find the vertex y so that $d(x,y)$ is the smallest among all y. Next, find the closest vertex to y that is not already in the tour, say vertex z, and add edge (y, z) to the tour. Repeat this process until the last vertex is added and then join the first and last vertices by the unique edge between them.

EXAMPLE 7

We will illustrate the nearest-neighbor heuristic using the problem in Example 2 shown again below.

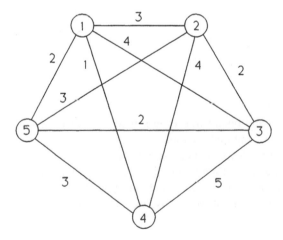

If we begin with vertex 1, we successively add vertex 4, then vertex 5, then vertex 3, and finally vertex 2. The final tour is 1-4-5-3-2. Only by coincidence is this the optimal solution. Generally, as the algorithm progresses, the edges selected become poor.

A second class of tour construction heuristics, called *insertion procedures*, was proposed by Rosenkrantz et al. (1974). We will describe *nearest insertion*. Select one vertex to start, say vertex i. Choose the nearest vertex, say j, and form the subtour i-j-i. At each iteration, find the vertex k not in the subtour that is closest to any vertex in the subtour. (This is reminiscent of Prim's minimum spanning tree algorithm.) Find the edge (i, j) in the subtour which minimizes $d(i, k) + d(k, j) - d(i, j)$. Insert vertex k between i and j. Repeat this process until a tour is constructed. Note that in the iterative step, we try to add the least amount of distance to the current subtour by removing edge (i, j) and adding edges (i, k) and (k, j).

EXAMPLE 8

We illustrate the nearest insertion procedure using the graph in Example 2. Suppose that we begin with vertex 1 and create the initial subtour 1-4-1. The vertex nearest to either 1 or 4 is vertex 5. We add vertex 5, creating the subtour 1-4-5-1. The closest vertex to any vertex in this subtour is vertex 3. We consider placing vertex 3 between vertices 1 and 4, 4 and 5, or 5 and 1. The incremental cost of placing vertex 3 between 1 and 4 is $d(1, 3) + d(3, 4) - d(1, 4) = 5 + 4 - 1 = 8$. Similarly, the incremental cost of placing vertex 3 between 4 and 5 is 4; the incremental cost of placing it between 5 and 1 is 4. Since we have a tie, we arbitrarily place it between 4 and 5, creating the subtour 1-4-3-5-1. Finally, we consider vertex 2. The calculations are summarized below.

Arc	Incremental Cost
(1, 4)	$3 + 4 - 1 = 6$
(4, 3)	$4 + 2 - 5 = 1$
(3, 5)	$3 + 2 - 2 = 3$
(5, 1)	$3 + 3 - 2 = 4$

We place vertex 2 between vertices 4 and 3, creating the final tour 1-4-2-3-5 having a total distance of 12.

Other variations of insertion procedures are *farthest insertion*, *cheapest insertion* (in which the incremental cost is computed for all vertices not in the tour, not just the closest one), and *arbitrary insertion*.

Christofides' Heuristic

A simple yet very clever heuristic was proposed by Christofides (1976). This method requires that the triangle inequality be satisfied and proceeds as follows.

1. Construct the minimum spanning tree of the graph.
2. Find the minimum-cost matching of the odd-degree vertices in the spanning tree. Add the edges from the optimal matching to the tree to create an Euler graph.
3. Find an Euler tour in this graph. Transform the Euler tour into a Hamiltonian cycle. This is done as follows. Let $x_1, x_2, \ldots, x_k$ be the sequence of vertices

visited in the Euler tour. Suppose that vertex x_i is the first vertex to be repeated. Let x_j be the first vertex after x_i that is not repeated in the Euler tour. Replace the path $x_{i-1}, x_i, \ldots, x_j$ by the single edge (x_{i-1}, x_j) as illustrated in Fig. 9.12.

EXAMPLE 9

Consider the TSP shown below.

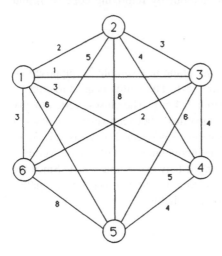

The minimal spanning tree is

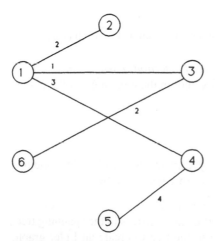

The odd-degree vertices are 1, 2, 5, and 6. The matching problem on these vertices follows.

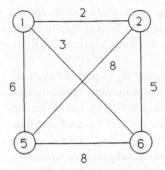

The optimal matching is (1, 2) and (5, 6). Adding these edges to the minimal spanning tree creates the Euler graph

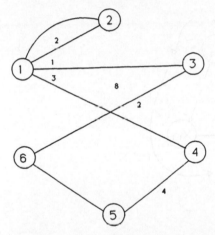

An Euler tour in this graph is 1-2-1-4-5-6-3-1. When vertex 1 is repeated, we replace the path 2-1-4 by the edge (2, 4) to create the Hamiltonian cycle 1-2-4-5-6-3-1.

Christofides' heuristic has a guaranteed worst-case bound of 1.5. This means that the value of the optimal solution will never be more than 1.5 times the value of the heuristic solution. To show this, we first prove the following.

Lemma 9.1. For an n-vertex TSP, n even, the value of the minimum-cost perfect matching, $C(M^*)$, is no greater than one-half the value of the minimum-cost Hamiltonian cycle, $C(H^*)$; that is, $C(M^*) \leq 0.5C(H^*)$.

Proof. Suppose that the minimum-cost Hamiltonian cycle H* is given by the sequence of vertices $x_1, x_2, \ldots, x_n$. Traverse this cycle beginning at vertex x_1 and allocate the edges alternately to two sets M_1 and M_2. Thus

$$M_1 = \{(x_1, x_2), (x_3, x_4), \ldots, (x_{n-1}, x_n)\}$$
$$M_2 = \{(x_2, x_3), (x_4, x_5), \ldots, (x_n, x_1)\}$$

Clearly, the union of M_1 and M_2 is the tour and the sum of the costs of M_1 and M_2 equals the cost of the tour. Then the smaller of the two matching costs must be less than or equal to one-half the cost of the tour. It follows that $C(M^*) \leq \min\{C(M_1), C(M_2)\} \leq 0.5C(H^*)$.   Q.E.D.

To prove the main result, we note that any Hamiltonian *path* H_p (a path through all vertices of the graph) is a spanning tree. Clearly, the cost of the minimum spanning tree $C(T^*)$ must be less than or equal to $C(H_p)$, which is strictly less than $C(H^*)$. Let M_T be the minimum-cost matching of all odd-degree vertices of T^*. Since $M_T \cup T^*$ is an Euler tour E, the cost of the Euler tour is simply the sum of the costs $C(T^*)$ and $C(M_T)$. From Lemma 1 we have $C(M_T) \leq 0.5C(H_T)$, where H_T is the optimal tour on the subgraph generated by the set of odd-degree vertices in the minimum spanning tree. Since $C(H_T) \leq C(H^*)$, we have $C(M_T) \leq 0.5C(H^*)$. Combining these results, we have that the cost of the Euler tour

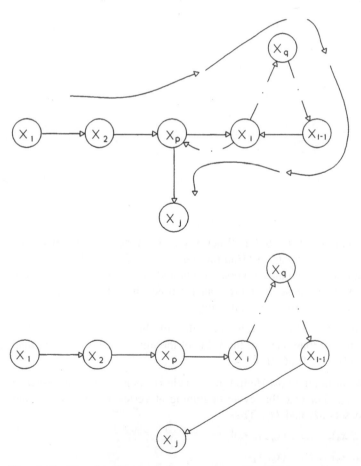

Fig. 9.12 Transforming an Euler tour into a Hamiltonian cycle.

must be strictly less than $1.5C(H^*)$. By the triangle inequality, it is easy to see that the cost of the Hamiltonian cycle created from the Euler tour can be no greater than the cost of the Euler tour. Therefore, the cost of the Hamiltonian cycle generated by the heuristic is less than 1.5 times the cost of the optimal tour.

Corneujols and Nemhauser (1978) tighten this bound somewhat and prove that the cost of the heuristic solution is no greater than $(3m - 1)/2m$ times the cost of the optimal tour, where m is the largest integer not greater than $0.5n$.

Tour Improvement Heuristics

The most popular tour improvement heuristics were introduced by Lin (1965) and extended by Lin and Kernighan (1973) and are called *k-opt* heuristics. A *k-change* of a tour consists of deleting k edges and replacing them by k other edges to form a new tour. An example of all possible 2-changes of a 5-vertex TSP is shown in Fig. 9.13. Figure 9.14 shows all possible 3-changes.

The heuristic procedure begins with any feasible tour. From this tour, all possible *k*-changes are examined. If a tour is found that has a lower cost than the current solution, it becomes the new solution. The process is repeated until no further *k*-change results in a better solution. When the algorithm stops, we have a *local optimal solution*. Of course, there is no guarantee that the resulting solution is globally optimal.

EXAMPLE 10

We will illustrate the 2-opt heuristic using the problem in Example 2. Let us begin with the arbitrary tour 1-2-3-4-5-1. All 2-changes and their costs are listed below:

Tour	Cost
1-3-2-4-5-1	15
1-4-3-2-5-1	13
1-2-4-3-5-1	16
1-2-5-4-3-1	18
1-2-3-5-4-1	11

Since the minimum cost of any 2-change (11) is less than the cost of the current solution (15) we select 1-2-3-5-4-1 as the new solution. Evaluating all 2-changes of this new solution, we have

Tour	Cost
1-3-2-5-4-1	13
1-5-3-2-4-1	11
1-2-5-3-4-1	14
1-2-4-5-3-1	16
1-2-3-4-5-1	15

No 2-change has a smaller cost than the current solution, so the algorithm stops.

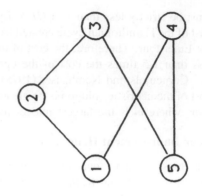

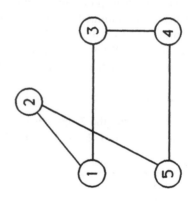

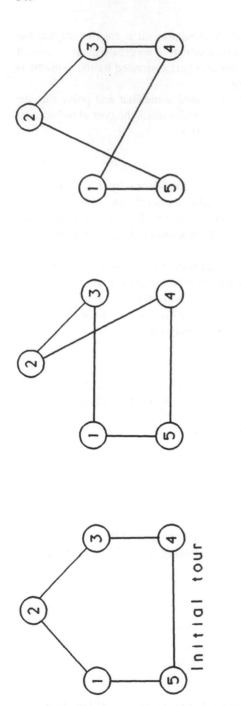

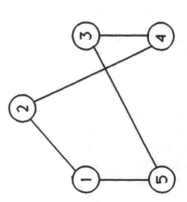

Fig. 9.13   2-changes of a 5-vertex TSP.

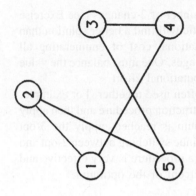
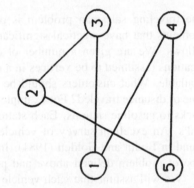
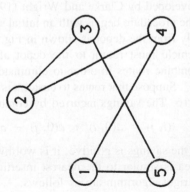
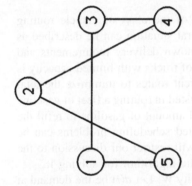
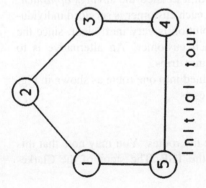

Initial tour

Fig. 9.14 3-changes of a 5-vertex TSP.

In general, the number of 3-changes exceeds the number of 2-changes (see Exercise 16); thus there is a greater chance that a 3-change heuristic will find a better solution than that a 2-change heuristic will. However, the computational cost of enumerating all 3-changes is larger than the cost of enumerating all 2-changes. One must balance the value of finding a better solution against the increased computational effort.

Tour construction and improvement heuristics are often used together. For example, one can begin with an initial solution using any tour construction procedure and then apply the 2-opt improvement heuristic. When a local optimum is reached, apply the 3-opt heuristic to attempt to improve the solution further. Continue switching between 2-opt and 3-opt until no further improvements can be made. Such a procedure is very effective and generally finds solutions that are within only a few percent of the optimum.

9.6 VEHICLE ROUTING PROBLEMS

The traveling salesman problem is only one example of the class of vehicle routing problems that have practical significance. A more general situation can be described as follows. We are given a number of customers with known delivery requirements and locations (assumed to be vertices in a network). A fleet of trucks with limited capacity is available. What customers should be assigned to different routes to minimize the total time or distance traveled? For example, we may be interested in routing a fleet of gasoline trucks to gasoline stations. Each station requires a fixed amount of gasoline to refill the tanks. An excellent survey of vehicle routing and related scheduling problems can be found in Bodin and Golden (1981). In this section we will restrict our discussion to the type of problem defined above and present one heuristic algorithm for solving it.

We will assume that each vehicle has a fixed capacity W. Let $d(i)$ be the demand at vertex i and $a(i, j)$ be the time or cost associated with traveling from vertex i to vertex j. We also assume that all vehicles are dispatched from a central location (depot) in the network, which will be defined to be vertex 0. An effective heuristic for this problem was developed by Clarke and Wright (1963) and is sometimes called the *savings approach*. The procedure begins with an initial solution in which each customer is serviced individually from the depot as shown in Fig. 9.15. Such a solution is very inefficient, since the vehicle must return to the depot after servicing each customer. An alternative is to combine routes in order to eliminate unnecessary return trips.

Suppose that routes to customers i and j are combined into one route as shown in Fig. 9.16. The savings incurred by combining these routes is

$$s(i, j) = a(0, i) + a(0, j) - a(i, j)$$

If the savings is positive, it is worthwhile to combine the routes. You may note that this idea is similar to the nearest insertion heuristic for the TSP. The steps of the Clarke-Wright algorithm are as follows:

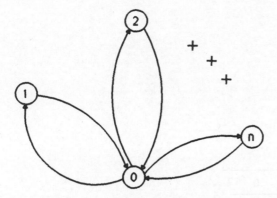

Fig. 9.15   Initial solution for the Clarke-Wright algorithm.

Step 1. Compute the savings $s(i, j)$ for all pairs of customers.

Step 2. Choose the pair of customers with the largest savings and determine if it is feasible to link them together. If so, construct a new route by joining them. If not, discard this possibility and choose the pair with the next largest savings.

Step 3. Continue with step 2 as long as the savings is positive. When all positive savings have been considered, stop.

EXAMPLE 11

Suppose that a beer distributorship has received orders from seven customers for delivery the next day. The number of cases required by each customer and travel times between each pair of customers are shown in the tables on the next page.

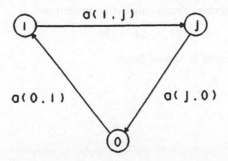

Fig. 9.16   Combining customers i and j.

Customer	Demand
1	46
2	55
3	33
4	30
5	24
6	75
7	30

Travel Times

From/To	0	1	2	3	4	5	6	7
0	—							
1	20	—						
2	57	51	—					
3	51	10	50	—				
4	50	55	20	50	—			
5	10	25	30	11	50	—		
6	15	30	10	60	60	20	—	
7	90	53	47	38	10	90	12	—

The delivery vehicle has capacity for 80 cases.

We begin with the initial solution consisting of seven routes.

0-1-0
0-2-0
0-3-0
0-4-0
0-5-0
0-6-0
0-7-0

If, for example, we link customers 1 and 2 together, the savings is computed as

$$s(1, 2) = a(0, 1) + a(0, 2) - a(1, 2) = 20 + 57 - 51 = 26$$

The savings associated with each pair of customers is shown below.

	1	2	3	4	5	6	7
1	—						
2	26	—					
3	61	58	—				
4	15	87	51	—			

	1	2	3	4	5	6	7
5	5	37	50	10	—		
6	5	62	6	5	5	—	
7	57	100	103	130	10	93	—

First, select the largest savings—customers 4 and 7. Joining customers 4 and 7 results in a feasible route because the demand is $30 + 30 = 60$, which is less than the vehicle's capacity. The new set of routes is

0-1-0
0-2-0
0-3-0
0-4-7-0
0-5-0
0-6-0

Next, try to join customers 3 and 7, since this represents the next-largest savings. However, the demand on this route would exceed the vehicle capacity, so this possibility is ignored.

Moving down the list of savings, combining customers 2 and 7, 7 and 6, 2 and 4, and 2 and 6 also violates the capacity constraint. The next-largest savings of 61 corresponding to customers 3 and 1 results in a feasible route. Continuing, we obtain the final set of routes:

0-4-7-0
0-3-1-0
0-2-5-0
0-6-0

APPENDIX: USING NETSOLVE TO FIND TRAVELING SALESMAN TOURS

The NETSOLVE traveling salesman algorithm finds a near-optimal tour in a directed or undirected network. The length of an edge is the value specified in the COST data field. The length of a tour is the sum of the lengths of the edges in the tour.

Five heuristics can be used to find an approximate solution to the TSP. If the keyword MIN is specified, a tour of minimum length is sought, whereas if MAX is specified, a tour of maximum length is sought. The heuristic used is selected from

1. CHEAPEST insertion
2. FARTHEST insertion
3. NEAREST insertion

FROM	TO	COST	LOWER	UPPER
1	2	2.00	0.00	999999.00
1	3	1.00	0.00	999999.00
1	4	3.00	0.00	999999.00
1	5	6.00	0.00	999999.00
1	6	3.00	0.00	999999.00
2	3	3.00	0.00	999999.00
2	4	4.00	0.00	999999.00
2	5	8.00	0.00	999999.00
2	6	5.00	0.00	999999.00
3	4	4.00	0.00	999999.00
3	5	6.00	0.00	999999.00
3	6	2.00	0.00	999999.00
4	5	4.00	0.00	999999.00
4	6	5.00	0.00	999999.00
5	6	8.00	0.00	999999.00

Fig. 9.17 NETSOLVE data for TSP example.

TRAVELING SALESMAN PROBLEM (CHEAPEST INSERTION ALGORITHM)

LENGTH OF SHORTEST TOUR FOUND: 21.00

EDGES IN THE TOUR:

FROM	TO	COST
1	3	1.00
3	6	2.00
6	5	8.00
5	4	4.00
4	2	4.00
2	1	2.00

TRAVELING SALESMAN PROBLEM (FARTHEST INSERTION ALGORITHM)

LENGTH OF SHORTEST TOUR FOUND: 21.00

EDGES IN THE TOUR:

FROM	TO	COST
1	2	2.00
2	4	4.00
4	5	4.00
5	6	8.00
6	3	2.00
3	1	1.00

Fig. 9.18 NETSOLVE heuristic solutions for TSP example.

```
TRAVELING SALESMAN PROBLEM   (NEAREST INSERTION ALGORITHM)

LENGTH OF SHORTEST TOUR FOUND:            21.00

    EDGES IN THE TOUR:
       FROM        TO               COST
       ----        --               --------
        1           2                2.00
        2           4                4.00
        4           5                4.00
        5           6                8.00
        6           3                2.00
        3           1                1.00
```

```
TRAVELING SALESMAN PROBLEM   (ORDINAL INSERTION ALGORITHM)

LENGTH OF SHORTEST TOUR FOUND:            21.00

    EDGES IN THE TOUR:
       FROM        TO               COST
       ----        --               --------
        1           2                2.00
        2           4                4.00
        4           5                4.00
        5           3                6.00
        3           6                2.00
        6           1                3.00
```

```
TRAVELING SALESMAN PROBLEM      (NEAREST NEIGHBOR ALGORITHM)

LENGTH OF SHORTEST TOUR FOUND:            21.00

    EDGES IN THE TOUR:
       FROM        TO               COST
       ----        --               --------
        1           2                2.00
        2           4                4.00
        4           5                4.00
        5           6                8.00
        6           3                2.00
        3           1                1.00
```

Fig. 9.18   (continued)

4. ORDINAL (arbitrary) insertion
5. or nearest NEIGHBOR

If none are specified, farthest insertion is used. For example, to find an approximate solution using cheapest insertion, the command line

    TSP CHEAPEST MIN

would be entered. The length of the shortest tour found is displayed at the terminal. If the terminal is ON, the edges defining the tour and their lengths are also displayed.

We will illustrate the use of this algorithm using the graph in Example 9 of the text. Figure 9.17 shows the edge list for this graph. Figure 9.18 gives the solution for each of the heuristics.

EXERCISES

1. Show that the set of all optimum solutions to the salesman problems is the same for all starting vertices.

2. Suppose that the salesman must visit vertex b immediately after visiting vertex a. How can this be incorporated into the salesman problem?

3. Suppose that the salesman must visit vertex b after (not necessarily immediate after) visiting vertex a. How can this be incorporated into the salesman problem?

4. Use the results of Section 9.2 to verify that the graphs shown in Figs. 9.19 and 9.20 possess Hamiltonian circuits.

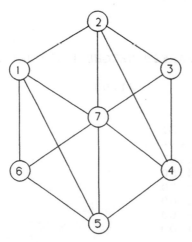

Fig. 9.19

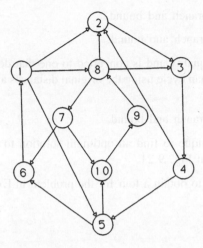

Fig. 9.20

5. For the graph in Fig. 9.5, we found that $L_1 < L_2$. Construct a graph for which $L_1 > L_2$. Construct a graph for which $L_1 = L_2$.

6. Compute an assignment-based lower bound for the following distance matrix:

	1	2	3	4	5	6
1	—	3	6	9	2	5
2	2	—	7	10	4	1
3	8	4	—	9	2	4
4	1	2	9	—	3	5
5	8	3	1	4	—	7
6	3	2	5	4	7	—

7. Compute a 1-tree lower bound for the following distance matrix:

	1	2	3	4	5	6
1	—	3	6	9	2	5
2		—	7	10	4	1
3			—	9	2	4
4				—	3	5
5					—	7

8. Solve the problem in Example 2 using branch and bound.

9. Solve the problem in Exercise 6 using branch and bound.

10. If a TSP does not satisfy the triangle inequality and is converted to one that does, how can you determine the optimal salesman cycle using the original distances after solving the transformed problem?

11. Solve the problem in Exercise 7 using branch and bound.

12. Use the branch-and-bound solution technique to find an optimum solution to the salesman problem for the graph shown in Fig. 9.21.

13. Use the following construction heuristics to obtain a tour for the problem in Exercise 7:
    (a) Nearest neighbor
    (b) Nearest insertion
    (c) Farthest insertion
    (d) Christofides' heuristic

14. For each initial tour in Exercise 13, apply the 2-opt improvement heuristic to find a local optimum.

15. For each initial tour in Exercise 13, apply the 3-opt improvement heuristic to find a local optimum.

16. For a complete graph on n vertices, determine the number of 2-changes and 3-changes that are possible as a function of n.

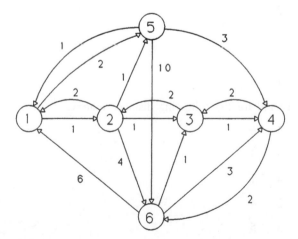

Fig. 9.21

17. Use the successive improvement technique to find a short salesman route for the graph shown in Fig. 9.21. Start this technique with the Hamiltonian cycle that visits the vertices in the order 4, 3, 2, 1, 5, 6.

18. In the past, all announcements for general circulation around the Operations Research Department of a large firm were photocopied and one copy was given to each person in the department. In the interest of economy, the department has decided that only one copy of the announcement will be made and that it will be routed throughout the department along some predetermined route. The distances between the desks of each of the six people in the department are given in the following table. What is the best routing for an announcement? (The announcement must always return to its sender so that he or she can be certain it was not lost somewhere along the route.)

From	To	Employee 1	2	3	4	5	6
Employee	1	0	9	8	7	6	10
	2		0	10	9	15	20
	3			0	5	15	25
	4				0	20	5
	5					0	20

(Distance in yards)

19. Four machine operations must be performed on every job entering a machine shop. These operations may be performed in any sequence; however, the setup time on a machine depends on the previous operation performed on the job. The setup times are given in the following table. What is the best sequence of operations?

From	To	Machine A	B	C	D
Machine	A	—	15	20	5
	B	30	—	30	15
	C	25	25	—	15
	D	20	35	10	—

20. Solve the following vehicle routing problem using the Clarke-Wright heuristic.

Customer	Demand
1	486
2	541
3	326
4	293

Customer	Demand
5	24
6	815
7	296

The vehicle capacity is 820, and the distance matrix is given below.

	1	2	3	4	5	6	7
0	—						
1	19	—					
2	57	51	—				
3	51	10	49	—			
4	49	53	18	50	—		
5	04	25	30	11	68	—	
6	12	80	06	91	62	48	—
7	92	53	47	38	09	94	09

REFERENCES

Bellmore, M., and G. L. Nemhauser, 1968. The Traveling Salesman Problem: A Survey, *Oper. Res.*, *16*, pp. 538–558.

Berge, C., 1973. *Graphs and Hypergraphs* (English translation by E. Minieka), North-Holland, Amsterdam, Chapter 10, pp. 186–227.

Bodin, L., and B. Golden, 1981. Classification in Vehicle Routing and Scheduling, *Networks*, *11*, pp. 97–108.

Christofides, N., 1972. Bounds for the Traveling-Salesman Problem, *Oper. Res.*, *20*(6), pp. 1044–1056.

Christofides, N., 1976. Worst-Case Analysis of a New Heuristic for the Traveling Salesman Problem, Management Sciences Research Group, Graduate School of Industrial Administration, Carnegie-Mellon University, Pittsburgh.

Chvátal, V., 1972. On Hamilton Ideals, *J. Combinatorial Theory Ser. B*, *12*(2), pp. 163–168.

Clarke, G. and J. W. Wright, 1963. Scheduling of Vehicles from a Central Depot to a Number of Delivery Points, *Oper. Res.*, *12*, pp. 568–581.

Cornuejols, G., and G. L. Nemhauser, 1978. Tight Bounds for Christofides' Traveling Salesman Heuristic, *Math. Program.*, *14*, pp. 116–121.

Garfinkel, R., and G. L. Nemhauser, 1972. *Integer Programming*, Wiley, New York, pp. 354–360.

Ghouila-Houri, A., 1960. Une condition suffisante d'existence d'un circuit hamiltonien, *C. R. Acad. Sci.*, *251*, pp. 494–497.

Held, M., and R. Karp, 1970. The Traveling-Salesman Problem and Minimum Spanning Trees, *Oper. Res.*, *18*, pp. 1138–1162.

Held, M., and R. Karp, 1971. The Traveling-Salesman Problem and Minimum Spanning Trees. Part II. *Math. Program.*, *1*(1), pp. 6–25.

Lenstra, J. K., and A. H. G. Rinnooy Kan, 1975. Some Simple Applications of the Traveling Salesman Problem, *Oper. Res. Q.*, *26*(4), pp. 717–733.

Lin, S., 1965. Computer Solutions of the Traveling Salesman Problem, *Bell Syst. Tech. J.*, *44*, pp. 2245–2269.

Lin, S., and B. W. Kernighan, 1973. An Effective Heuristic Algorithm for the Traveling-Salesman Problem, *Oper. Res.*, *21*(2), pp. 498–516.

Magirou, V. F., 1986. The Efficient Drilling of Printed Circuit Boards, *Interfaces*, *16*(4), pp. 13–23.

Ratliff, H. D., and A. S. Rosenthal, 1983. Order Picking in a Rectangular Warehouse: A Solvable Case of the Traveling Salesman Problem, *Oper. Res.*, *31*(3), pp. 507–521.

Rosenkrantz, D. J., R. E. Stearns, and P. M. Lewis, 1974. Approximate Algorithms for the Traveling Salesman Problem, *Proc. 15th Ann. IEEE Symp. on Switching and Automatic Theory*, pp. 33–42.

Steckhan, H., 1970. A Theorem on Symmetric Traveling Salesman Problems, *Oper. Res.*, *18*, pp. 1163–1167.

10
LOCATION PROBLEMS

10.1 INTRODUCTION AND EXAMPLES

Location theory is concerned with the problem of selecting the best location in a specified region for a service facility such as a shopping center, fire station, factory, airport, warehouse, etc. The mathematical structure of a location problem depends on the region available for the location and on how we judge the quality of a location. Consequently, there are many different kinds of location problems, and the literature is filled with a variety of solution techniques.

In this chapter, we shall confine ourselves to location problems in which the region in which the facility is to be located is a graph; i.e., the facility must be located somewhere on an arc or at a vertex of a graph. Moreover, we shall confine ourselves to location problems that do not take us too far afield mathematically. Let us first describe some practical applications of location theory.

Distribution System Design

Warehousing plays a crucial role in the physical distribution of products. Consider, for instance, a large national grocery chain that manufactures many products under its own name, maintains regional distribution centers, and owns hundreds of retail stores. This firm would have control over the location of all intermediate components in the logistics system.

Suppose that the firm does not own any warehouses. Then shipments of goods would have to be made directly from plants to retail stores. If the factory is located far from its supplies of raw materials, premiums must be paid for transporting these materials to the plant. Also, longer delivery times increase the chances of material shortages for production. On the other hand, if the plant is located far from clusters of retail stores, transportation costs incurred in shipping from the plant to the retail stores are higher and it would take longer to deliver an order. The use of warehouses placed close to markets can provide quick and efficient delivery to retail stores, while still allowing factories to be near suppliers. Warehouses also are useful for consolidating many small orders from retail stores to reduce transportation costs.

Problems of locating warehouses often can be modeled on road networks since common carriers typically are the mode of transportation. Warehouses would be located either at intersections (nodes) or along the roads (arcs) of the network. A typical objective would be to minimize the total cost of shipping goods to and from the warehouse.

Bank Account and Lockbox Location (Cornuejols et al., 1977)

The number of days required to clear a check drawn on a bank in one city depends on the city in which the check is cashed. This time is called ''float,'' and the payer will continue to earn interest on the funds until the check clears. For large corporations, the difference of even a few days can have a large economic significance. Thus, to maximize its available funds, a company that pays bills to numerous clients in various locations may find it advantageous to maintain accounts in several strategically located banks. It would then pay bills to clients in one city from a bank in some other city that had the largest clearing time or float.

A related problem is called the lockbox problem. With regard to accounts receivable, corporations want to collect funds due them as quickly as possible. This can be done by locating check collection centers or lockboxes at strategic locations so that the float is minimized.

If the cities are represented by vertices of a graph and edges correspond to the information links, both of these problems can be cast as location problems on a graph. We seek to locate the bank accounts or lockboxes at the vertices of the graph to optimize the appropriate objective.

Emergency Facility Location

Many location problems involve emergency facilities such as hospitals, fire stations, civil defense, or accident rescue. As in the distribution situation, emergency vehicles must travel along a road network and the facility may be located either at an intersection or along some road segment. The usual objective is to minimize the response time from the notification of an emergency to the delivery of the service. Often, the goal is to locate the facility so that the maximum response time to any point of demand is minimized.

Other applications of location problems include locating switching centers in communication networks, computer facilities, bus stops, mailboxes, public facilities such as parks and shopping centers, and military supply points.

10.2   CLASSIFYING LOCATION PROBLEMS

The examples we have discussed highlight some of the different types of location problems that exist. We will classify location problems according to three characteristics (for a more complete classification and comprehensive survey of network location problems, see Tansel et al., 1983a, 1983b):

1. The potential location of the facility to be located—either at a vertex or anywhere on the network. Thus, while a fire station can be located anywhere on a road network, the logical network model of the lockbox problem requires the location to be at a vertex.
2. The location of demands—either at vertices or anywhere on the network. For example, houses exist along roads, while demands for lockbox services are restricted to vertices in the network.
3. Objective function—either to minimize the total cost to all demand points or to minimize the maximum cost to any demand point. In the distribution system and bank account examples, total cost was the appropriate objective; in emergency service location, we seek to minimize the maximum travel cost to any demand point.

This classification scheme is summarized in Fig. 10.1. Each combination of problem characteristics has a unique name as noted in the figure. These are defined below.

1. A *center* of a graph is any vertex whose farthest vertex is as close as possible. In this case, both the facility and demands occur only at vertices.
2. A *general center* of a graph is any vertex whose farthest point in the graph is as close as possible. Note that while the facility is located at a vertex, demand points lie along edges of the graph as well as vertices.
3. An *absolute center* of a graph is any point whose farthest vertex is as close as possible. In this case, the facility is located anywhere along the graph, but demands occur only at vertices.
4. A *general absolute center* of a graph is any point whose farthest point is as close as possible. Here, both facilities and demands are located anywhere on the graph.

By analogy to each of these four types of location problems, we can define the *median*, *general median*, *absolute median*, and *general absolute median* simply by changing the objective function from minimizing the maximum distance from the facility to a demand to minimizing the sum of the distances from the facility to all demand points. Let us consider some examples.

A. A county has decided to build a new fire station, which must serve all six townships in the county. The fire station is to be located somewhere along one of the highways in the county so as to minimize the distance to the township farthest from the fire station.

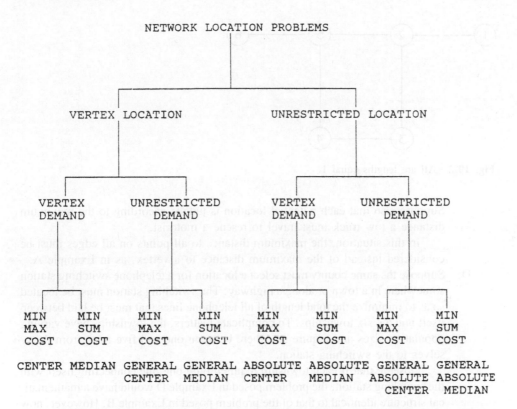

Fig. 10.1 Classification of network location problems.

If the highways of the county are depicted as the edges of a graph, this fire station location problem becomes the problem of locating the point on an edge with the property that the distance along the edges (highways) from this point to the farthest vertex (township) is as small as possible.

Consider the graph in Fig. 10.2. If vertex 3 is selected as the location, the most distant vertex from 3 is 6, which is 3 units away. Better still is vertex 2, which is at most 2 units away from any vertex. Even better is the midpoint of edge (2, 5), which is 1½ units away from vertices 1, 3, 4, and 6.

B. Suppose that the same county must locate a post office so that the total distance from the post office to all the townships is minimized. In this case, locating the post office at the midpoint of edge (2, 5) would give a total distance of 1½ + ½ + 1½ + 1½ + ½ + 1½ = 7 units since the distance from the midpoint of edge (2, 5) to vertices 1, 2, 3, 4, 5, 6 is, respectively, 1½, ½, 1½, 1½, ½, 1½ units.

C. Suppose that the same county must locate a station for tow trucks to rescue motorists who have become stranded somewhere on the county's highways.

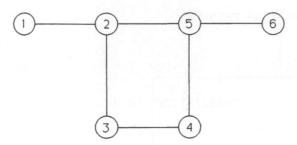

Fig. 10.2 All arc lengths equal 1.

Suppose also that each potential location is judged according to the maximum distance a tow truck must travel to rescue a motorist.

In this situation, the maximum distance to all points on all edges must be considered instead of the maximum distance to a vertex, as in Example A.

D. Suppose the same county must select a location for a telephone switching station somewhere in a town or along a highway. The switching station must be located so as to minimize the total length of all telephone lines that must be laid between itself and the six townships. To complicate matters, the townships have varying population sizes and require anywhere between one and five lines from themselves to the switching station.

Observe that if each township required only one line connecting itself with the switching station, the problem posed in Example D would have a mathematical structure identical to that of the problem posed in Example B. However, now certain townships must be considered more heavily and the location of the switching station must be influenced by the juxtaposition of the more populous townships.

Observe that Example A calls for an absolute center, Example B calls for an absolute median, Example C calls for a general absolute center, and Example D calls for a "weighted absolute median," which is discussed in Section 10.6.

10.3   MATHEMATICS OF LOCATION THEORY

Before making more rigorous definitions for the various types of locations to be considered, some definitions to describe the points on arcs and the various distances in a graph are needed.

Number the vertices of the graph G that is under consideration 1 through n. Consider any arc (i, j) whose length is given by $a(i, j) > 0$. Let the *f-point* of arc (i, j) denote the point on arc (i, j) that is $f\,a(i, j)$ units from vertex i and $(1 - f)\,a(i, j)$ units from vertex j for all f, $0 \le f \le 1$. Thus, the ¼-point of arc (i, j) is one-fourth of the way along arc (i, j) from vertex i toward vertex j.

The 0-point of arc (i, j) is vertex i, and the 1-point of arc (i, j) is vertex j. Thus, the vertices may also be regarded as points. Points that are not vertices are called *interior points*. A point must be either an interior point or a vertex. As before, let X denote the set of all vertices. Let P denote the set of all points. Thus, $P - X$ is the set of all interior points.

Let $d(i, j)$ denote the length of a shortest path from vertex i to vertex j. Let D denote the $n \times n$ matrix whose i,jth element is $d(i, j)$. The elements in matrix D are called *vertex-vertex distances*. Recall from Chapter 4 that either the Floyd or the Dantzig algorithm can be used to calculate matrix D.

Let $d(f - (r, s), j)$ denote the length of a shortest path from the f-point on arc (r, s) to vertex j. This is called a *point-vertex distance*. If arc (r, s) is undirected, i.e., allows travel in both directions, this distance must be the smaller of the two following distances:

(a) The distance from the f-point to vertex r plus the distance from vertex r to vertex j

(b) The distance from the f-point to vertex s plus the distance from vertex s to vertex j

Thus,

$$d(f - (r, s), j) = \min\{fa(r, s) + d(r, j), (1 - f)a(r, s) + d(s, j)\} \tag{1a}$$

If (r, s) is a directed arc, i.e., travel is allowed only from r to s, the first term in the above minimization is eliminated and

$$d(f - (r, s), j) = (1 - f)a(r, s) + d(s, j) \tag{1b}$$

Observe that only the arc lengths and the D matrix are needed to compute all the point-vertex distances.

When plotted as a function of f, the point-vertex distance for a given arc (r, s) and given vertex j must take one of the three forms shown in Fig. 10.3. Note that the slope of this piecewise linear curve is either $+a(r, s)$ or $-a(r, s)$, and the slope makes at most one change from $+a(r, s)$ to $-a(r, s)$.

Next, consider the shortest distance from vertex j to each point on arc (r, s). For some point on arc (r, s), this distance takes its maximum value. This maximum distance from vertex j to any point on arc (r, s) is denoted by $d'(j,(r, s))$ and is called a *vertex-arc distance*.

If arc (r, s) is undirected, there are two ways to travel from vertex j to the f-point on (r, s): via vertex r or via vertex s. Naturally, we select the shorter of the two routes. If these two routes have unequal distances, then some neighboring points of the f-point of arc (r, s) are even farther away from vertex j. For example, in Fig. 10.1, the 0.25-point of edge $(3, 4)$ is 1.25 units or 2.75 units away from vertex 2, depending on whether you travel via vertex 3 or via vertex 4. If f is increased from 0.25 to 0.26, the shortest distance from vertex 2 to the 0.26-point of edge $(3, 4)$ is $\min\{1.26, 2.74\} = 1.26$. These two

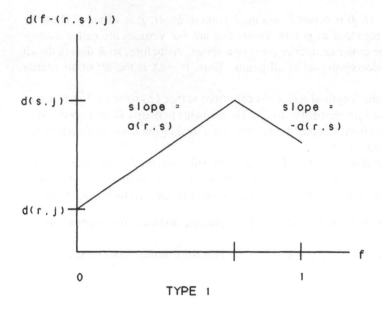

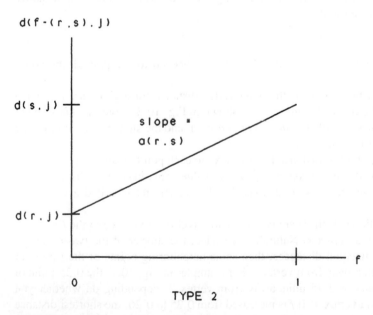

Fig. 10.3 Plots of point-vertex distances.

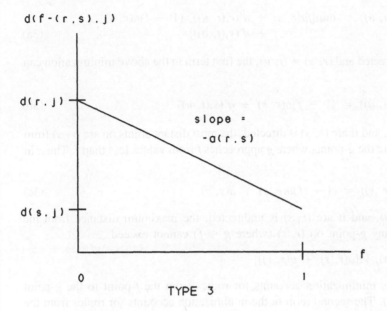

$$d(f-(r,s),j)$$

slope =
-a(r,s)

d(r,j)

d(s,j)

0   1   f

TYPE 3

distances are equal at the most distant point. Observe that these two distances always sum to

$$d(j, r) + fa(r, s) + d(r, s) + (1 - f)a(r, s) = d(j, r) + d(j, s) + a(r, s)$$

Thus, it follows that

$$d'(j,(r, s)) = \frac{d(j, r) + d(j, s) + a(r, s)}{2} \tag{2a}$$

If, on the other hand, arc (r, s) is directed, then a point on arc (r, s) can be reached only via vertex r. Consequently, the most distant points on (r, s) from any vertex are the points closest to vertex s, that is, the f-points for which f approaches 1. In this case

$$d'(j,(r, s)) = d(j, r) + a(r, s) \tag{2b}$$

Number the arcs in graph G 1 through m. Let D' denote the $n \times m$ matrix whose j,kth element is the vertex-arc distance from vertex j to arc k. Observe that the vertex-arc distance matrix D' can be computed from the vertex-vertex distance matrix D and the arc lengths by using equations (2a) and (2b).

Let $d'(f - (r, s), (t, u))$ denote the maximum distance from the f-point of arc (r, s) to the points on arc (t, u). This distance is called a *point-arc distance*.

If arc (r, s) is undirected and if $(r, s) \neq (t, u)$, then the route from the f-point on (r, s) to the most distant point on (t, u) must be either via vertex r or via vertex s. Thus, it follows that

$$d'(f - (r, s), (t, u)) = \min\{fa(r, s) + d'(r,(t, u)), (1 - f)a(r, s)$$
$$+ d'(s,(t, u))\} \tag{3a}$$

If arc (r, s) is directed and $(r, s) \neq (t, u)$, the first term in the above minimization can be eliminated and

$$d'(f - (r, s), (t, u)) = (1 - f)a(r, s) + d'(s,(t, u)) \tag{3b}$$

If $(r, s) = (t, u)$, and if arc (r, s) is directed, the most distant points on arc (r, s) from the f-point on (r, s) are the g-points where g approaches f from values less than f. Thus, in this case,

$$d'(f - (r, s), (r ,s)) = (1 - f)a(r, s) + d(s, r) \tag{3c}$$

If $(r, s) = (t, u)$, and if arc (r, s) is undirected, the maximum distance from the f-point on (r, s) to any g-point on (r, s) (where $g < f$) cannot exceed

$$A = \min\{fa(r, s), \tfrac{1}{2}[a(r, s) + d(s, r)]\}$$

The first term in this minimization accounts for routes from the f-point to the g-point restricted to arc (r, s). The second term in the minimization accounts for routes from the f-point on (r, s) to the g-point on (r, s) that traverse vertex s.

Similarly, the maximum distance from the f-point on (r, s) to any g-point on (r, s) (where $g > f$) cannot exceed

$$B = \min\{(1 - f)a(r, s), \tfrac{1}{2}[a(r, s) + d(r, s)]\}$$

The first term in the preceding minimization accounts for routes from the f-point to the g-point restricted to arc (r, s). The second term in the preceding minimization accounts for routes from the f-point on (r, s) to the g-point on (r, s) that traverse vertex r.

Consequently, if arc (r, s) is undirected,

$$d'(f - (r, s), (r, s)) = \max\{A, B\}$$

or, equivalently,

$$d'(f - (r, s),(r, s)) = \max \begin{cases} \min\{fa(r, s), \tfrac{1}{2}[a(r, s) + d(s, r)]\}, \\ \min\{(1 - f)a(r, s), \tfrac{1}{2}[a(r, s) + d(r, s)]\} \end{cases} \tag{3d}$$

When $d'(f - (r, s), (t, u))$ is plotted as a function of f for all $(r, s) \neq (t, u)$, the curve takes the same form as the point-vertex distances shown in Fig. 10.3, since equation (3a) has the same form as equation (1a) and equation (3b) has the same form as equation (1b). Only the constants are different; the equational forms are the same.

On the other hand, when $d'(f - (r, s),(r, s))$ is plotted as a function of f for any undirected arc (r, s), the curve takes the form shown in Fig. 10.4. This follows from equation (3d). This discussion is summarized as follows:

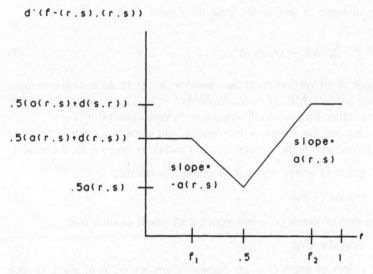

Fig. 10.4 Plot of point-arc distance $[d'(f - (r, s),(r, s))]$.

Symbol	Name	Equation
$a(i, j)$	Arc length	Given
$d(i, j)$	Vertex-vertex distance, VV	Floyd or Dantzig algorithm
$d(f - (r, s), j)$	Point-vertex distance, PV	(1a), (1b)
$d'(j,(r, s))$	Vertex-arc distance, VA	(2a), (2b)
$d'(f - (r, s),(t, u))$	Point-arc distance, PA	(3a), (3b), (3c), (3d)

Let

$$MVV(i) = \max_j \{d(i, j)\} \tag{4}$$

denote the maximum distance of any vertex from vertex i.
Let

$$SVV(i) = \sum_j d(i, j) \tag{5}$$

denote the total distance of all vertices from vertex i.
Similarly, let

$$MPV(f - (r, s)) = \max_j \{d(f - (r, s), j)\} \tag{6}$$

denote the maximum distance of any vertex from the f-point on arc (r,s).

Similarly, let

$$\text{SPV}(f - (r, s)) = \sum_j d(f - (r, s), j) \tag{7}$$

denote the total distance of all vertices from the f-point on arc (r,s). In a similar manner we can define $\text{MVA}(i)$, $\text{SVA}(i)$, $\text{MPA}(f - (r, s))$, $\text{SPA}(f - (r, s))$ by taking maximums or sums over all arcs, rather than over all vertices as in equations (4)–(7).

With all these definitions for distances, maximums, and sums, we are now ready to state rigorously the definitions of the various types of locations that we shall consider.

1. A *center* of graph G is any vertex x of graph G such that

$$\text{MVV}(x) = \min_i \{\text{MVV}(i)\} \tag{8}$$

2. A *general center* of graph G is any vertex x of graph G such that

$$\text{MVA}(x) = \min_i \{\text{MVA}(i)\} \tag{9}$$

3. An *absolute center* of graph G is any f-point on any arc (r, s) of graph G such that

$$\text{MPV}(f - (r, s)) = \min_{f-(t, u)\epsilon P} \{\text{MPV}(f - (t, u))\} \tag{10}$$

4. A *general absolute center* of graph G is any f-point on any arc (r, s) of G such that

$$\text{MPA}(f - (r, s)) = \min_{f-(t, u)\epsilon P} \{\text{MPA}(f - (t, u))\} \tag{11}$$

The definitions of (5) *median*, (6) *general median*, (7) *absolute median*, and (8) a *general absolute median* are analogous to the preceding definitions except that everywhere the maximization operation [that is, $\text{MVV}(i)$, $\text{MVA}(i)$, $\text{MPV}(f - (t ,u))$, $\text{MPA}(f - (t, u))$] is replaced by the summation operation [that is, $\text{SVV}(i)$, $\text{SVA}(i)$, $\text{SPV}(f - (t, u))$, $\text{SPA}(f - (t, u))$].

In Section 10.4, methods are developed for finding the four kinds of centers defined above. In Section 10.5, methods are developed for finding the four kinds of medians defined in the preceding discussion.

10.4 CENTER PROBLEMS

This section presents a method for finding each of the four types of centers defined in Section 10.3.

Center

Recall that a center is any vertex x with the smallest possible value of $\text{MVV}(x)$; that is, a center is any vertex x with the property that the most distant vertex from x is as close as possible.

The Floyd algorithm or the Dantzig algorithm of Chapter 4 can be used to calculate the vertex-vertex distance matrix D whose i,jth element is $d(i,j)$, the length of a shortest path from vertex i to vertex j. The maximum distance MVV(i) of any vertex from vertex i is the largest entry in the ith row of matrix D. A center is any vertex x with the smallest possible value of MVV(x); that is, a center is any vertex whose row in the D matrix has the smallest maximum entry.

EXAMPLE 1

Let us find a center for the graph shown in Fig. 10.5. It is left to the reader to verify by using either the Floyd or Dantzig algorithm that

$$D = \begin{bmatrix} 0 & 2 & 3 & 3 \\ 4 & 0 & 2 & 1 \\ 6 & 2 & 0 & 3 \\ 3 & 5 & 4 & 0 \end{bmatrix}$$

Thus,

$$\text{MVV}(1) = \max\{0, 2, 3, 3\} = 3$$
$$\text{MVV}(2) = \max\{4, 0, 2, 1\} = 4$$
$$\text{MVV}(3) = \max\{6, 2, 0, 3\} = 6$$
$$\text{MVV}(4) = \max\{3, 5, 4, 0\} = 5$$

Thus, $\min_i \text{MVV}(i) = \min\{3, 4, 6, 5\} = 3 = \text{MVV}(1)$. Consequently, vertex 1 is a center of this graph. The farthest vertex from vertex 1 is 3 units away. No other vertex can do better than 3 units.

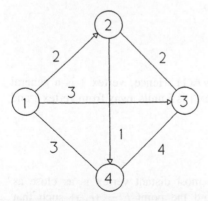

Arc	Order
1	(1, 2)
2	(1, 3)
3	(1, 4)
4	(2, 4)
5	(2, 3)
6	(3, 4)

Fig. 10.5 Computational example.

General Center

Recall that a general center is any vertex x with the smallest possible value of $MVA(x)$, that is, any vertex such that the most distant point from vertex x is as close as possible.

A general center can be found by finding the row of D' with the smallest maximum entry. This row corresponds to a vertex that is a general center. This follows since $MVA(i)$ equals the largest entry in the ith row of the vertex-arc distance matrix D'.

EXAMPLE 2

Let us find a general center for the graph in Fig. 10.5. The arcs of this graph are numbered 1 through 6. Using the vertex-vertex distance matrix D given in the preceding example and the arc lengths given in Fig. 10.5, we can use equations (2) to calculate

$$D' = \begin{bmatrix} 2 & 3 & 3 & 3 & 3.5 & 5 \\ 6 & 7 & 4 & 1 & 2 & 3.5 \\ 8 & 9 & 6 & 3 & 2 & 3.5 \\ 5 & 6 & 3 & 6 & 5.5 & 4 \end{bmatrix}$$

For example, from equation (2a),

$$d'(1,(3, 4)) \quad = \tfrac{1}{2}[d(1, 3) + d(1, 4) + a(3, 4)]$$
$$= \tfrac{1}{2}(3 + 3 + 4) = 5$$

From equation (2b),

$$d'(1,(2, 4)) = d(1, 2) + a(2, 4) = 2 + 1 = 3$$

Thus,

$$MVA(1) = \max\{2, 3, 3, 3, 3\tfrac{1}{2}, 5\} = 5$$
$$MVA(2) = \max\{6, 7, 4 ,1, 2, 3\tfrac{1}{2}\} = 7$$
$$MVA(3) = \max\{8, 9, 6, 3, 2, 3\tfrac{1}{2}\} = 9$$
$$MVA(4) = \max\{5, 6, 3, 6, 5\tfrac{1}{2}, 4\} = 6$$

Thus, min $MVA(i) = \min\{5, 7, 9, 6\} = 5 = MVA(1)$. Hence, vertex 1 is a general center of the graph. The most distant point from vertex 1 is 5 units away from vertex 1 and lies on arc (3,4).

Absolute Center

Recall that an absolute center is any point whose most distant vertex is as close as possible. To find an absolute center, we must find the point $f - (r, s)$ such that $MPV(f - (r, s)) = \min_{f-(t,u)\in P} MPV(f - (t, u))$.

The absolute center problem is more difficult than the center problem or general center problem because all points, not only the vertices, must be considered.

First, an observation: *No interior point of a directed arc can be an absolute center.* Since all travel on a directed arc is in one direction, it follows that the terminal vertex of a directed arc is closer to each vertex in the graph than is any interior point of the directed arc. Consequently, we need consider only vertices and interior points of undirected arcs in our search for an absolute center.

Consider any undirected arc (r, s). The distance $d(f - (r, s), j)$ from the f-point on (r, s) to vertex j is given by equation (1a) and plotted in Fig. 10.3. This distance is easy to plot as a function of f since its plot is a piecewise linear curve with at most two pieces.

Plot $d(f - (r, s), j)$ for all f, $0 \leq f \leq 1$, for all vertices j. The uppermost portion of all these plots represents $\max_j d(f - (r, s), j)$. The value f^* of f at which this uppermost portion of all these plots takes its minimum value is the best candidate for absolute center on edge (r, s).

The best candidate on each undirected arc must be located by this method. The absolute center is any candidate $f^* - (r, s)^*$ with the minimum distance to its farthest vertex, i.e.,

$$\max_j \{d(f^* - (r, s)^*, j)\} = \min_{(t,u)} \{\max_j d(f^* - (t, u), j)\}$$

To summarize, a candidate for absolute center is found by selecting the point on each undirected arc whose most distant vertex is as close as possible. The candidate with smallest distance between itself and its most distant vertex is selected as an absolute center. The selection of the candidate on each edge requires the plotting of all point-vertex distances as a function of the points on the edge. This is relatively uncomplicated since these distance functions are piecewise linear curves with at most two pieces. Unfortunately, there seems to be no way to avoid plotting the point-vertex distances. This method is due to Hakimi (1964).

EXAMPLE 3

Let us find an absolute center for the graph in Fig. 10.5. We know that all absolute centers (there may be ties and, consequently, more than one absolute center) must be either vertices or interior points of undirected arcs. The best vertex candidate for absolute center would be the vertex selected as center. In the example of calculating the center of this graph, we found that vertex 1 was the center and all vertices were within 3 units of vertex 1. Thus, vertex 1 is the best vertex candidate with a range of 3 units.

It remains to examine the interiors of the three undirected arcs $(1, 4)$, $(2, 3)$, and $(3, 4)$.

First, let us examine edge $(3, 4)$. From equation (1a),

$$d(f - (3, 4), 1) = \min\{fa(3, 4) + d(3, 1), (1 - f)a(3, 4) + d(4, 1)\}$$
$$= \min\{4f + 6, 4(1 - f) + 3\}$$

$$= \begin{cases} 4f + 6 & \text{for } f \leq \frac{1}{8} \\ 7 - 4f & \text{for } f \geq \frac{1}{8} \end{cases}$$

$$d(f - (3, 4), 2) = \min\{fa(3, 4) + d(3, 2), (1 - f)a(3, 4) + d(4,2)\}$$
$$= \min\{4f + 2, 4(1 - f) + 5\}$$
$$= \begin{cases} 4f + 2 & \text{for } f \leq \frac{7}{8} \\ 9 - 4f & \text{for } f \geq \frac{7}{8} \end{cases}$$

$$d(f - (3, 4), 3) = \min\{fa(3, 4) + d(3, 3), (1 - f)a(3, 4) + d(4, 3)\}$$
$$= \min\{4f, 4(1 - f) + 4\}$$
$$= 4f \quad \text{(for all } f, 0 \leq f \leq 1)$$

$$d(f - (3, 4), 4) = \min\{fa(3, 4) + d(3, 4), (1 - f)a(3, 4) + d(4, 4)\}$$
$$= \min\{4f + 3, 4(1 - f) + 0\}$$
$$= \begin{cases} 4f + 3 & \text{for } f \leq \frac{1}{8} \\ 4 - 4f & \text{for } f \geq \frac{1}{8} \end{cases}$$

These point-vertex distances are plotted in Fig. 10.6. The lowest value taken by the uppermost portion of these curves occurs when $d(f - (3, 4), 1) = d(f - (3, 4), 2)$. Thus,

$$7 - 4f^* = 4f^* + 2, \quad f^* = \frac{5}{8}$$

$$d(\tfrac{5}{8} - (3,4), 1) = d(\tfrac{5}{8} - (3, 4), 2) = 4\frac{1}{2}$$

Consequently, the $\frac{5}{8}$-point is the best candidate for absolute center on edge (3, 4), and no vertex is more than $4\frac{1}{2}$ units away from the $\frac{5}{8}$-point of edge (3, 4).

Figure 10.7 shows the same result for edge (1, 4). Note that the best candidate for absolute center on edge (1, 4) is the 0-point which is vertex 1. From before, we know that every vertex is within 3 units of vertex 1.

Figure 10.8 shows the same result for edge (2, 3). Note that the best candidate for absolute center on edge (2, 3) is the 0-point which is vertex 2. From before, we know that every vertex is within 4 units of vertex 2.

Consequently, the best interior point candidate is the $\frac{5}{8}$-point on edge (3, 4) with a maximum distance of $4\frac{1}{2}$ units. The best vertex candidate is vertex 1 with a maximum distance of 3 units. Hence, vertex 1 is the absolute center of this graph.

An improved algorithm having a polynomial time bound is given by Minieka (1981).

General Absolute Center

Recall that a general absolute center is any point x such that the farthest point from point x is as close as possible. To find a general absolute center we must find a point $f - (r, s)$ such that

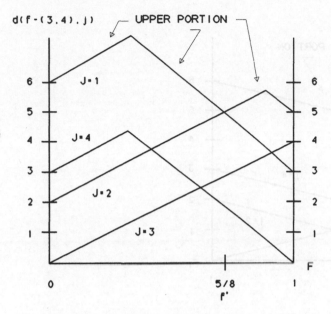

Fig. 10.6 Plot of point-vertex distances $d(f - (3, 4), j)$.

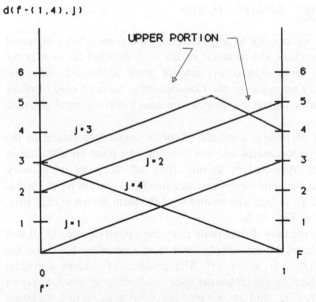

Fig. 10.7 Plot of point-vertex distances $d(f - (1, 4), j)$.

d(f -(2 . 3) . j)

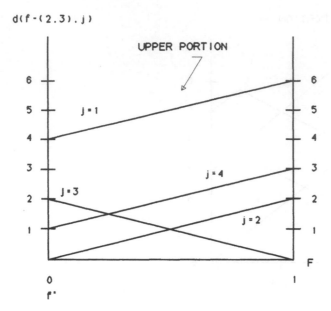

Fig. 10.8 Plot of point-vertex distances $d(f - (2, 3), j)$.

$$MPA(f - (r, s)) = \min_{f - (t, u) \in P} \{MPA(f - (t, u))\}$$

No interior point of a directed arc can be a general absolute center. Since all travel on a directed arc is in one direction, the terminal vertex of a directed arc is a better candidate for general absolute center than is any interior point of this arc, since the terminal vertex is closer to every arc in the graph. Consequently, we need only consider vertices and the interior points of undirected arcs in our search for a general absolute center.

Observe that the problem of finding a general absolute center is identical to the problem of finding an absolute center, except we must now consider point-arc distances in place of point-vertex distances. As noted in Section 10.3, all the point-arc distance functions have the same form as the point-vertex distance functions, except for the point-arc distance function $d'(f - (r, s), (r, s))$. The former have the form shown in Fig. 10.3; the latter has the form shown in Fig. 10.4.

In most realistic problems, the most distant point from the f-point on edge (r, s) will not lie on edge (r, s). In this case, we can simply omit from further consideration the point-arc distance function $d'(f - (r, s), (r, s))$. The problem of finding a general absolute center can now be solved by the technique used for finding an absolute center above. The only difference is that the point-arc distance functions must replace the point-vertex distance functions. As there are more arcs than vertices, more plotting is required to find the general absolute center.

However, if there is a possibility that the most distant point from the f-point on edge (r, s) also lies on edge (r, s), then the plot of the point-arc distance function $d'(f - (r, s), (r, s))$ must be included in the calculations for the best candidate on edge (r, s). Equation (3d) can be used to construct this plot. Happily, this plot is also piecewise linear with at most four pieces. See Fig. 10.4.

In summary, the technique for finding a general absolute center is the same as the technique for finding an absolute center except that the point-vertex distances are replaced by the point-arc distances.

10.5 MEDIAN PROBLEMS

This section presents methods for finding the four types of medians described in Section 10.3.

Median

Recall that a median is any vertex x with the smallest possible total distance from x to all other vertices. Thus, a median is any vertex x such that

$$SVV(x) = \min_i \{SVV(i)\}$$

The sum of the entries in the ith row of the vertex-vertex distance matrix D equals the sum of the distances from vertex i to all other vertices, that is, $SSV(i)$. Hence, a median corresponds to any row of D with the smallest sum.

EXAMPLE 4

Find a median of the graph in Fig. 10.5. From the previous examples, we know that the vertex-vertex distance matrix for this graph is

$$D = \begin{bmatrix} 0 & 2 & 3 & 3 \\ 4 & 0 & 2 & 1 \\ 6 & 2 & 0 & 3 \\ 3 & 5 & 4 & 0 \end{bmatrix}$$

Thus,

$$SVV(1) = 0 + 2 + 3 + 3 = 8$$
$$SVV(2) = 4 + 0 + 2 + 1 = 7$$

$$SVV(3) = 6 + 2 + 0 + 3 = 11$$
$$SVV(4) = 3 + 5 + 4 + 0 = 12$$

Hence, $\min_i\{SVV(i)\} = \min\{8, 7, 11, 12\} = 7 = SVV(2)$, and vertex 2 is the median of this graph. The total distance from vertex 2 to all other vertices is 7 units.

General Median

A general median is any vertex x with the smallest total distance to each arc, where the distance from a vertex to an arc is taken to be the maximum distance from the vertex to the points on the arc. Thus, a general median is any vertex x such that

$$SVA(x) = \min_i\{SVA(i)\}$$

The sum of the entries in the ith row of the vertex-arc distance matrix D' equals the sum of the distances from vertex i to all arcs, that is, $SVA(i)$. Hence, a median corresponds to any row of D' with the smallest sum.

EXAMPLE 5

Find a general median for the graph in Fig. 10.5. From previous examples, we know that the vertex-arc distance matrix for this graph is

$$D' = \begin{bmatrix} 2 & 3 & 3 & 3 & 3\frac{1}{2} & 5 \\ 6 & 7 & 4 & 1 & 2 & 3\frac{1}{2} \\ 8 & 9 & 6 & 3 & 2 & 3\frac{1}{2} \\ 5 & 6 & 3 & 6 & 5\frac{1}{2} & 4 \end{bmatrix}$$

Thus,

$$SVA(1) = 2 + 3 + 3 + 3 + 3\frac{1}{2} + 5 = 19$$
$$SVA(2) = 6 + 7 + 4 + 1 + 2 + 3\frac{1}{2} = 23\frac{1}{2}$$
$$SVA(3) = 8 + 9 + 6 + 3 + 2 + 3\frac{1}{2} = 31\frac{1}{2}$$
$$SVA(4) = 5 + 6 + 3 + 6 + 5\frac{1}{2} + 4 = 29\frac{1}{2}$$

Hence, $\min_i\{SVA(i)\} = \min\{19\frac{1}{2}, 23\frac{1}{2}, 31\frac{1}{2}, 29\frac{1}{2}\} = 19\frac{1}{2} = SVA(1)$. Thus, vertex 1 is the general median of this graph. The total distance from vertex 1 to all arcs is $19\frac{1}{2}$ units.

Absolute Median

An absolute median is any point whose total distance to all vertices is as small as possible.

Theorem 10.1.   There is always a vertex that is an absolute median.

Proof: Consider each point-vertex distance $d(f - (r, s), j)$ as a function of f. When plotted, this function takes the forms shown in Fig. 10.3. Observe that each function of this form has the property that if any two points on the curve are connected by a straight line (see Fig. 10.9), then this straight line always lies on or below the curve. Any function with this property is called a *concave function*. Moreover, the minimum value of a concave function always occurs at one of its boundary points, i.e., either at $f = 0$ or at $f = 1$.

Next, consider $SPV(f - (r, s)) = \Sigma_j \, d(f - (r, s), j)$ as a function of f. Since this function is the sum of concave functions, it too must be concave. Thus, $SPV(f - (r, s))$ is minimized at either $f = 0$ or $f = 1$. Consequently, no interior point of edge (r, s) is a better candidate for absolute median than one of its end vertices.   Q.E.D.

Observe that the proof of Theorem 10.1 remains valid not only for the point-vertex distances defined by equations (1) but also for any concave point-vertex distance function of f.

As a consequence of Theorem 10.1, we need only consider the vertices in our search for an absolute median. Thus, *any median is also an absolute median*, and no new solution techniques are needed.

General Absolute Median

A general absolute median is any point with the property that the total distance from it to all arcs is as small as possible. Again, the distance from a point to an arc is taken as the

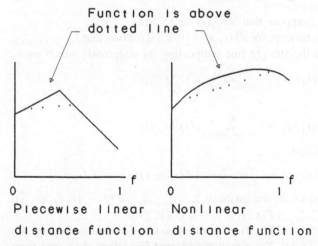

Function is above
dotted line

0   1   f   0   1   f

Piecewise linear   Nonlinear

distance function   distance function

Fig. 10.9   Concave distance functions.

maximum distance from the point to all points on the arc. Thus, a general absolute median is any point $f - (r, s)$ such that

$$\text{SPA}(f - (r, s)) = \min_{f-(t, u)\in P} \{\text{SPA}(f - (t, u))\}$$

Theorem 10.1 stated that there always is a vertex that is an absolute median. The proof of Theorem 10.1 rested on the fact that all the point-vertex distance functions were concave. If all the point-arc distance functions were also concave, an analogous theorem could be proved for the general absolute median. Unfortunately, this is not the case, since the point-arc distance function $d'(f - (r, s), (r, s))$ is not concave, as can be seen in Fig. 10.4. Otherwise, all point-arc distance functions are concave as seen from Fig. 10.3. Consequently, it is possible that all general absolute medians are interior points. (An example of this is given later.) It is possible to eliminate from consideration the interiors of some arcs. First, observe that *no interior point* of a directed arc can be a general absolute median. This follows because the terminal vertex of a directed arc is a better candidate for general absolute median than any interior point of this directed arc. Moreover,

Theorem 10.2. No interior point of undirected arc (r, s) is a general absolute median if

$$|\text{SVA}(r) - \text{SVA}(s)| > \tfrac{1}{2}[d(r, s) + d(s, r)] \tag{12}$$

Proof: The total distance from vertex r to all arcs is

$$\text{SVA}(r) = d'(r,(r,s)) + \sum_{(x,y):(x,y)\neq(r,s)} d'(r,(x,y)) \tag{13}$$

Likewise,

$$\text{SVA}(s) = d'(s,(r,s)) + \sum_{(x,y):(x,y)\neq(r,s)} d'(s,(x,y)) \tag{14}$$

Without loss of generality, suppose that $\text{SVA}(r) \leq \text{SVA}(s)$.

Since $d'(f - (r, s), (t, u))$ is concave for all $(r, s) \neq (t, u)$, it follows that $\sum_{(x,y):(x,y)\neq(r,s)} d'(f - (r, s),(x, y))$ lies above the straight line connecting its endpoints, which are

$$\sum_{(x,y):(x,y)\neq(r, s)} d'(0 - (r, s), (x, y)) = \sum_{(x,y):(x,y)\neq(r,s)} d'(r,(x, y))$$

and

$$\sum_{(x,y):(x,y)\neq(r,s)} d'(1 - (r, s), (x, y)) = \sum_{(x,y):(x,y)\neq(r,s)} d'(s,(x, y))$$

Thus, when $f = \tfrac{1}{2}$, it follows that

$$\sum_{(x,y):(x,y)\neq(r,s)} d'(\tfrac{1}{2} - (r, s),(x, y)) \geq \tfrac{1}{2}\sum_{(x,y):(x,y)\neq(r,s)} [d'(s,(x, y)) + d'(r,(x, y))]$$

Hence, as f increases from zero to $\tfrac{1}{2}$, the quantity $\sum_{(x,y):(x,y)\neq(r,s)} d'(f - (r, s), (x, y))$ increases by at least $\tfrac{1}{2} \sum_{(x,y):(x,y)\neq(r,s)} [d'(s,(x, y)) - d'(r,(x, y))]$.

Next, let us examine $d'(f - (r, s), (r, s))$. This function takes its minimum value at $f = \tfrac{1}{2}$, which is $d'(\tfrac{1}{2} - (r, s), (r, s))$. See equation (3d) and Fig. 10.4. As f goes from zero to $\tfrac{1}{2}$, $d'(f - (r, s), (r, s))$ experiences a decrease equal to $d'(r,(r, s)) - \tfrac{1}{2}a(r, s)$.

If $\Sigma_{(x,y)}\, d'(f - (r, s), (x, y)) < SVA(r)$ for some value of f, $0 < f < 1$, then it is necessary that the maximum decrease of $d'(f - (r, s), (r, s))$ at $f = \frac{1}{2}$ equal or exceed the minimum increase of $\Sigma_{(x,y):(x,y)\neq(r,s)}\, d'(f - (r, s), (x, y))$ at $f = \frac{1}{2}$. In other words, if an interior point of edge (r, s) is to be a better candidate for general absolute median than vertex r, it is necessary that

$$d'(r,(r, s)) - \frac{1}{2}a(r, s) \geq \frac{1}{2}\sum_{(x,y):(x,y)\neq(r,s)} [d'(s,(x, y)) - d'(r,(x, y))] \tag{15}$$

Equivalently, it is necessary that

$$d'(s,(r, s)) + d'(r,(r, s)) - a(r, s) \geq SVA(s) - SVA(r) \tag{16}$$

Substituting equation (2a) into inequality (16) yields

$$\frac{1}{2}[d(s, r) + a(r, s)] + \frac{1}{2}[d(r, s) + a(r, s)] - a(r, s)$$
$$\geq SVA(s) - SVA(r) \tag{17}$$

Simplifying inequality (17) yields

$$\frac{1}{2}[d(r, s) + d(s, r)] \geq SVA(s) - SVA(r) \tag{18}$$

If we had initially assumed that $SVA(s) \leq SVA(r)$, then inequality (18) would become

$$\frac{1}{2}[d(r, s) + d(s, r)] \geq SVA(r) - SVA(s) \tag{19}$$

Combining inequalities (18) and (19) yields inequality (12).   Q.E.D.

Theorem 10.2 is useful because it provides an easy way to eliminate edges from further consideration in our search for a general absolute median. To check condition (12), only the vertex-vertex distance matrix D and the vertex-arc distance matrix D' are needed.

If not all edges are eliminated by Theorem 10.2, are some further eliminations possible? Yes:

Lemma 10.1.   For any interior point f on any edge (r,s),

$$SPA(f - (r, s)) \geq SVA(r) - \frac{1}{2}d(r, s) \tag{20}$$

and

$$SPA(f - (r, s)) \geq SVA(s) - \frac{1}{2}d(s, r) \tag{21}$$

Proof: From the proof of Theorem 10.2, we know that as f increases from zero to $\frac{1}{2}$, the distance $d'(f - (r, s), (r, s))$ decreases by $d'(r,(r, s)) - \frac{1}{2}a(r, s)$, which by equation (2a) equals $\frac{1}{2}d(r, s)$. Thus, condition (20) follows.

If f is decreased from 1 to $\frac{1}{2}$, then $d'(f - (r, s), (r, s))$ decreases by $d'(s,(r, s) - \frac{1}{2}a(r, s) = \frac{1}{2}d(s, r)$, and condition (21) follows.   Q.E.D.

Lemma 10.1 can be used to generate a lower bound on the total distance for every interior point on any edge that was not eliminated by Theorem 10.2. Each of these edge lower bounds can be compared to the least total distance from a vertex, namely $\min_i$ $\{SVA(i)\}$. If the lower bound for an edge is greater the least total distance from a vertex, this edge can be eliminated.

Each remaining, noneliminated edge (r, s) must then be examined completely by evaluating $SPA(f - (r, s))$ for all f. Hopefully, the best candidate for general absolute median on the interior of the examined edge (r, s) will have a total distance that will be less than the lower bound of some nonexamined edges. In this case, these nonexamined edges can also be eliminated.

Ultimately, all edges must be either eliminated or completely examined. A general absolute median is selected from the set of vertices and interior point candidates.

EXAMPLE 6

Find a general absolute median for the graph in Fig. 10.2. Using either the Floyd algorithm or the Dantzig algorithm of Chapter 4, the vertex-vertex distance matrix D is found to be

$$D = \begin{bmatrix} 0 & 1 & 2 & 3 & 2 & 3 \\ 1 & 0 & 1 & 2 & 1 & 2 \\ 2 & 1 & 0 & 1 & 2 & 3 \\ 3 & 2 & 1 & 0 & 1 & 2 \\ 2 & 1 & 2 & 1 & 0 & 1 \\ 3 & 2 & 3 & 2 & 1 & 0 \end{bmatrix}$$

Next, order the edges as follows:

1. $(1, 2)$
2. $(2, 3)$
3. $(3, 4)$
4. $(4, 5)$
5. $(5, 6)$
6. $(2, 5)$

From equation (2a), the vertex-arc distance matrix D' can be calculated, yielding

$$D' = \begin{bmatrix} 1 & 2 & 3 & 3 & 3 & 2 \\ 1 & 1 & 2 & 2 & 2 & 1 \\ 2 & 1 & 1 & 2 & 3 & 2 \\ 3 & 2 & 1 & 1 & 2 & 2 \\ 2 & 2 & 2 & 1 & 1 & 1 \\ 3 & 3 & 3 & 2 & 1 & 2 \end{bmatrix}$$

Thus,

$$SVA(1) = 1 + 2 + 3 + 3 + 3 + 2 = 14$$
$$SVA(2) = 1 + 1 + 2 + 2 + 2 + 1 = 9$$
$$SVA(3) = 2 + 1 + 1 + 2 + 3 + 2 = 11$$
$$SVA(4) = 3 + 2 + 1 + 1 + 2 + 2 = 11$$
$$SVA(5) = 2 + 2 + 2 + 1 + 1 + 1 = 9$$
$$SVA(6) = 3 + 3 + 3 + 2 + 1 + 2 = 14$$

Consequently, vertices 2 and 5 are the best vertex candidates for general absolute median since each of these vertices has a total distance to all arcs equal to 9 units.

Next, try to eliminate the interiors of some of the edges by applying condition (12) of Theorem 10.2. Observe that in this graph all arcs are undirected, and consequently the right side of condition (12) becomes $d(r, s)$.

1. Edge (1, 2) is eliminated because

   $$|SVA(1) - SVA(2)| = |14 - 9| = 5 > 1 = d(1, 2)$$

2. Edge (2, 3) is eliminated because

   $$|SVA(2) - SVA(3)| = |9 - 11| = 2 > 1 = d(2, 3)$$

3. Edge (3, 4) is not eliminated because

   $$|SVA(3) - SVA(4)| = |11 - 11| = 0 < 1 = d(3, 4)$$

4. Edge (4, 5) is eliminated because

   $$|SVA(4) - SVA(5)| = |11 - 9| = 2 > 1 = d(4, 5)$$

5. Edge (5, 6) is eliminated because

   $$|SVA(5) - SVA(6)| = |9 - 14| = 5 > 1 = d(5, 6)$$

6. Edge (2, 5) is not eliminated because

   $$|SVA(2) - SVA(5)| = |0 - 0| = 0 < 1 = d(2, 5)$$

Thus, only edges (3, 4) and (2, 5) remain under consideration. Next, let us apply conditions (20) and (21) of Lemma 10.1 to see if any further edge eliminations are possible.

1. Edge (3, 4) can be eliminated by condition (20) because

   $$SPA(f - (3, 4)) \geq SVA(3) - \tfrac{1}{2}d(3, 4) = 11 - \tfrac{1}{2} = 10\tfrac{1}{2}$$

   which is greater than the 9 units achieved by selecting a vertex as general absolute median.

2. Edge (2, 5) cannot be eliminated by condition (20) or (21) because

$$\mathrm{SPA}(f - (2, 5)) \geq \mathrm{SPA}(2) - \tfrac{1}{2}d(2, 5) = 9 - \tfrac{1}{2} < 9$$

and

$$\mathrm{SPA}(f - (2, 5)) \geq \mathrm{SPA}(5) - \tfrac{1}{2}d(5, 2) = 9 - \tfrac{1}{2} < 9$$

Only edge (2, 5) remains under consideration. Equations (3a) and (3d) can be used to generate the point-arc distances for edge (2, 5):

$$d'(f - (2, 5),(1, 2)) = 1 + f$$
$$d'(f - (2, 5),(2, 3)) = 1 + f$$
$$d'(f - (2, 5),(3, 4)) = 2$$
$$d'(f - (2, 5),(4, 5)) = 1 + (1 - f)$$
$$d'(f - (2, 5),(5, 6)) = 1 + (1 - f)$$
$$d'(f - (2, 5),(2, 5)) = \max\{f, (1 - f)\}$$

Adding these point-arc distances yields

$$\mathrm{SPA}(f - (2, 5)) = (1 + f) + (1 + f) + 2 + (2 - f) + (2 - f) + \max\{f, (1 - f)\}$$
$$= 8 + \max\{f, (1 - f)\}$$

Since $\min_{0 \leq f \leq 1} \max\{f, (1 - f)\}$ occurs at $f = \tfrac{1}{2}$, the best candidate for general absolute median on edge (2, 5) is the $\tfrac{1}{2}$-point because $\mathrm{SPA}(\tfrac{1}{2} - (2, 5)) = 8\tfrac{1}{2}$. Since 9 is the best possible total distance of a vertex, we conclude that the $\tfrac{1}{2}$-point of edge (2, 5) is the general absolute median of this graph with a total distance to all arcs equal to $8\tfrac{1}{2}$ units.

10.6   EXTENSIONS

Weighted Locations

In the preceding sections, every vertex carried the same weight in the selection of a location. Every arc carried the same weight in the selection of a location. However, as shown in Section 10.2, Example D, there are practical reasons for associating a different weight to each vertex and multiplying the distance to a vertex by the vertex's weight. Similarly, there are situations in which the distance to an arc should be multiplied by the arc's weight. For example, if the arcs represent highway segments that must be served from a central emergency station, each segment should be weighted with regard to the amount of traffic it carries.

The methods of Sections 10.4 and 10.5 can be easily extended to find the various weighted locations—weighted center, weighted general center, weighted absolute center, etc. Merely multiply each distance (i.e., each vertex-vertex distance, each vertex-arc distance, each point-vertex distance, each point-arc distance) by the weight associated with its destination vertex or destination arc. If these weighted distances replace the original nonweighted distances, the various methods of Sections 10.4 and 10.5 will generate the corresponding weighted locations.

Multicenters and Multimedians

Sections 10.4 and 10.5 were concerned with the problem of selecting exactly one location to serve as a center or median. Suppose, instead, that we are allowed to select several facility locations. Each vertex (or arc) would then be associated with the location closest to it.

These multilocation problems are very complicated in that they consist of two stages: (a) partition of the vertices (or arcs) and (b) selection of the best location to serve all members of each subset of vertices (or arcs). Unfortunately, the techniques that are available for these problems ultimately rely on integer programming for a final solution. Any detailed discussion of these techniques would take us too far afield and would best be treated in a text on integer programming. The reader is referred to the book by Handler and Mirchandani (1979) and surveys by Tansel et al. (1983a, 1983b), Dearing (1985), and Halpern and Maimon (1982).

However, one result concerning multiabsolute medians is worth noting (Hakimi, 1965; Goldman, 1969). Suppose we are searching for a set of p locations, $p > 1$, such that each vertex is associated with the location closest to it and the total distance from each location to the vertices associated with it is minimized. Such a set of points is called a p-absolute median.

Theorem 10.3.   There is a p-absolute median that consists entirely of vertices.

Proof: Theorem 10.1 proved this result for $p = 1$. For $p > 1$, the vertices are partitioned in p sets such that each set of vertices is served by the same median. Since *any* sum of point-vertex distance functions is a concave function, we know that each vertex set is optimally served by a median that is a vertex.   Q.E.D.

EXERCISES

1.   The four towns 1, 2, 3, 4 in our county are connected by roads as shown in Fig. 10.10. Construct the D and D' matrices for this graph. Next, calculate a
   (a)   Center
   (b)   Absolute center
   (c)   General center
   (d)   General absolute center
   (e)   Median
   (f)   Absolute mean
   (g)   General median
   (h)   General absolute median

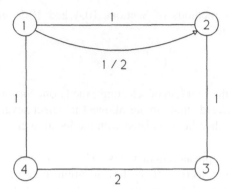

Fig. 10.10

2. Suppose that an additional road is built connecting cities 2 and 4 in Exercise 1. This road has length equal to 1½. Repeat Exercise 1. (Assume the new road is a two-way road.)

3. Suppose that an edge can be traveled in both directions, but that due to wind direction the distance from x to y is 5 and the distance from y to x is 7. How can an edge such as (x, y) be incorporated into our models for finding centers and medians? [Note that you cannot replace this edge with two arcs (x, y) and (y, x) in median problems. Why?]

4. Suppose that town 1 has twice the population of town 2, which has twice the population of town 3, which has the same population as town 4. Repeat Exercise 1 using the weights described.

5. Simplify the method of Section 10.3 for finding an absolute center for the special case when the graph under consideration is a tree.

6. Show that the interior of a directed arc can never contain an absolute center or an absolute median.

7. How much can the length of edge (1, 4) in Fig. 10.5 increase without changing the location of the center of this graph? Median? How much can the length of edge (2, 3) increase without changing the location of the absolute canter of this graph? Absolute median? Develop a general theory of sensitivity analysis.

REFERENCES

Christofides, N., and P. Viola, 1971. The Optimum Location of Multicentres of a Graph, *Oper. Res. Quart.*, 22, pp. 45–54.

Cornuejols, G., M. L. Fisher, and G. L. Nemhauser, 1977. Location of Bank Accounts to Optimize Float: An Analytic Study of Exact and Approximate Algorithms, *Manage. Sci.*, *23*(8), pp. 789–810.

Dearing, P. M., 1985. Review of Recent Developments—Location Problems, *Oper. Res. Lett.*, *4*(3), pp. 95–98.

Goldman, A. J., 1969. Optimum Locations for Centers in a Network, *Transp. Sci.*, *4*, pp. 352–360.

Hakimi, S. L., 1964. Optimum Locations of Switching Centers and the Absolute Centers and Medians of a Graph, *Oper. Res.*, *12*, pp. 450–459.

Hakimi, S. L., 1965. Optimum Distribution of Switching Centers in a Communications Network and Some Graph Theoretic Problems, *ORSA*, *13*, pp. 462–475.

Halpern, J., and O. Maimon, 1982. Algorithms for the *m*-Center Problems: A Survey, *Eur. J. Oper. Res.*, *10*, pp. 90–99.

Handler, G., and P. Mirchandani, 1979. *Location on Networks: Theory and Algorithms*, MIT Press, Cambridge, Massachusetts.

Levy, J., 1967. An Extended Theorem for Location in a Network, *Oper. Res. Quart.*, *18*, pp. 433–442.

Marsten, R., 1972. An Algorithm for Finding Almost All of the Medians of a Network, DP #23, Center for Mathematical Studies in Economics and Management Science, Northwestern University, Evanston, Illinois, November.

Minieka, E., 1970. The *m*-Center Problem, *SIAM Rev.*, *12*, pp. 138–139.

Minieka, E., 1977. The General Centers and Medians of a Graph, *Oper. Res.*, *25*, pp. 641–650.

Minieka, E., 1981. A Polynomial Time Algorithm for Finding the Absolute Center of a Network, *Networks*, *11*, pp. 351–355.

ReVelle, G., and R. Swain, 1970. Central Facility Location, *Geogr. Anal.* 2(1), pp. 30–42.

Tansel, B. C., R. L. Francis, and T. J. Lowe, 1983a. Location on Networks: A Survey. Part I: The *p*-Center and *p*-Median Problems, *Manage. Sci.*, *29*(4), pp. 482–497.

Tansel, B. C., R. L. Francis, and T. J. Lowe, 1983b. Location on Networks: A Survey. Part II: Exploiting Tree Network Structure, *Manage. Sci.*, *29*(4), pp. 498–511.

Toregas, C., C. ReVelle, R. Swain, and L. Bergman, 1971. The Location of Emergency Facilities, *Oper. Res.*, *19*, pp. 1363–1373.

Wendell, R. E., and A. P. Hurter, Jr., 1973. Optimal Locations on a Network, *Transp. Sci.*, *7*, pp. 18–33.

11
PROJECT NETWORKS

11.1 INTRODUCTION AND EXAMPLES

Large projects, such as the construction of a building, development of an accounting information system, graduation from college in four years, or even preparation of a dinner party, involve a large number of different activities. Some of these activities can be performed at the same time; others can be performed only after certain other activities have been completed. For example, the landscaping and dry wall installation of a building under construction can be performed simultaneously; however, the walls cannot be erected until after the foundation has been laid. In a four-year college program, each course may be regarded as an activity. Some courses may be taken concurrently with others, while other courses serve as prerequisites to more advanced courses. When preparing a dinner, you can set the table while the roast is in the oven. However, you cannot cook the potatoes until after they have been washed.

Since we must contend with precedence relations between activities, as described above, and since each activity in a project requires a certain amount of time, the project manager is confronted with the dilemma of carrying out the activities in the best possible way so that the project finishes on time. For example, a contractor who delays too long in laying the building foundation will fail to meet the construction deadline. If a student delays taking basic courses, he or she will not graduate on time. If you spend all your time setting the table and forget to put the roast in the oven, dinner will be late. In other words, the project manager must determine which activities are *critical* to the on-time completion of the project.

A project may be represented by a graph, called a *project network*. In a project network, each activity is represented by a directed arc. Nodes represent the start and finish of activities; these are called *events*. An event is said to have been completed when all the activities directed into it have been completely performed. Arcs which are directed into a node x are called the *immediate predecessors* of the arcs that are directed out of node x. We shall first illustrate some examples of project networks.

Construction

Project networks are used routinely in the construction industry to help manage projects consisting of many activities. Suppose that the construction of a new home requires the following activities:

Activity		Immediate Predecessors	Time
A	Clear land	None	1
B	Lay foundation	Clear land	4
C	Frame walls	Lay foundation	4
D	Wire	Frame walls	3
E	Install dry wall	Wire, roof	4
F	Landscape	Lay foundation	6
G	Interior work	Install dry wall	4
H	Roof	Frame walls	5

Observe that the dry wall installation activity requires as prerequisite activities both wiring and roofing. The project network for this example is shown in Fig. 11.1. In this network, node 1 represents the start of activity A; node 2 represents the start of activity B and the end of activity A; node 4 represents the start of both activities D and H as well as the end of activity C, and so forth. If you check each node, you will see that all immediate predecessors are represented correctly.

Note that there are two parallel activities joining nodes 4 and 5. In practice, parallel arcs are discouraged and often replaced by a composite arc representing a composite activity (in this case, wiring and roofing), or one of the parallel arcs is replaced by two arcs in series. Many computer packages, such as NETSOLVE, do not allow parallel arcs.

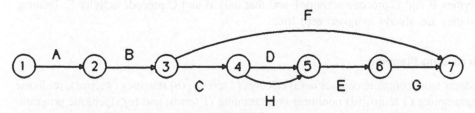

Fig. 11.1 Project network for building a house.

If we replaced activity H by two arcs in series, we would have the following:

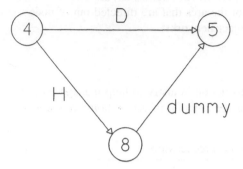

The new arc (8, 5) is a fictitious arc that we call a *dummy arc*. It is used simply to show precedence and has no meaning in the actual set of project activities.

Occasionally it is not possible to represent the correct precedence relations in a network using only arcs that correspond to the activities. In such situations, we use dummy arcs also to ensure that precedence relations are met correctly. The following abstract project is a simple example where dummy arcs are necessary.

Activity	Predecessors
A	None
B	A
C	A
D	A
E	B, C
F	B, C, D
G	E, F

Observe that the prerequisites of activity E are a proper subset of the predecessors of activity F. In this situation, we need a dummy activity to depict these precedence relationships (see Fig. 11.2). Arc (3, 4) is a dummy arc; it is needed to ensure that activities B and C precede activity F and that *only* B and C precede activity E. Dummy activities are always assigned zero time.

Curriculum Planning

Students must complete courses in (a) calculus (2 terms), (b) statistics (3 terms), (c) linear programming (1 term), (d) nonlinear programming (1 term), and (e) stochastic programming (1 term) before they can graduate in operations research. Needless to say, they cannot enroll in Calculus II until they have successfully completed Calculus I; they cannot enroll in Statistics III until they have completed Statistics II and Calculus II, and they

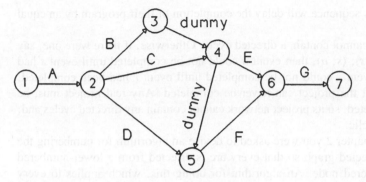

Fig. 11.2 Project network with dummy arcs.

cannot enroll in Statistics II until they have successfully completed Statistics I, which requires Calculus I. There are no prerequisites for linear programming; however, the prerequisites for nonlinear programming are Calculus II and linear programming. The prerequisites for stochastic programming are Calculus II, Statistics III, and linear programming.

The project network for this study program is shown in Fig. 11.3. Note that a dummy arc (3, 4) is required since Statistics II requires both Statistics I and Calculus II, but nonlinear programming requires Calculus II and linear programming.

Students can complete this program in as little as five terms if they pass Calculus I, Statistics I, Statistics II, Statistics III, and stochastic programming in five successive

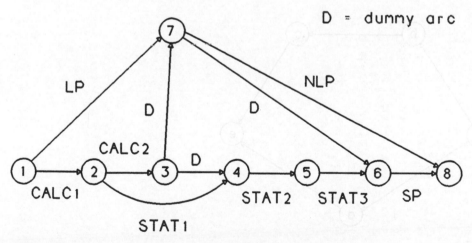

Fig. 11.3 Project network for operations research course program.

terms. Any delay in this sequence will delay the completion of their program by an equal amount of time.

A project network cannot contain a directed cycle. Otherwise, if there were one, say (a, b), (b, c), . . ., (r, s), (s, a), then event a could not be completed until event s had been completed, and event s could not be completed until event r had been completed, etc., which implies that the project can never be completed. Any real project must be capable of being completed, so its project network cannot contain any directed cycles and, therefore, must be acyclic.

In Exercise 6 of Chapter 2 you were asked to devise an algorithm for numbering the nodes of an acyclic directed graph so that every arc is directed from a lower-numbered node to a higher-numbered node. An algorithm for doing this, which applies to every project network, is given below.

Event-Numbering Algorithm

Step 1. Give the start event number 1.
Step 2. Give the next number to any unnumbered event whose predecessor events are each already numbered. (Such an event exists, since there are no cycles in the network.) Repeat step 2 until all events have been numbered. [Note that the finish event will always receive the last (highest) number.]

EXAMPLE 1 (Event-Numbering Algorithm)

Number the events in the project network in Fig. 11.4. Event a is the start event and according to step 1 of the algorithm is given number 1. Proceeding to step 2, event b has

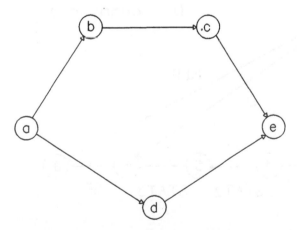

Fig. 11.4 Project network with three possible event numberings.

only numbered preceding events (i.e., event a); hence, event b is given number 2. Next, event c is given number 3, event d is given number 4, and finally the last event e is given number 5. Observe that in this example there was no choice as to which event would receive the next number.

The events in the project network in Fig. 11.4 can be numbered in a variety of different ways:

Event	First Numbering	Second Numbering	Third Numbering
a	1	1	1
b	2	2	3
c	3	4	4
d	4	3	2
e	5	5	5

11.2 CONSTRUCTING PROJECT NETWORKS

Constructing a project network correctly is something of an art, particularly when dummy activities must be used. In this section we shall work through an example which provides some guidelines for doing this.

Suppose that a project involving the selection and installation of an automated warehouse system is represented by the following activities:

	Activity	Immediate Predecessors
A	Determine equipment needs	None
B	Obtain vendor proposals	None
C	Select vendor	A, B
D	Order system	C
E	Design new warehouse layout	C
F	Lay out warehouse	E
G	Design computer interface	C
H	Interface computer	D, F, G
I	Install system	D, F
J	Train system operators	H
K	Test system	I, J

We begin with a node corresponding to the start event. The activities that leave this node are those with no immediate predecessors. From the project description, these are activities A and B. The beginning of the network would appear as follows:

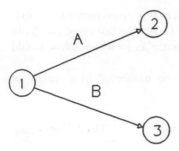

Node 2 represents the event "activity A ends," and node 3 represents the event "activity B ends."

Next, we see that activity C must follow the completion of both activities A and B. We represent this by a dummy activity from node 2 to node 3 and draw activity C from node 4 to node 4 as follows:

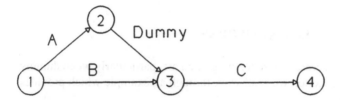

Next, we see that activities D, E, and G all have activity C as their immediate predecessor. This results in the following network:

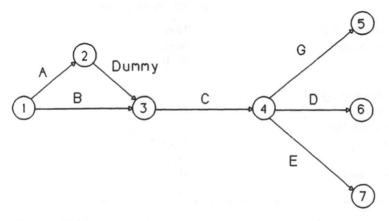

We continue to add activities, making sure that the proper predecessor relationships are maintained. Activity F follows activity E, and both D and F are predecessors of activity I. This results in the following:

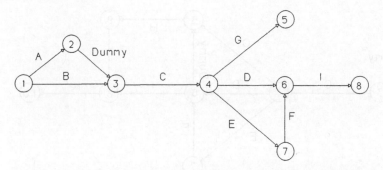

Now, activity H has immediate predecessors D, F, and G. If we attempt to draw G and D in parallel and extend H from the end of these activities as shown in the subnetwork below, we will have *added* G as a predecessor to activity I, which is incorrect.

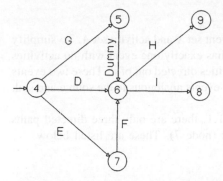

The only way to represent the precedence relations correctly is to use a dummy activity as shown below.

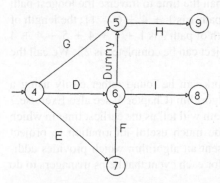

The complete project network is given in Fig. 11.5. It is useful to check each event in the "final" network to make sure that all predecessors are correctly specified.

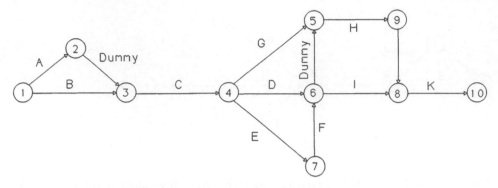

Fig. 11.5 Project network for automated warehouse design.

11.3 CRITICAL PATH METHOD

Consider any project network $G = (X, A)$ with event set X and activity set A. To simplify all future developments, assume that network G has exactly one event with no activities directed into it and exactly one event with no activities directed out of it. These two events can be regarded respectfully as the *start* and *finish* events analogous to the source and sink in a flow network.

In the house construction network in Fig. 11.1, there are only three directed paths from the start event (node 1) to the finish event (node 7). These are listed below.

    Path 1: A-B-F
    Path 2: A-B-C-D-E-G
    Path 3: A-B-C-H-E-G

Each activity along every path from start to finish must be performed *in sequence*. Thus, the time to complete all activities cannot be less than the time to traverse the longest path from start to finish. In this example, the length of path 1 is $1 + 4 + 6 = 11$; the length of path 2 is $1 + 4 + 4 + 3 + 4 + 4 = 20$; the length of path 3 is $1 + 4 + 4 + 5 + 4 + 4 = 22$. Therefore the earliest time in which the project can be completed is 22. We call the longest path in the network the *critical path*.

The longest path in a directed acyclic network can be found rather easily using a suitable modification of Dijkstra's shortest-path algorithm (Chapter 4; see also Exercise 7 in Chapter 4). Although the length of the longest path will tell us the earliest time in which the project can be completed, it does not provide much useful information to project managers for scheduling purposes. We shall present an algorithm which provides additional information about starting and finish times for each event that helps managers to do a better job scheduling project activities.

Denote the amount of time required to perform activity (x, y) by $t(x, y) \geq 0$. If activity (x, y) is a dummy activity, then let $t(x, y) = 0$.

For each event $x \in X$, let $E(x)$ denote the *earliest time* at which event x can possibly be completed. Let $L(x)$ denote the *latest time* at which event x can be completed such that the project will still be completed on time.

For example, in Fig. 11.3, $E(2) = 1$ since event 2 can be completed as easily as the end of the first term. If the completion deadline for this project is the end of the fifth term, $L(2) = 1$ since the latest that event 2 can be completed is the end of the first term; otherwise, Statistics I, II, and III and stochastic programming could not be completed by the end of the fifth term.

As another example, suppose that event 14 has only three immediate predecessor events, namely events 5, 8, and 9, where $E(5) = 4$, $E(8) = 7$, and $E(9) = 6$ (see Fig. 11.6). Event 14 cannot be completed earlier than time 10 since $E(5) + t(5, 14) = 4 + 6 = 10$. Also, event 14 cannot be completed earlier than time 11 since $E(8) + t(8, 14) = 7 + 4 = 11$. Moreover, event 14 cannot be completed earlier than time 9 since $E(9) + t(9, 14) = 6 + 3 = 9$. Thus, $E(14) = \max\{10, 11, 9\} = 11$.

In general we see that

$$E(j) = \max_{i:(i, j)\in A} \{E(i) + t(i, j)\} \tag{1}$$

With equation (1) as motivation, we may now state the *earliest event time algorithm*.

Earliest Event Time Algorithm

Step 1. Number the events $1, 2, \ldots, n = |X|$ so that all activities (i, j) have $i < j$. Use the event-numbering algorithm to accomplish this. Let $E(1) = 0$.

Step 2. For $j = 2, 3, \ldots, n$, let

$$E(j) = \max_{i:(i, j)\in A} \{E(i) + t(i, j)\}$$

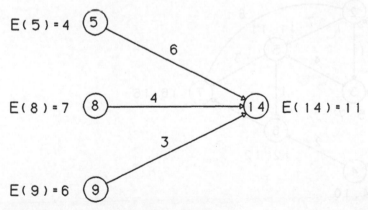

Fig. 11.6 Calculating earliest event times.

EXAMPLE 2 (Earliest Event Time Algorithm)

Calculate the earliest event times for the project network shown in Fig. 11.7. The events are already numbered 1 through 7.

$Step\ 1.$ $E(1) = 0$

$Step\ 2.$ $E(2) = E(1) + t(1, 2) = 0 + 4 = 4$

$E(3) = \max[E(1) + t(1, 3), E(2) + t(2, 3)]$

$\qquad = \max[4 + 1, 0 + 3] = 5$

$E(4) = E(1) + t(1, 4) = 0 + 4 = 4$

$E(5) = \max[E(2) + t(2, 5), E(3) + t(3, 5)]$

$\qquad = \max[4 + 7, 5 + 4] = 11$

$E(6) = \max[E(4) + t(4, 6), E(5) + t(5, 6)]$

$\qquad = \max[4 + 2, 11 + 1] = 12$

$E(7) = \max [E(2) + t(2, 7), E(5) + t(5, 7), E(6) + t(6, 7)]$

$\qquad = \max[4 + 8, 11 + 3, 12 + 4] = 16$

Consequently, the project cannot be completed any earlier than time 16.

Next, let us calculate the latest event times. For example, consider event 17 in Fig. 11.8. Event 17 is a predecessor to exactly three events, namely events 20, 24 and 29. Event 17 cannot be completed later than time 11; otherwise, event 20 would be delayed past its latest time, which is 16. Similarly, event 17 cannot be completed later than time

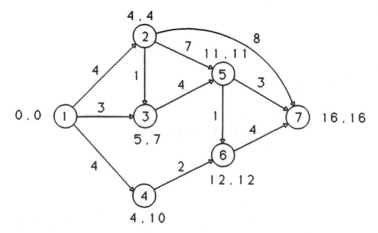

Fig. 11.7   Example of event time algorithms. Node labels are E(x), L(x).

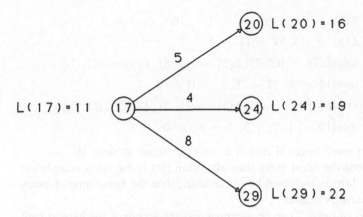

Fig. 11.8 Calculating latest event times.

15, since $L(24) - t(17, 24) = 19 - 4 = 15$. Also, event 17 cannot be completed later than time 14, since $L(29) - t(17, 29) = 22 - 8 = 14$.

In general, we see that

$$L(i) = \min_{j:(i,j) \in A} \{L(j) - t(i, j)\} \qquad (2)$$

With equation (2) as motivation, we may now state the *latest event time algorithm*.

Latest Event Time Algorithm

> *Step 1.* Number the events $1, 2, \ldots, n = |X|$ so that all activities (i, j) have $i < j$. Use the event-numbering algorithm to accomplish this.
>
> Let $L(n)$ equal the time at which the project must be completed. [In any realistic situation $L(n) \geq E(n)$.]
>
> *Step 2.* For $i = n - 1, n - 2, \ldots, 1$, let
>
> $$L(i) = \min_{j:(i,j) \in A} \{L(j) - t(i, j)\}$$

EXAMPLE 3 (Latest Event Time Algorithm)

Compute the latest event times for the project network shown in Fig. 11.7. Let $L(7) = E(7) = 16$, which indicates that the project must be finished as early as possible, i.e., time 16.

> *Step 1.* $L(7) = 16$
>
> *Step 2.* $L(6) = L(7) - t(6, 7) = 16 - 4 = 12$
>
> $\quad\ L(5) = \min\{L(7) - t(5, 7), L(6) - t(6, 5)\}$
>
> $\qquad\quad = \min\{16 - 3, 12 - 1\} = 11$

$$L(4) \;=\; L(6) \,-\, t(4,\,6) \,=\, 12 \,-\, 2 \,=\, 10$$

$$L(3) \;=\; L(5) \,-\, t(3,\,5) \,=\, 11 \,-\, 4 \,=\, 7$$

$$L(2) \;=\; \min\{L(7) \,-\, t(2,\,7),\; L(5) \,-\, t(2,\,5),\; L(3) \,-\, t(2,\,3)\}$$

$$\;=\; \min\{16 \,-\, 8,\; 11 \,-\, 7,\; 7 \,-\, 1\} \,=\, 4$$

$$L(1) \;=\; \min\{L(4) \,-\, t(1,\,4),\; L(3) \,-\, t(1,\,3),\; L(2) \,-\, t(1,\,2)\}$$

$$\;=\; \min\{10 \,-\, 4,\; 7 \,-\, 3,\; 4 \,-\, 4\} \,=\, 0$$

Consequently, the project must begin at time 0 in order to finish at time 16.

It follows directly from the latest event time algorithm that if the latest completion time of the entire project $L(n)$ is increased by t time units, then the latest time of every event will also be increased by t units.

The earliest time $E(x)$ of event x can be interpreted as the length of the longest path from the start event to event x. Similarly, $L(n) - L(x)$ can be interpreted as the length of the longest path from event x to the finish event. Lastly, observe that if $L(n) \geq E(n)$, then $L(x) \geq E(x)$ for all events x.

The earliest event time algorithm requires only one addition for each activity in the project and one maximization for each event in the project, except event 1. Similarly, the latest event time algorithm requires only one subtraction for each activity in the project and one minimization for each event in the project, except for event n.

Consider any activity $(x,\, y)$. What is the maximum amount of time that can be allotted to activity $(x,\, y)$ without delaying the on-time completion of the entire project? Activity $(x,\, y)$ may start as early as time $E(x)$ and may finish as late as time $L(y)$. Hence, at most $L(y) - E(x)$ time periods may be allotted to the performance of activity $(x,\, y)$ without delaying the on-time completion of the entire project. Consequently, the maximum delay that can be tolerated in activity $(x,\, y)$ is $L(y) - E(x) - t(x,\, y) \geq 0$. The quantity

$$L(y) \,-\, E(x) \,-\, t(x,\, y) \tag{3}$$

is called the *total float* of activity $(x,\, y)$. Obviously, if the total float of an activity equals zero, then any delay in the performance of this activity will delay the on-time completion of the entire project by an equal amount.

How much time can be allotted to the performance of activity $(x,\, y)$ without imposing any additional time constraints on the activities that are performed after $(x,\, y)$? In this case, activity $(x,\, y)$ must be completed by time $E(y)$. Since activity $(x,\, y)$ may begin as early as time $E(x)$, it follows that at most $E(y) - E(x)$ time periods may be allotted to the performance of activity $(x,\, y)$ without imposing any additional time constraints on the activities that follow $(x,\, y)$. The quantity

$$E(y) \,-\, E(x) \,-\, t(x,\, y) \tag{4}$$

is called the *free float* of activity (x, y). The free float of activity $(x,\, y)$ equals the maximum delay that can occur in the performance of activity $(x,\, y)$ without affecting any activity that follows (x, y). From equation (1), it follows that the free float is always nonnegative.

How much time can be allotted to the performance of activity (x, y) without imposing any additional time constraint on any other activity in the project? In order not to impose any additional requirements on any other activity in the project, activity (x, y) must begin as late as possible and be completed as easily as possible. Thus, activity (x, y) would have to begin at time $L(x)$ and end at time $E(y)$. Thus, at most $E(y) - L(x)$ time periods can be allotted to the performance of activity (x, y). The quantity

$$E(y) - L(x) - t(x, y) \tag{5}$$

is called the *independent float* of activity (x, y). The independent float of activity (x, y) can be interpreted as the maximum delay that can occur in the performance of activity (x, y) without imposing any additional time restriction on any other activity in the project. A negative value for an independent float indicates that any delay will affect the flexibility of other activities in the project.

How are the three kinds of floats related? Since $L(x) \geq E(x)$ for all events x, it follows from statements (3)–(5) that for each activity (x, y)

$$\text{Total float} \geq \text{free float} \geq \text{independent float} \tag{6}$$

The following table gives the three floats for each activity in the project network in Fig. 11.7.

Activity	Total Float	Free Float	Independent Float	
(1, 2)	$4 - 0 - 4 = 0$	$4 - 0 - 4 = 0$	$4 - 0 - 4 = 0$	(critical)
(1, 3)	$7 - 0 - 3 = 4$	$5 - 0 - 3 = 2$	$5 - 0 - 3 = 2$	
(1, 4)	$10 - 0 - 4 = 6$	$4 - 0 - 4 = 0$	$4 - 0 - 4 = 0$	
(2, 3)	$7 - 4 - 1 = 2$	$5 - 4 - 1 = 0$	$5 - 4 - 1 = 0$	
(2, 5)	$11 - 4 - 7 = 0$	$11 - 4 - 7 = 0$	$11 - 4 - 7 = 0$	(critical)
(2, 7)	$16 - 4 - 8 = 4$	$16 - 4 - 8 = 4$	$16 - 4 - 8 = 4$	
(3, 5)	$11 - 5 - 4 = 2$	$11 - 5 - 4 = 2$	$11 - 7 - 4 = 0$	
(4, 6)	$12 - 4 - 2 = 6$	$12 - 4 - 2 = 6$	$12 - 10 - 2 = 0$	
(5, 6)	$12 - 11 - 1 = 0$	$12 - 11 - 1 = 0$	$12 - 11 - 1 = 0$	(critical)
(5, 7)	$16 - 11 - 3 = 2$	$16 - 11 - 3 = 2$	$16 - 11 - 3 = 2$	
(6, 7)	$16 - 12 - 4 = 0$	$16 - 12 - 4 = 0$	$16 - 12 - 4 = 0$	(critical)

An activity is called *critical* if any delay in the performance of the activity delays the on-time completion of the entire project. In other words, a critical activity is any activity whose total float equals zero.

It is vital for the project manager to identify all critical activities in order to guard against any delays in these activities, as these delays would delay the on-time completion of the entire project. Delays less than the total float may occur in noncritical activities without delaying the on-time completion of the entire project.

Recall that $E(n)$ equals the length of a longest path from the start event to the finish event of the project. If $E(n) = L(n)$, then every activity on a longest path from the start event to the finish event is critical. A path consisting entirely of critical activities is called a *critical path*.

For example, in the project network shown in Fig. 11.7, activities (1, 2), (2, 5), (5, 6), and (6, 7) each have total float equal to zero. Thus, each of these activities is a critical activity. Notice that these activities form a path from 1 to 7 whose total length is 4 + 7 + 1 + 4 = 16. This is the longest path in the project network. Any delays along the activities of this path will result in an equal delay in the on-time completion of the entire project.

As a further example, consider the project network shown in Fig. 11.9. Notice that path (1, 2), (2, 5) and path (1, 3), (3, 5) each has total length equal to 8 times units. Thus, both of these paths are critical.

Activities (1, 4) and (4, 5) are not critical as limited delays in both of these activities can be tolerated without delaying the completion of the entire project by time 8. How much can activity (1, 4) be delayed without delaying the completion of the project beyond time period 8?

Up to now, we have assumed that $t(x, y)$ is known with certainty for all activities (x, y). Obviously, this is hardly realistic. Can we ever be certain how much time a machine shop job will require? Can we be certain that we shall complete a course in one term? Can we be certain how long the landscaping will require? Can we be certain how much cooking time a 5-lb roast will require?

To deal with the problem of uncertainty in activity times, a technique known as the *program evaluation review technique (PERT)* was developed. In essence, PERT is the same as the critical path method (CPM) described except that activity times are replaced

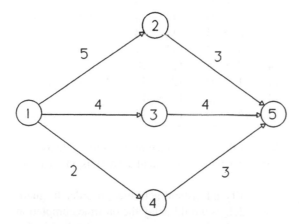

Fig. 11.9 Project with multiple critical paths.

with expected activity times. To calculate the expected time of an activity, PERT requires that three time estimates be made. They are

A, the *optimistic activity time*
B, the *realistic activity time*
C, the *pessimistic activity time*

The expected activity time is estimated to be

$$\frac{A}{6} + \frac{4}{6}B + \frac{C}{6} \tag{7}$$

In other words, a weighted average of the three time estimates is taken, where A and C have weight ⅙ and B has weight ⅘.

The variance of each activity time is taken as

$$\left(\frac{C - A}{6}\right)^2 \tag{8}$$

The critical path method can now be applied to the project network with activity times replaced by expected activity times as computed in expression (7). Now $E(x)$ denotes the *expected* earliest time of event x and $L(x)$ denotes the *expected* latest time of event x. Hence, $E(n)$ denotes the *expected* earliest completion time of the entire project.

The *actual* earliest completion time of the entire project is assumed to be a normally distributed random variable whose mean equals $E(n)$ and whose variance equals the sum of the variances of the activities in the longest path from event 1 to event n in the project network. (If there is more than one such path as in Fig. 11.9, then the largest variance of any such path is used.) This assumption about the mean, variance, and normal distribution of the actual completion time permits us to make probability statements about the actual completion time. See any introductory statistics text for the details of how to make probability statements about a normally distributed random variable.

The PERT user should bear in mind that this normal assumption is weak if the activity times tend not to be statistically independent of one another. Moreover, the theoretical justifications for expressions (7) and (8) rest on some tenuous connections between activity times and the beta distribution. Nonetheless, PERT has received widespread industrial acceptance.

11.4   GENERALIZED PROJECT NETWORKS

Up to now, we assumed that (a) all activities preceding an event must be completed before any activities emanating from the event could be performed and (b) all activities in the project must be performed.

Assumption (a) would be unnecessary when any one of several courses is the prerequisite for another course. Also, the arrival of any one of a number of checks would be sufficient for you to begin your shopping activity. Similarly, the success of any one of several grant proposals would suffice to finance a research project.

Assumption (b) would be unnecessary in a university program that allowed elective courses. Or, a milling job may have to undergo one, two, or three drillings depending on the result of quality control tests. The results of a market survey may determine which type of advertising policy should be pursued.

Thus, we can see that many projects cannot be realistically described in terms of the confines of the project networks that we have studied. For this reason, generalized project networks that avoid the above assumptions have been developed. A detailed description is available in Eisner (1962), Elmaghraby (1964), Pritsker and Happ (1966), Pritsker and Whitehouse (1966), and Pritsker (1977).

Unlike project networks, which have only one kind of vertex called an event, generalized project networks have a variety of vertices, all commonly called *decision boxes*. A decision box or db is characterized by the conditions placed on the activities entering it and by the condition placed on the activities emanating from it.

Three different conditions can be placed on the activities entering a db:

(a) "And input": All activities entering the db must be performed before the db is considered completed.
(b) "Inclusive input": At least one activity entering the db must be performed before the db is considered completed.
(c) "Exclusive input": Exactly one of the activities entering the db must be performed before the db is considered completed.

Two different conditions can be placed on the activities emanating from a db:

(a) "Deterministic output": All activities emanating from the db are to be performed once the db has been completed.
(b) "Probabilistic output": Exactly one of the activities emanating from the db is performed after the db has been completed.

Consequently, there are $3 \times 2 = 6$ different types of db's. Their pictorial representations are given in Fig. 11.10.

In a project network, a time $t(x, y)$ was specified for each activity (x, y). In a generalized activity network, both a time $t(x, y)$ and a probability $p(x, y)$ must be specified for each activity (x, y). Probability $p(x, y)$ denotes the chance that activity (x, y) will actually be performed once db x has been reached. If db x has a deterministic output, then $p(x, y)$ must equal one and activity (x, y) is certainly performed. Moreover, the sum of the probabilities of the activities emanating from a probabilistic output db cannot exceed one.

EXAMPLE 4

Consider the project whose generalized project network is shown in Fig. 11.11. Decision box 1 has a deterministic output; hence, both activities (1, 2) and (1, 3) will be per-

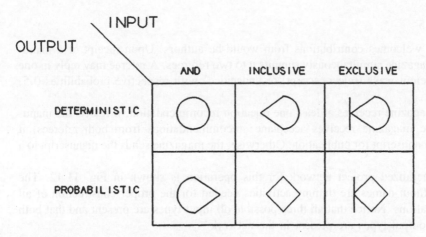

Fig. 11.10 Six types of decision boxes.

formed. Thus, $p(1, 2) = 1$ and $p(1, 3) = 1$. Decision box 2 has an "and" input and is completed as soon as activity $(1, 2)$ has been performed. After db 2 has been completed, activity $(2, 5)$ will be performed with 60% probability and activity $(2, 4)$ will be performed with 40% probability. Only one of the two activities $(2, 4)$ and $(2, 5)$ will occur. Decision box 4 is reached only if activity $(2, 4)$ or activity $(3, 4)$ is completed but not if both are completed. If neither $(2, 4)$ nor $(3, 4)$ is performed, decision box 4 is never reached.

Moreover, it is possible that both activities $(2, 5)$ and $(3, 5)$ will be performed. In this case, db 5 is never reached.

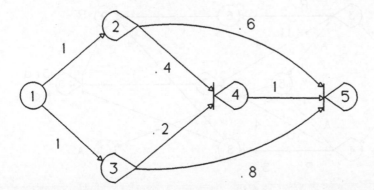

Fig. 11.11 Generalized project network. Activity numbers are probabilities.

EXAMPLE 5

A magazine welcomes contributions from would-be authors. Upon receipt of a manuscript, the magazine simultaneously submits it to two referees. A referee may reply in one of three different ways (reject, accept, or undecided) with respective probabilities 0.5, 0.4, and 0.1.

If the magazine receives at least one rejection recommendation, it rejects the manuscript. If the magazine receives acceptance recommendations from both referees, it accepts the manuscript for publication. Otherwise, the magazine sends the manuscript to a third referee.

The generalized project network for this operation is shown in Fig. 11.12. The activities without names are dummy activities needed for the proper construction of all possible situations. Notice that all three possible db input types are present and that both possible db output types are present in the network.

Next, let us alter the situation. Suppose, instead, that the editor does not send the manuscript to the second referee unless the first referee returns an acceptance or undecided verdict. For this situation, the generalized project network is shown in Fig. 11.13.

In a project network, all events are eventually reached; it is merely a matter of time. The same is not necessarily true for a generalized project network. Since not all activities need be performed, not all db's need be reached. For example, in Fig. 11.12, the project will eventually terminate at either db 9, 10, or 11.

Also, it can happen that the project will terminate, not with a db, but with an activity. For example, in Fig. 11.11, if both activities (2, 4) and (3, 4) are performed, the project terminates without reaching db 4. However, this can occur only if exclusive input db's are present in the network.

Before proceeding into further analysis of generalized activity networks, let us note that it is often possible to simplify a network into an equivalent network with fewer

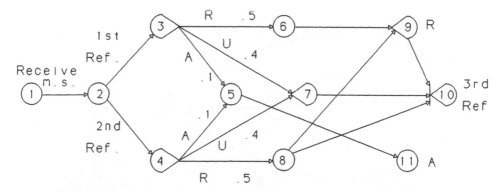

Fig. 11.12 Generalized project network for simultaneous referees. R = reject, A = accept, U = undecided.

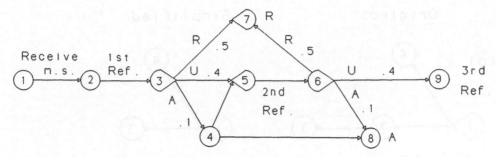

Fig. 11.13 Generalized project network for sequential referees. R = reject, A = accept, U = undecided.

activities. Such simplifications are possible for activities that occur in series or parallel combinations. These simplifications are shown in Fig. 11.14.

Because of the variety of possibilities for termination of a generalized project network, the project manager understandably wishes to know the probability that any db will in fact be reached and that any particular activity will in fact be performed. Moreover, it is important to know the expected time at which a db will be reached (supposing it is reached at all). (Note that in PERT, the actual time required to perform an activity was random, but we were certain that the activity would sooner or later be performed. In generalized activity networks the opposite case occurs: the time required to perform an activity is assumed to be nonrandom, but the activities that are performed are randomly selected.)

These probabilities and expected times are not easy to calculate. The difficulties in these calculations arise from the presence of probabilistic output db's. Only one of the set of activities emanating from a probabilistic output db may occur. Hence, the probabilities of actually completing activities and db's following a probabilistic output db are not statistically independent, and consequently we cannot use the convenient traditional probability rules, which assume statistical independence.

For example, in Fig. 11.15, both activities (2, 4) and (3, 4) must be performed before db 4 is reached. There is a $0.6 \times 0.5 = 0.3$ probability that activity (2, 4) is performed. There is a $0.4 \times 0.4 = 0.16$ probability that activity (3, 4) is performed. Hence, we might hastily conclude that there is a 0.3×0.16 probability that db 4 is reached. However, upon closer inspection, we see that activities (2, 4) and (3, 4) are not statistically independent of one another. Both (2, 4) and (3, 4) cannot occur, since otherwise both (1, 2) and (1, 3) must occur, which is impossible. (Recall that only one activity emanating from a probabilistic output db may occur.) Thus, db 4 can never be reached.

Thus, the lack of statistical independence between activity probabilities leaves large networks virtually intractable. Even if the definition of the probabilistic output db were changed so that any number of activities could emanate from it (each with its own

Original   Simplified

(a)

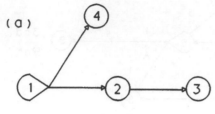
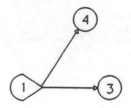

$p(1, 3) = p(1, 2)p(2, 3)$
$\qquad = p(1, 2) \times 1 = p(1, 2)$
$t(1, 3) = t(1, 2) + t(2, 3)$

(b)

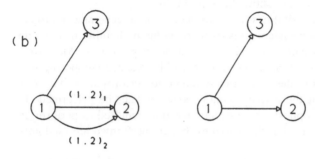

$p(1, 2) = 1$
$t(1, 2) = \max\{t(1, 2)_1, t(1, 2)_2\}$

(c)

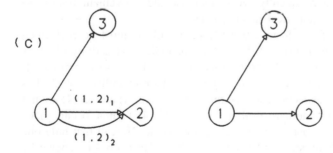

$p(1, 2) = 1$
$t(1, 2) = \min\{t(1, 2)_1, t(1, 2)_2\}$

Fig. 11.14 Network simplifications.

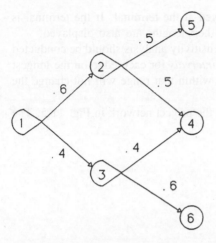

Fig. 11.15 Statistical dependence.

probability of occurrence), the same computational difficulties would arise. For example, several activities may all be descendents of one particular activity emanating from a probabilistic output db. Then the occurrence of all these descendent activities would be statistically dependent on one another, and we would encounter the same computational difficulties as shown above.

As if the computational situation is not bad enough, many projects generate generalized project networks that contain cycles. For example, there might be an activity in the project that must be repeated until it is performed correctly, such as a required course in a university program. The presence of cycles complicates the calculations even further. Some methods for simplifying cycles into equivalent cycleless configurations can be found in Elmaghraby (1964), Pritsker and Happ (1966), and Pritsker and Whitehouse (1966).

APPENDIX: USING NETSOLVE TO FIND LONGEST PATHS

The longest-path algorithm finds a path of longest length between a given pair of nodes in a directed acyclic network. The "length" of an edge is the value specified by the COST data field, and the length of a path is the sum of the lengths of the edges on the path. Although the input graph should be acyclic, the existence of a directed cycle will be detected by the algorithm.

For example, to determine a longest-length path from the node named ATL to the node named PHX, the command line

    LPATH ATL PHX

is entered. The length of a longest path is displayed at the terminal. If the terminal is currently ON, the edges on the longest path and their lengths are also displayed.

In addition, the user will be asked whether a sensitivity analysis should be conducted on the problem. This analysis produces *tolerance intervals* for each edge on the longest path, such that varying the individual edge length within that range will not change the longest path.

We will illustrate the use of this algorithm for the project network in Fig. 11.5. The following table gives the time for each activity.

Activity	Time
A	3
B	5
C	2
D	6
E	5
F	3
G	4
H	3
I	4
J	2
K	2

Figure 11.16 shows the edge data for the network created with NETSOLVE. Figure 11.17 shows the solution to the longest-path problem. The critical path has length 22 and consists of activities B, C, E, F, H, J, K, and the dummy activity.

FROM	TO	COST	LOWER	UPPER
1	2	3.00	0.00	999999.00
1	3	5.00	0.00	999999.00
2	3	0.00	0.00	999999.00
3	4	2.00	0.00	999999.00
4	5	5.00	0.00	999999.00
4	6	6.00	0.00	999999.00
4	7	4.00	0.00	999999.00
5	6	3.00	0.00	999999.00
6	7	0.00	0.00	999999.00
6	9	4.00	0.00	999999.00
7	8	3.00	0.00	999999.00
8	9	2.00	0.00	999999.00
9	10	2.00	0.00	999999.00

Fig. 11.16 NETSOLVE data for longest path problem.

```
LONGEST PATH LENGTH FROM 1          TO 10        :       22.00

   EDGES IN THE LONGEST PATH
       FROM       TO              COST
       ----       --          --------
       1          3              5.00
       3          4              2.00
       4          5              5.00
       5          6              3.00
       6          7              0.00
       7          8              3.00
       8          9              2.00
       9          10             2.00


SENSITIVITY ANALYSIS FOR EDGE COSTS

                                EDGE             TOLERANCE
       FROM       TO            COST        LOWER          UPPER
       ----       --            ----        -----          -----
       1          3             5.00         3.00      999999.00
       3          4             2.00   -999999.00      999999.00
       4          5             5.00         3.00      999999.00
       5          6             3.00         1.00      999999.00
       6          7             0.00        -1.00      999999.00
       7          8             3.00         2.00      999999.00
       8          9             2.00         1.00      999999.00
       9          10            2.00   -999999.00      999999.00
```

Fig. 11.17 NETSOLVE solution to longest path problem.

EXERCISES

1. Construct the project network for the following set of activities.

Activity	Immediate Predecessors
A	None
B	None
C	A
D	A, B
E	A, B
F	C
G	D, F
H	E, G

2. Construct a project network for the following project.

Activity	Immediate Predecessors
A	None
B	None
C	A
D	A
E	C, B
F	C, B
G	D, E

3. A project consists of seven activities called A through G. The precedence relationships between these activities are

Activity	Immediate Predecessors
A	None
B	A
C	None
D	B
E	A, C
F	A, D
G	E, D

Construct the network corresponding to this project.

4. Given the following project network, find the critical path.

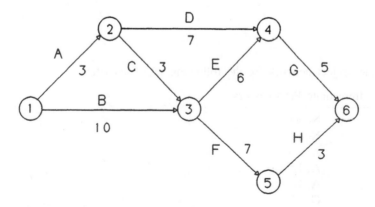

5. For the project network in Fig. 11.18:
   (a) Calculate the earliest event times.

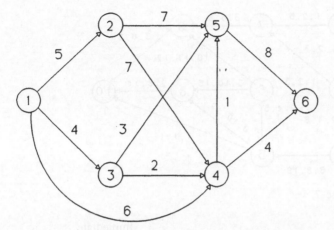

Fig. 11.18 Network for Exercise 5. Numbers indicate activity times.

(b) Calculate the latest event times so that the project finishes as early as possible.
(c) Find the critical path.
(d) Calculate the total float of each activity.
(e) Calculate the free float of each activity.
(f) Calculate the independent float of each activity.

6. You discover that the time given for activity (3, 5) in Fig. 11.18 should have been 5. Update the results of Exercise 5 without repeating all calculations. How much can the performance time of activity (3, 5) increase without altering the earliest completion time of this project?

7. In Fig. 11.19 the three numbers next to each activity are, respectively, the activity's optimistic time, realistic time, and pessimistic time. Use PERT to calculate
(a) The expected time for each activity
(b) The variance of each activity time
(c) The earliest expected event times
(d) The latest expected event times so that the project finishes at the earliest expected time
(e) The critical path
(f) The variance of the earliest completion time of the entire project

8. A manufacturing company is planning to install a new flexible manufacturing system. The activities that must be performed, along with their immediate predecessors and estimated activity times, are shown below.

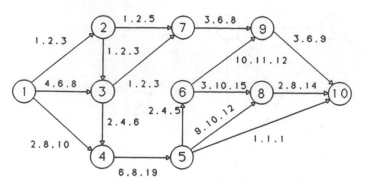

Fig. 11.19 Exercise 7.

Activity	Description	Immediate Predecessors	Time
A	Analyze current performance	None	3
B	Identify goals	A	1
C	Conduct study of existing system	A	7
D	Define new system capability	B	6
E	Study existing technologies	None	2
F	Determine specifications	D	9
G	Conduct equipment analyses	C, F	13
H	Identify implementation activities	C	3
I	Determine organizational impacts	H	4
J	Prepare report	E, G, I	2
K	Establish audit procedure	H	2

Draw the project network and find the critical path.

9. Consider the following project network:

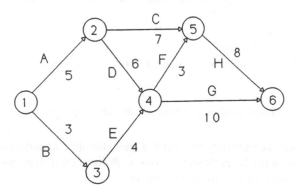

(a) Identify the critical path.

(b) Can activity D be delayed without delaying the entire project? If so, how much?

(c) Can activity C be delayed without delaying the entire project? If so, how much?

(d) What are the starting and finish times for activity E?

10. Show that the earliest event time algorithm and the latest event time algorithm both fail when the network contains a directed cycle.

11. Construct a project network in which some activity has an independent float that has a negative value. Interpret this value.

12. A drilling machine must drill two holes into an automobile part. If both holes are of the standard size, the part is considered finished. If the first hole drilled is below standard size, the second hole is drilled with extreme care on a special setting of the drilling machine. If the first hole is above standard size, the driller must check whether any of the pistons in stock fit this hole. If so, the second hole is drilled at the usual setting. If not, there is a 60% chance that the quality control engineer will decide to discard the part; there is a 40% chance that the quality control engineer will decide to have the second hole drilled on the special setting. Construct a generalized project network for this operation.

REFERENCES

Eisner, H., 1962. A Generalized Network Approach to the Planning and Scheduling of a Research Project, *Oper. Res., 10*, pp. 115–125.

Elmaghraby, S., 1964. An Algebra for the Analysis of Generalized Activity Networks, *Manage. Sci., 10*(3), pp. 494–514.

Evans, J. R., D. R. Anderson, D. J. Sweeney, and T. A. Williams, 1990. *Applied Production and Operations Management*, 3rd ed., West, St. Paul.

Ford, L. R., and D. R. Fulkerson, 1962. *Flows in Networks*, Princeton University Press, Princeton, pp. 151–161.

Kelley, J. E., Jr., 1969. Critical-Path Planning and Scheduling: Mathematical Basis, *Oper. Res., 9*, pp. 296–320.

Moder, J., and C. Phillips, 1970. *Project Management with CPM and PERT*, 2nd ed., Van Nostrand Reinhold, New York, (This is an excellent comprehensive introductory treatment of project networks.)

Pritsker, A. B., 1977. *Modeling and Analysis Using Q-GERT Networks*, Halsted Press, New York.

Pritsker, A. B., and W. W. Happ, 1966. GERT: Graphical Evaluation Review Technique; Part I. Fundamentals, *J. Ind. Eng., 17*, pp. 267–274.

Pritsker, A. B., and G. Whitehouse, 1966. GERT: Graphical Evaluation Review Technique; Part II. Probabilistic and Industrial Engineering Applications, *J. Ind. Eng., 17*, pp. 293–301.

USER'S GUIDE

to

NETSOLVE

Version 1.3.1

Interactive Software for Network Analysis
for the IBM PC and Compatibles

James P. Jarvis
Douglas R. Shier
Clemson University
Clemson, South Carolina

to accompany the text
Optimization Algorithms for Networks and Graphs,
Second Edition, Revised and Expanded
by J. Evans and E. Minieka
© 1992 by Marcel Dekker, Inc.

Published and distributed by: Upstate Resources, Inc.
                                 PO Box 152
                                 Six Mile, SC 29682

Getting Started with NETSOLVE

Requirements:

Use of NETSOLVE requires an IBM PC or compatible running under PC-DOS or MS-DOS (version 2.1 or later) with a minimum 512K of RAM. A numeric coprocessor is not required, but is utilized if present.

Installation:

First, please complete and return the registration card in order to be informed of future updates. Second, make a copy of the enclosed diskette and use the copy for installation. NETSOLVE is not copy protected as a convenience to the user. Please do not give copies away. The only file required for using the software is NETSOLVE.EXE. Installation on a hard disk is accomplished by copying that file to the hard disk. NETSOLVE can also be run directly from the copy of the enclosed diskette.

Starting NETSOLVE:

Execution of NETSOLVE is initiated by entering NETSOLVE from the DOS prompt (or A:NETSOLVE if the diskette is in drive A). After a greeting message, NETSOLVE will then display its command prompt, the character ">". The command prompt indicates that NETSOLVE is ready to accept another command. The commands likely to be of most interest to the first-time user are HELP, COMMANDS, and QUIT. The command HELP introduces the built-in help system; COMMANDS produces a list of all commands grouped by function; and QUIT exits the software and returns to DOS. In the enclosed version, networks are limited to a maximum of 50 nodes and 200 edges. (An extended version is also available as described below.)

Sample files:

The NETSOLVE diskette includes two sample networks, NET1 and NET2, which are referenced in the User's Guide. The networks can be loaded by entering GET NET1 (or GET NET2) when NETSOLVE displays its command prompt. Also included are a README file similar to this page and the PIPELINE network used in the appendix to Chapter 2 of the text.

License:

NETSOLVE is copyright © 1985–1992 by J.P. Jarvis and D.R. Shier. The copy of NETSOLVE on the enclosed diskette is licensed for use by a single user in conjunction with the purchase of *Optimization Algorithms for Networks and Graphs, Second Edition, Revised and Expanded,* by J. Evans and E. Minieka, © 1992 by Marcel Dekker, Inc. Additional copies and site licenses can be purchased by contacting:

> Upstate Resources, Inc.
> P.O. Box 152
> Six Mile, SC 29682

TABLE OF CONTENTS

Getting Started with NETSOLVE 421

1.0 Overview 423

2.0 Introducing the NETSOLVE System 424
    2.1 Networks 424
    2.2 The NETSOLVE Command Language 424

3.0 Creating Networks 427

4.0 Listing Information 433
    4.1 Information Commands 433
    4.2 Listing Commands 434
    4.3 Controlling Output 437

5.0 Modifying Networks 440

6.0 Network Algorithms 442
    6.1 Assignment 442
    6.2 Longest Path 444
    6.3 Weighted Matching 444
    6.4 Maximum Flow 445
    6.5 Minimum Cost Flow 446
    6.6 Minimum Spanning Tree 446
    6.7 Shortest Path 447
    6.8 Transportation 448
    6.9 Traveling Salesman 449

7.0 External Data Files 451

8.0 Advanced Features 454
    8.1 Defaults 454
    8.2 Representation Types 456
    8.3 Multiple Journals 458

Appendix A. Commands and Command Syntax 459
    A.1 Commands 459
    A.2 Command Syntax 459

Appendix B. NETSOLVE Keyword Aliases 463

1.0 Overview

Network models provide an extremely useful representation for describing a wide variety of interconnected systems. In view of the applicability of network models to diverse situations (transportation planning, project scheduling, distribution system design), it is not surprising that a number of techniques have been developed to analyze networks. Currently available software for carrying out network analysis tends, however, to be fairly specialized and often requires a substantial effort by the novice user to gain familiarity with the conventions (and idiosyncrasies) of the particular computer code. Moreover, there are rarely facilities available within an optimization code for modifying or correcting network data, so that conducting sensitivity analyses or exploring "what if" scenarios is fairly tedious.

By contrast, the NETSOLVE system has been designed to be

- *Interactive*
- *Integrated*
- *Flexible*
- *Easy-to-use*

In effect, NETSOLVE provides a "workbench" upon which a network can be assembled, validated, modified, and analyzed using various editing and optimization tools. Because the system is an *interactive* one, the user is given immediate feedback and can thus work more effectively. The *integrated* nature of the system means that the user who learns the commands once has immediate access to a number of editing and optimization tools. Several other features enhance the *flexibility* and *user-friendliness* of the system. For example, networks can be entered in a "free format" manner, descriptive names can be used (CHICAGO versus node 1072), an on-line library of "help" documents is available, and output can be selectively routed to a number of external data files for subsequent printing.

The purpose of this manual is to provide a step-by-step introduction to the NETSOLVE system and its many capabilities. Needed terminology and concepts are first described in Section 2. Subsequent sections give a guided tour through the various facilities available in NETSOLVE. In addition, several appendices provide the "seasoned" user with easily accessed reference material.

2.0 Introducing the NETSOLVE System

In this section we introduce some basic network terminology and discuss various conventions used in NETSOLVE for representing networks. The command language structure used by the NETSOLVE system to perform various tasks is described as well.

2.1 Networks

Because the NETSOLVE system is used to construct, modify, and analyze networks, we first remind the user of some basic network terminology. A *network* consists of a number of *nodes* (or *vertices*) joined by *edges* (or *arcs*). Various data items may in addition be associated with the nodes and edges. In *directed* networks, the edge (a,b) is considered to be oriented from a to b, while in *undirected* networks there is no implied orientation to the edge. Moderately large networks (with up to 350 nodes and 1000 edges) that are either directed or undirected can be handled by the professional version of NETSOLVE. (The educational version allows networks with up to 50 nodes and 200 edges.)

Each network in the NETSOLVE system is assigned a descriptive *name* consisting of up to 8 alphanumeric characters in length; an underscore (_) is also considered to be a valid character for a name. This naming convention also applies to the names of nodes in the network. Thus, the network named RAILNET might include nodes with the names DETROIT, NYC_2, 108, and so forth. Edges are represented by pairs of node names indicating the two endpoints and separated by *delimiters* (blanks or commas). For example, either of

```
CHICAGO DETROIT
CHICAGO,DETROIT
```

specifies the directed edge from CHICAGO to DETROIT (if RAILNET is directed) or the undirected edge joining CHICAGO and DETROIT (if RAILNET is undirected).

More generally, each node or edge in the network has a number of *data fields* associated with it. Nodes have the associated data fields NAME (required) and SUPPLY (optional). The supply field is used in the execution of certain network algorithms (assignment, matching, minimum cost flow, transportation). Edges have the required data fields FROM, TO (indicating the endpoints of the edge) and the optional data fields LOWER (lower bound), UPPER (upper bound), and COST. Again, these optional fields are needed to store input values for certain of the algorithms described in Section 6.

2.2 The NETSOLVE Command Language

The NETSOLVE system utilizes a command language, in contrast to a menu-driven interface. That is, the user is not constantly confronted with a series of yes/no questions or prompted to select a series of actions. Rather, a single command line, expressed in a fairly natural and self-evident way, is entered in order to carry out the desired series of actions. In this way, the user who has gained some familiarity with the command language of NETSOLVE is able to operate with a minimum of interference from the package. However, when some information is needed concerning the use of NETSOLVE, help is but a few keystrokes away. For example, by entering the word COMMANDS, followed by a carriage return, you can obtain a listing of all commands available in NETSOLVE. (Try it!)

```
NETSOLVE COMMANDS:

        EDITING:    ADD        APPEND      CHANGE      CREATE
                    DELETE     LIST

         SYSTEM:    CLEAR      DECIMALS    DEFAULT     EFORMAT
                    NFORMAT    PATH        QUIT

    INFORMATION:    ALIAS      COMMANDS    HELP        STATUS

   INPUT/OUTPUT:    GET        JOURNAL     MERGE       NOTE
                    SAVE       TERMINAL

     ALGORITHMS:    ASSIGN     LPATH       MATCH       MAXFLOW
                    MINFLOW    MST         SPATH       TRANS
                    TSP
```

Sample output from COMMANDS.

We will be taking a closer look at each of these commands in subsequent sections. By way of introduction, suppose we would like to delete the node named ALBANY (along with its incident edges) from the current network. Then the command line

    DELETE NODE ALBANY

(again followed by a carriage return) is entered. At this point, it is worth emphasizing that

• Any command line entered applies to the network currently residing on the "workbench". (Section 7 discusses how to transfer networks to and from the workbench.)

• In order to enter a command, the typed line must be followed by a carriage return. (We won't bother to remind you again about this!)

In addition, many of the command names (e.g. DELETE) and keywords (e.g. NODE) have aliases or abbreviations for greater convenience. The previous action can therefore be carried out using either of the alternative forms:

    DEL VERTEX ALBANY
    DEL N ALBANY

Some of the important aliases in this package reflect the diversity of existing nomenclature for networks. Thus, the terms *network* and *graph* are synonymous, as are *node/vertex* and *edge/arc*. A complete listing of all aliases can be obtained by simply typing

        ALIAS   .

   The previous example, DELETE NODE ALBANY, also indicates that the command name (DELETE), keyword (NODE), and the user-supplied input (ALBANY) should be separated by at least one blank. Commas can likewise be used to serve the same purpose. We reiterate:

- A command line consists of a command name and possibly several keywords and user-supplied inputs.

- Aliases for command names or keywords can be freely used in place of their (longer) counterparts.

- Delimiters (blanks and commas) are used to separate portions of the command line.

   One final note: in order to terminate a NETSOLVE session, simply use the command line

        QUIT   .

Be warned, however, that unless you "save" the current network (see Section 7), this network will be lost when you exit from NETSOLVE. As a precaution, the system provides the user with one additional opportunity to save the current network prior to terminating the session.

3.0 Creating Networks

The first step in using NETSOLVE is, of course, entering a network into the system. While there are several ways to do this, the most straightforward method is through the use of the CREATE command, which we now explore. (Other methods will be indicated later when discussing the ADD and GET commands.)

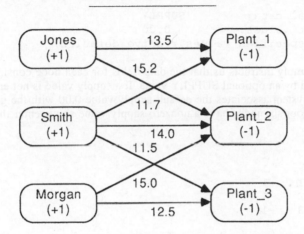

Figure 1. Example network composed of 6 nodes and 7 edges.

Consider, for example, the directed network in Figure 1, composed of 6 nodes (with associated supplies) and 7 edges (with associated costs). Such a network might be useful in assigning certain managers to plants in an optimal fashion. The simplest way to enter the network is to issue the command

    CREATE .

The system then asks for the name of the network to be created, which we provide by entering JOBNET. This exchange is summarized by the following transaction line.

    Enter name of network: JOBNET

In a similar way, the system requests the network orientation, and we respond with the designation D for directed.

    Directed or undirected? (D/U): D

Because our network has supply values associated with the nodes, we answer "yes" to the next prompt

```
Any node data? (Y/N): Y
```

after which the system issues the reminder

```
Enter node data in order: NAME, SUPPLY.  (Null line to end)

Required data:          NAME

Optional data:          SUPPLY
Default values:         0.00
   (Unspecified values receive defaults)
```

This message simply instructs us that the data fields for each node consist of a required NAME followed by an optional SUPPLY value. If a supply value is not entered for some node, then the system associates the default supply value 0.00 with the given node. In this case, all of our six nodes have a nonzero supply value so we enter the node data:

```
JONES       1
SMITH       1
MORGAN      1
PLANT_1    -1
PLANT_2    -1
PLANT_3    -1
```

The node data above can be entered in a "free format" manner: namely, the data items need not be aligned in particular columns but need only be separated by delimiters (blanks or commas). Furthermore, the supply values could equally have been entered with decimal points (such as 1.0 or -1.) if desired. Each line of node data is terminated by entering a carriage return. In order to signify that we have finished entering the node data for this first phase, a null line (another carriage return) is entered.

At this point, we are then asked if there will be any edge data entered, to which we reply "yes":

```
Any edge data? (Y/N): Y
```

The screen then contains the message:

```
Enter edge data in order: FROM, TO, COST, LOWER, UPPER.

Required data:          FROM            TO

Optional data:          COST          LOWER         UPPER
Default values:         1.00           0.00      999999.00
   (Unspecified values receive defaults)
```

This serves as a reminder of the order in which we should enter the edge data for our problem (FROM, TO, COST). Since there are no LOWER or UPPER edge data for this network, nothing need be entered for those fields. Therefore we enter the edge data:

```
JONES     PLANT_1   13.5
JONES     PLANT_2   15.2
SMITH     PLANT_2   14.0
SMITH     PLANT_1   11.7
MORGAN    PLANT_2   15
MORGAN    PLANT_3   12.5
SMITH     PLANT_3   11.5
```

Again note that the edge data can be entered using a free format, with items on a line separated by delimiters. Edge costs can be written either with or without decimal points. If some of edge costs had been equal to 1, we could have entered simply the FROM and TO fields, since the unspecified COST field would then be assigned the default value 1. Moreover, there is no particular order in which the various edges need to be entered. Entering an edge with an endpoint not specified in the first phase (node data) automatically results in the addition of that node to the network. To signify the end of this second phase (entering edge data), a null line is entered and the network has been successfully created. The system responds with the *prompt character* ">" to indicate that NETSOLVE is ready to accept the next command.

To allay any lingering doubts, we can check that the network has really been created by issuing the command line

```
STATUS NETWORK
```

and seeing that in fact a directed network called JOBNET now exists with 6 nodes and 7 edges. Later sections will show how to list portions of this network, correct any input errors, and run various algorithms on this network.

```
CURRENT NETWORK
        NAME           JOBNET
        ORIENTATION    DIRECTED
        NODES          6
        EDGES          7
```

Sample output from STATUS NETWORK.

The above procedure has led us through the "interactive" form of the CREATE command, initiated by issuing the simple command line CREATE. Once the user has become more familiar with the NETSOLVE interface, the number of interactions necessary to construct a network can be reduced by using the "extended" form of the CREATE command. In this example, this command line becomes

```
CREATE JOBNET DIR NF=(NAME SUPPLY) EF=(FROM TO COST)
```

This single line instructs the system that we wish to begin constructing a directed network named JOBNET. The NF keyword signifies the node format to be assumed for the node data fields. In this example, the node data fields are NAME and SUPPLY and they will be entered in precisely that order. Similarly, the EF (edge format) specification means that the edge data fields are to be entered in the order FROM, TO, COST. As before, the network data are entered using two phases (node data followed by edge data) exactly as described earlier.

The specific command line we have entered:

```
CREATE JOBNET DIR NF=(NAME SUPPLY) EF=(FROM TO COST)
```

represents only one of several forms that are useful in creating networks. To enter an undirected network, the keyword UNDIR is specified instead of DIR. In this case each undirected edge is entered once, not in both directions. The node and edge formats can also be changed to accommodate the preferences of the user. For instance, if it were more convenient to enter the edge data in the form COST, FROM, TO we would simply use the specification EF=(COST FROM TO) in the command line. Similarly, the node data fields could be permuted as in the specification NF=(SUPPLY NAME) or, if the given network does not require supplies, the specification NF=(NAME) can be used.

In order to check on the specific actions and syntax of any command, we can make use of the on-line library of help documents resident in NETSOLVE. Specifically, by typing

```
HELP CREATE
```

we obtain a description of what the CREATE command accomplishes, its aliases, the syntax of the command, and some examples of valid command lines. (Now is a good time to try this!)

The description of the syntax for this command says, in effect, that a valid CREATE command line may consist of simply the command name CREATE, which signifies that the user will be prompted using the interactive mode for creating a network. Alternatively, the extended mode of network creation can be invoked by issuing the command name CREATE, followed by a network name, an orientation (DIR or UNDIR), and the node/edge formats. If the node format is not specified, the form NF=(NAME) is assumed; if the edge format is not specified, the form EF=(FROM TO) is assumed. In this way, the command line

```
CREATE HIGHWAY UNDIR
```

defines an undirected network called HIGHWAY, with the assumed node format of NAME and the assumed edge format of FROM, TO.

CREATE

Used to enter a new network from the terminal. In an undirected
network, each edge need only be entered once. When using the
interactive mode, the user will be prompted for the network name
and orientation. In the extended command line mode, the user can
specify certain node and edge formats which give the order in
which node and edge data will be entered.

ALIAS: CR

SYNTAX:
  CREATE
  CREATE "NETNAME" "TYPE" <NF=(<"NFMT">*)> <EF=(<"EFMT">*)>

    "NETNAME" : Name of new network (at most 8 characters)
    "TYPE"   : Selected from DIR,UNDIR
    "NFMT"   : Selected from NAME,SUPPLY (must include NAME)
    "EFMT"   : Selected from COST,FROM,LOWER,TO,UPPER (must
              include FROM,TO)

EXAMPLES: CREATE
  CREATE RAILNET DIR
  CREATE RAILNET UNDIR EF=(FROM TO COST)

Sample output from HELP CREATE.

It is important to note that whenever the node format consists only of NAME, then
it is not necessary to enter during the first phase of CREATE the names of nodes that will
already be parts of edges entered during the second phase. Thus, the first input to the
"Any node data?" prompt can be a "no", thereby ending the first phase; any node not
already in the network will automatically be placed in the network when an edge incident
with that node is processed during the second phase. All of this is to say that normally
one need not enter network nodes at all, unless (1) there are supply values to be input, or
(2) there are isolated nodes in the network to be input.

In summary, when the CREATE command is used to enter a network from the
terminal, the following conventions should be respected:

• Data for a particular node or edge is entered by using a carriage return at the
  end of the line.

• Each phase of data entry is terminated by using a null line.

In addition, if the extended mode is used for the CREATE command line, then

- The command line must consist of the command name CREATE, followed by the network name, the orientation (DIR, UNDIR), and possible node/edge formats.

- The node format, if specified, must contain NAME and may contain SUPPLY. Any order of data fields is acceptable.

- The edge format, if specified, must contain FROM, TO and may contain any of LOWER, UPPER, COST. Any order of data fields is acceptable.

- If not specified, the node format becomes NAME and the edge format becomes FROM, TO.

4.0 Listing Information

Now that we have expended some effort entering a network, it seems only fair that NETSOLVE should reciprocate by producing some output of its own. In this section, we will show how to obtain information about the NETSOLVE system and about the particular network resident on the workbench.

4.1 Information Commands

There are four commands that provide the user information about the NETSOLVE system:

- COMMANDS
- ALIAS
- HELP
- STATUS

We have already observed that entering COMMANDS generates a listing of the available commands, while entering ALIAS generates a listing of aliases (or abbreviations) for various commands and keywords in NETSOLVE.

The HELP command can be used to obtain more detailed information about the function and usage of any particular command. As we have seen in Section 3, typing

```
HELP CREATE
```

produces a description, a list of aliases, the syntax, and examples of the CREATE command. The syntax given for the extended form of CREATE is

```
CREATE "NETNAME" "TYPE" <NF=(<"NFMT">*)> <EF=(<"EFMT">*)>
```

In this and other syntax descriptions, any user-supplied inputs are enclosed in quotes. For example, in a previous illustration we used JOBNET for "NETNAME" and DIR for "TYPE". Any expression enclosed in pointed brackets <...> is optional: it need not be supplied. In this example, the node format and edge format specifications are both enclosed in pointed brackets, so either or both may be omitted from the command line. Finally, any expression enclosed within <...>* can be repeated any number of times, including none. This is just a concise way of stating, in the present instance, that any number of edge data fields can comprise the edge format. A summary of these conventions for describing the syntax of command lines can be obtained by simply typing HELP from the terminal. The collection of on-line help documents thus allows the novice user to learn rapidly about the system capabilities, without interfering with the experienced user (who only on occasion needs to recheck the function and syntax of commands).

```
GENERAL INFORMATION

The following commands provide information about NETSOLVE:

   ALIAS           : Lists aliases for NETSOLVE keywords
   COMMANDS        : Lists the available commands
   HELP "COMMAND"  : Provides information on specific commands
   STATUS          : Lists NETSOLVE system parameters

All descriptions of command syntax use
      " ... "     to denote a user-supplied expression
      < ... >     to denote an optional expression
      < ... >*    when an expression can be repeated

A NETSOLVE session is ended by the QUIT command.
```

Sample output from HELP.

The STATUS command will be discussed at the end of Section 4.3, after certain other commands have been illustrated.

4.2 Listing Commands

The LIST command is used to list node and edge information about the entire network or portions of the network. Let us return to JOBNET, created in Section 3, and list out various items about this network. The command line

```
LIST NODES
```

produces a listing of the names and supplies of all 6 nodes. The command lines

```
EF FROM TO COST
LIST EDGES
```

produce a listing of only the FROM, TO, and COST data fields for all 7 edges. The use of the edge format command EF to insure that only certain relevant fields are listed will be discussed later in this section. Such listing facilities allow the user to detect data entry errors present in the network. The editing commands discussed in Section 5 will be useful in actually correcting such data errors.

```
NAME              SUPPLY
----              ------
JONES               1.00
MORGAN              1.00
PLANT_1            -1.00
PLANT_2            -1.00
PLANT_3            -1.00
SMITH               1.00

FROM       TO                 COST
----       --                 -----
JONES      PLANT_1           13.50
JONES      PLANT_2           15.20
MORGAN     PLANT_2           15.00
MORGAN     PLANT_3           12.50
SMITH      PLANT_1           11.70
SMITH      PLANT_2           14.00
SMITH      PLANT_3           11.50
```

Sample output from LIST NODES and LIST EDGES.

When the network is reasonably large, it is clearly desirable to list only selected portions of the network. This can be accomplished by using certain additional qualifiers on the LIST command. For example, entering

```
LIST NODE JONES
```

will simply list out the various data fields associated with that node, while entering

```
LIST EDGE JONES PLANT_2
```

will list out the various data fields for that edge. Entering

```
LIST FROM JONES
```
or
```
LIST TO PLANT_2
```

will list the data fields for all edges leaving JONES, or all edges entering PLANT_2, respectively.

FROM	TO	COST
JONES	PLANT_1	13.50
JONES	PLANT_2	15.20

FROM	TO	COST
JONES	PLANT_2	15.20
MORGAN	PLANT_2	15.00
SMITH	PLANT_2	14.00

Sample output from LIST FROM JONES and LIST TO PLANT_2.

In addition, the ADJACENT keyword allows selective listing of all edges entering or leaving a given node as in

```
LIST ADJACENT JONES
```

It is also possible to change the way data fields appear in such listings. For example, the command line

```
DECIMALS 1
```

will change the number of decimal digits (to the right of the decimal point) that are displayed by subsequent LIST commands. Enter now the line

```
LIST FROM SMITH
```

and observe how the display of cost data has been changed.

FROM	TO	COST
SMITH	PLANT_1	11.7
SMITH	PLANT_2	14.0
SMITH	PLANT_3	11.5

Sample output from LIST FROM SMITH.

Another way of modifying the displayed results from LIST is through the NFORMAT and EFORMAT commands, usually abbreviated using NF and EF. For example, the command line

```
   NF NAME
```

suppresses the output of the supply field in subsequent listings of node data. Similarly, the command

```
   EF TO FROM
```

suppresses the cost field and rearranges the order in which the other two fields are displayed. In general,

- The NFORMAT command can select and rearrange the various node data fields (NAME cannot be suppressed however).

- The EFORMAT command can select and rearrange the various edge data fields (FROM and TO cannot be suppressed however).

4.3 Controlling Output

NETSOLVE allows the user to obtain a log of selected portions of a session's activity by using the JOURNAL command. When the journal is "on", all output produced by the system is reproduced in a text file, which may later be printed off-line to obtain a permanent record of the session. The command lines to turn the journal on and to turn the journal off are simply

```
   JOURNAL ON
   JOURNAL OFF  .
```

Because the journal can be switched on and off several times during a session, a selective record of the session's work can be obtained. The journal recording is saved in the file path.JRN, where "path" is the current DOS path associated with the session. (See Section 7 for a more complete discussion of paths and external data files, and Section 8.3 for the ability to route output to several different journals.)

In a similar way, the terminal can be turned on and off. Normally the terminal remains on throughout a session and all output will be listed at the terminal. However, in cases where a great deal of output is to be generated, either by listing a large network in its entirety or running an algorithm on such a network, it is desirable to turn the journal on and the terminal off. Only summary information will then appear at the terminal (e.g. in the case of output from algorithms), while the full set of output information will be routed to the journal. As you might have guessed, the command lines to turn the terminal on or off are just

```
   TERMINAL ON
   TERMINAL OFF  .
```

An additional command useful for enhancing the readability of the journal output is the NOTE command. In essence, the user can add documentation to the output in the journal by sending a message to the journal directly. For example, the command line

```
NOTE What follows is a listing of JOBNET
```

will produce "What follows is a listing of JOBNET" as the next line in the journal output.

The final command discussed here is the STATUS command, which produces a listing of information about the currently configured system and network. There are four categories of information that can obtained using this command. Entering

```
STATUS NETWORK
```

produces a listing of the name, orientation, and size of the current network. Entering

```
STATUS SIZE
```

produces a listing of the maximum size of network that can be processed in NETSOLVE. The command line

```
STATUS DEFAULT
```

lists the "system defaults" (see Section 8.1) while

```
STATUS OUTPUT
```

lists the status of the terminal and journal as well as the values of additional output-related parameters. Entering STATUS alone generates a listing of all four categories of information.

```
SYSTEM DEFAULTS
      SUPPLY                                  0.0
      LOWER                                   0.0
      UPPER                              999999.0
      COST                                    1.0

OUTPUT PARAMETERS
      TERMINAL                      ON
      JOURNAL             NETSOLVE  ON
      PATH                          .\
      DECIMALS                       1
      NODE FORMAT                   NAME      SUPPLY
      EDGE FORMAT                   FROM      TO        COST

STORAGE CAPACITIES
      NODES                                  50
      EDGES                                 200

CURRENT NETWORK
      NAME                          JOBNET
      ORIENTATION                   DIRECTED
      NODES                            6
      EDGES                            7
```

Sample output from STATUS.

5.0 Modifying Networks

Now that we have seen how to enter a network and how to list out portions of a network, it is appropriate to consider how to modify the network currently on the workbench. The listing facilities of NETSOLVE may well have revealed an error in entering some data for the network, and it is then imperative to be able to correct such errors. Even after a network has been verified and has been analyzed using one of the network algorithms, it is frequently desired to study the sensitivity of the reported solution to changes in the problem data. The ability to modify networks in a painless way thus permits rapid sensitivity analyses to be performed and various "what if" scenarios to be evaluated. In this section, a number of *editing* features in NETSOLVE will be explained.

Perhaps the simplest facility is that of adding nodes or edges to the current network using the ADD command. Entering

```
ADD NODE ATLANTA
```

will add the new (isolated) node ATLANTA to the network. The supply field can also be specified by using, for example

```
ADD NODE ATLANTA SUPPLY=3.5
```

Similarly, a new edge from ATLANTA to BOSTON having an edge cost of 1321 can be added, using

```
ADD EDGE ATLANTA BOSTON COST=1321
```

Any of the optional edge data fields (LOWER, UPPER, COST) can be specified when adding a new edge, if desired.

Of course, since NETSOLVE initializes the workbench at the very start to consist of a "null network" (no nodes or edges), it is possible to bypass the CREATE command and build up a network by issuing successive ADD commands. In this case, the user should recognize that the null network has the default name of NETWORK and is assumed directed, so any network constructed by successive additions of nodes/edges to the null network will likewise have this name and orientation.

A more efficient way to add several new nodes or new edges to an existing network is by means of the APPEND command. The simplest form of this command is

```
APPEND
```

This will allow the entering of node and edge data just as if we were back in CREATE – except in this case, there may already be nodes and edges present in the network. The name and orientation of the network, of course, stay the same. The node/edge formats are those specified when the original network was created (or if NFORMAT or EFORMAT had been used subsequently, the latest definitions for the node/edge formats are used).

The current node and edge formats can be viewed by using the STATUS OUTPUT command line.

It is also possible to override the current node/edge formats (and define new ones) explicitly in the APPEND command; for example, entering

    APPEND NF=(NAME SUPPLY) EF=(FROM TO LOWER)

will redefine the current node/edge formats and will allow the user to enter node and edge data (just as in CREATE) according to these new node/edge formats. Normally one would not be so devious as this, but there are circumstances (see the material on defaults in Section 8.1) where it may be advantageous to do so.

The DELETE command removes nodes or edges from the network. For example,

    DELETE NODE ATLANTA

will remove the node named ATLANTA from the network, as well as any edges incident with this node. Similarly, the command line

    DELETE EDGE ATLANTA BOSTON

will remove the edge between ATLANTA and BOSTON from the network. These two nodes will still remain in the network, however.

The CHANGE command allows the user to modify the data fields associated with a particular node or edge. For instance, entering

    CHANGE NODE ATLANTA NAME=ATL SUPPLY=3.0

will change the current name of that node from ATLANTA to ATL and will change the current supply value associated with this node. In a similar way,

    CHANGE EDGE ATL PHX COST=1421 LOWER=10.0

will change the cost and lower bound data fields associated with this edge to those values specified. More generally, any of the optional data fields for edges can be changed in a single command line and can be listed in any order on the command line.

A final way to modify the current network (rather drastically, in fact) is by use of the CLEAR command, which completely removes the current network. The user is given a second chance to reconsider this option before proceeding. It should also be mentioned that the successful CLEAR will reset the node and edge formats to NAME and FROM, TO respectively.

6.0 Network Algorithms

This section discusses the nine network analysis algorithms available within NETSOLVE. Most of these are exact optimization routines: namely, they are guaranteed to produce an optimal solution to the problem. One of these (the TSP procedure) is a heuristic algorithm for generating a good, but not necessarily optimal, solution to the problem. All algorithms are efficient, well tested, state-of-the-art procedures that have been implemented using sophisticated data structures (e.g. in order to exploit network sparsity). Table 1 summarizes the input requirements and restrictions relative to the use of such algorithms. Each algorithm is described separately in the following subsections. It should be noted that the presentation of numerical output from any algorithm can be modified by use of the DECIMALS command prior to executing the algorithm.

6.1 Assignment

This algorithm determines a minimum cost (or maximum value) assignment of supply nodes to demand nodes in a directed network. Supply nodes have a value of +1 in the SUPPLY data field; demand nodes have a value of -1. All nodes must be either supply or demand nodes and each edge must be directed from a supply node to a demand node. The total cost of an assignment is given by the sum of the COST data fields over those edges used to assign a distinct supply node to each demand node. (Note that there must be at least as many supply nodes as demand nodes.)

This algorithm is executed on the current network by issuing the command line

```
     ASSIGN
or
     ASSIGN MIN   .
```

If the keyword MAX is used in the second form above, then a maximum value assignment (based on the COST field) will be produced. In either case, output consists of the total cost (or value) of the optimal assignment and (if the terminal is ON) a list of the edges used to assign supplies to demands. Assignments are indicated by a FLOW value of 1 in the edge list display.

In addition, the user will be asked whether a sensitivity analysis should be conducted on the problem. This analysis produces *dual variables* for each node and *reduced costs* for each edge. Also, *tolerance* intervals are produced for each relevant edge cost, such that varying the individual edge cost within that range will not change the optimality of the current solution.

Particular assignments can be forced or prohibited by appropriate specification of the LOWER and UPPER data fields on individual edges. Note that for any edge we must have UPPER at least as large as LOWER. The LOWER value can be 0 (no restriction) or 1 (forcing an assignment). The UPPER value can be 0 (prohibiting an assignment) or greater than or equal to 1 (no restriction). The normal default values assigned to LOWER and UPPER (see Section 8.1) are such that no restrictions are placed on the solution.

Table 1. Network Algorithms Available in NETSOLVE

Command Name	Optimization Problem	Data Fields Used	Restrictions
ASSIGN	Assignment	SUPPLY, LOWER, UPPER, COST	Directed networks with all edges directed from supply nodes (SUPPLY = 1) to demand nodes (SUPPLY = -1). UPPER ≥ LOWER = 0,1 for each edge.
LPATH	Longest Path	COST	Directed, acyclic networks.
MATCH	Weighted Matching	SUPPLY, COST	Orientation of edges ignored. Supply and cost data rounded to precision specified by DECIMALS command.
MAXFLOW	Maximum Flow	LOWER, UPPER	Directed networks. UPPER ≥ LOWER ≥ 0 for each edge.
MINFLOW	Minimum Cost Flow	SUPPLY, LOWER, UPPER, COST	Directed networks. UPPER ≥ LOWER ≥ 0 for each edge. Total supply ≥ total demand.
MST	Minimum Spanning Tree	COST	Directed or undirected network. Orientation of edges is ignored.
SPATH	Shortest Path	COST	Directed or undirected networks. No directed cycles of negative length.
TRANS	Transportation	SUPPLY, LOWER, UPPER, COST	Directed networks with all directed from supply nodes to demand nodes. UPPER ≥ LOWER ≥ 0 for each edge. Total supply ≥ total demand.
TSP	Traveling Salesman	COST	Directed or undirected networks. Approximate solution produced.

Thus, the user who wishes to solve an assignment problem having neither required nor prohibited assignments need only specify values for the SUPPLY and COST data fields.

6.2 Longest Path

This algorithm finds a longest length path between a given pair of nodes in a directed acyclic network. The "length" of an edge is the value specified by the COST data field, and the length of a path is the sum of the lengths of the edges on the path. The existence of a directed cycle will be detected by the algorithm.

For example, to determine a longest length path from the node named ATL to the node named PHX, the command line

```
LPATH ATL PHX
```

is entered. The length of a longest path is displayed at the terminal. If the terminal is currently ON, then the edges on the longest path and their lengths are also displayed.

In addition, the user will be asked whether a sensitivity analysis should be conducted on the problem. This analysis produces *tolerance* intervals for each edge on the longest path, such that varying the individual edge length within that range will not change the longest path.

This algorithm is useful for doing PERT-type calculations in networks representing activities and their precedences. In this case, the first node specified (ATL in the above example) would correspond to the beginning of the project and the second node (PHX) to the completion of the project. The length of an edge (i.e. the value in the COST data field) could be an estimate of the time required to complete an activity. A longest path then identifies the activities that are most crucial for completing the project on time.

6.3 Weighted Matching

This algorithm determines an optimal matching in a network without self-loops. (If self-loops are present in the network, they are in fact ignored.) The network can be directed or undirected, but edge direction is ignored. A matching consists of a set of nonadjacent edges, each of which has a "weight" specified by the COST data field. In addition, a matching may contain isolated nodes, which are considered to be matched to themselves, incurring a certain "self-loop weight" defined below. An optimal matching is a matching that has minimum (or maximum) total weight, where the weight of a matching is simply the sum of all constituent self-loop and edge weights. Because the matching algorithm requires integer data, all weights are rounded to the precision set by the current DECIMALS value and then scaled to integer values. The output will list the actual weights.

There are two ways in which self-loop weights can be defined. In the first, the self-loop weight of a node is given by the associated SUPPLY data field for that node. In the second, signified by the presence of the keyword SLCOST, a common self-loop weight value is incurred for matching any node to itself. In calculating a minimum weight matching (signified by the keyword MIN), the user should normally define the common self-loop weight as a large positive value; for a maximum weight matching (signified by the keyword MAX), the user should employ a large negative value for the common self-loop weight.

As an example, to determine a minimum weight matching for the current network, the command line

```
MATCH MIN SLCOST=1000
```

can be used. Here the weight associated with matching any node to itself is 1000. If the command line

```
MATCH MIN
```

were used instead, then the values found in the SUPPLY data field would be used as weights for matching nodes to themselves. In a similar way, the keyword MAX when used in place of MIN in the above examples will generate a maximum weight matching for the network.

The first line of output displays the total weight of an optimal matching. If the terminal is ON, then the edges and nodes comprising this matching are displayed, together with the associated edge/self-loop weights.

6.4 Maximum Flow

This algorithm determines a maximum flow from an origin node to a destination node in a capacitated, directed network. Edge flows must lie between a nonnegative lower bound and an upper bound specified by the LOWER and UPPER data fields respectively. The value of a flow is the net flow *leaving* the origin node (or, equivalently, *entering* the destination node). Flows are required to satisfy *flow conservation* at every node other than the origin and destination. That is, the total edge flow out of a node must exactly equal the total edge flow into that node.

This algorithm is executed by entering the command line

```
MAXFLOW ORIGIN DEST
```

which would find the maximum flow from the node named ORIGIN to the node named DEST. If an optimal solution is found, the first line of output contains the value of the optimal flow. If the terminal is ON, a listing of the edge data fields and edge flows for edges having nonzero flow is produced. If an optimal solution does not exist, then an indication is given of whether the problem is infeasible or the solution is unbounded. Additional information is produced if the terminal is ON and the formulation is infeasible. In that case, a list of nodes where flow conservation cannot be attained is displayed.

6.5 Minimum Cost Flow

This algorithm determines a minimum cost (or maximum value) flow from supply nodes to demand nodes in a capacitated, directed network. Supply nodes have a positive value in the SUPPLY data field. Demand nodes have a negative value and transshipment nodes have a zero value. Total supply must be greater than or equal to total demand (i.e. the sum of the SUPPLY values over all nodes is nonnegative). Edge flows must lie between a nonnegative lower bound and an upper bound specified by the LOWER and UPPER data fields respectively. The total cost (or value) of a flow is the sum (over all edges) of the edge flow multiplied by the unit edge weight given in the COST data field. Flows are required to satisfy *flow conservation* at every node. That is, the total edge flow out of a node minus the total edge flow into that node must exactly equal the node supply for demand and transshipment nodes, and must not exceed node supply for supply nodes.

This algorithm is executed by entering the command line

```
    MINFLOW
```
or
```
    MINFLOW MIN
```

An alternative form with the keyword MAX instead of MIN is used to find a maximum value flow in the network. If an optimal solution is found, the first line of output consists of the cost (value) for the optimal flow. If the terminal is ON, a listing of the edge data fields and edge flows for edges having nonzero flow is produced. If an optimal solution does not exist, then an indication is given of whether the problem is infeasible or the solution is unbounded. Additional information is produced if the terminal is ON. In the case of an infeasible formulation, a list of nodes where flow conservation cannot be attained is displayed. For an unbounded solution, a list of edges having a negative unit cost (positive unit value) and infinite flow is exhibited for a minimum cost (maximum value) problem.

In addition, the user will be asked whether a sensitivity analysis should be conducted on the problem. This analysis produces *dual variables* for each node and *reduced costs* for each edge. Also, *tolerance* intervals are produced for each relevant edge cost, such that varying the individual edge cost within that range will not change the optimality of the current solution.

6.6 Minimum Spanning Tree

This algorithm determines a minimum cost (or maximum value) spanning tree in a directed or undirected network. The "cost" associated with an edge is the value specified in the COST data field, and the cost of a spanning tree is simply the sum of all edge costs comprising the tree. This problem is most naturally formulated for undirected networks; when the given network is directed, the orientation of edges is ignored and the algorithm produces an optimal spanning tree for the underlying undirected network. If the network is connected, then an optimal spanning *tree* is determined; otherwise, an optimal spanning *forest* (spanning each connected component) is determined. The command line to implement this algorithm is simply

```
        MST
or
        MST MIN
```

which will produce the total cost of a minimum spanning tree (or forest) for the network. On the other hand using

```
        MST MAX
```

will produce the total value of a maximum value spanning tree (or forest). If the terminal is currently ON, then in addition the (undirected) edges comprising an optimal spanning tree (or forest) will be displayed, together with the costs of the associated edges. Isolated nodes in the network will also be displayed.

Additionally, the user will be asked whether a sensitivity analysis should be conducted on the problem. This analysis produces *tolerance* intervals for each relevant edge cost, such that varying the individual edge cost within that range will not change the optimality of the current tree.

The MST command can also be used to determine the connected components of an undirected graph. In this case any values defined for the COST data field (such as the default settings given in Section 8.1) will suffice. The output from MST will group together nodes comprising the same connected component.

6.7 Shortest Path

This algorithm determines shortest length paths between various node pairs in either a directed or undirected network. The "length" of an edge is the value specified by the COST data field, and the length of a path is simply the sum of all edge lengths comprising the path. The edge lengths can have any algebraic sign, but it is necessary for all directed cycles to have nonnegative length. (However, any negative length cycle will be detected and reported by the algorithm.)

Several variants of the SPATH command are available for performing shortest path analyses, depending on whether one needs to determine shortest paths

- from a given source node to a given destination node
- from a given source node to all network nodes
- from all network nodes to a given destination node
- from any network node to any other network node.

For example, to determine a shortest path from the node named ATL to the node named PHX, the command line

```
        SPATH ATL PHX
```

is entered. The length of a shortest path is displayed at the terminal. If the terminal is currently ON, then in addition the edges comprising a shortest path are displayed, together with the length of each such edge. Also, the user will be asked whether a sensitivity analysis should be conducted on the problem. This analysis produces *tolerance* intervals for each edge on the shortest path, such that varying the individual edge length within that range will not change the shortest path.

If the command line

```
SPATH ATL *
```

is entered, then shortest paths are calculated from ATL to all other nodes in the network. If the terminal is ON, then the "distance" (shortest path length) to each accessible node is displayed, together with its "predecessor" node. The predecessor of any node A is the node immediately preceding A on a shortest path from the source node to A. By repeatedly consulting the predecessor list, a shortest path from the source node to A can be easily obtained. In addition, the user will be asked whether a sensitivity analysis should be conducted on the problem. This analysis produces *dual variables* for each (accessible) node and *reduced costs* for each edge. Also, *tolerance* intervals are produced for each relevant edge length, such that varying the individual edge length within that range will not change the optimality of the current shortest path tree.

In a similar way, the command line

```
SPATH * ATL
```

generates information on shortest paths from all network nodes to ATL. In this case, however, a "successor" node is displayed, indicating the node immediately following a given node on a shortest path from that given node to the destination node. Finally, the command line

```
SPATH * *
```

produces information on shortest paths from any network node to any other network node. Because the output generated for this "all pairs" problem can be rather substantial, the user may wish to turn the terminal OFF and turn the journal ON when solving this type of problem for large networks.

6.8 Transportation

This algorithm determines a minimum cost (or maximum value) shipping pattern from supply nodes to demand nodes in a directed network. All edges must be directed from supply nodes (having a positive value in the SUPPLY data field) to demand nodes (having a negative SUPPLY value). All nodes must be either supply or demand nodes, and total supply must meet or exceed total demand. Shipments (or flows) between particular supply and demand nodes must lie between a nonnegative lower bound (specified by the LOWER data field) and an upper bound (specified by the UPPER data field).

The algorithm is executed by issuing the command line

or
```
TRANS

TRANS MIN   .
```

A maximum value shipment can be determined by using the keyword MAX in the second form above. If an optimal solution is determined, the cost (value) of the associated shipping pattern is displayed. If the terminal is ON, an additional listing of all edges used in the optimal solution is displayed along with the associated edge data fields. If no optimal solution exists then the system reports an infeasible formulation and (if the terminal is ON) a list is produced of nodes at which supply is insufficient or demand is excessive.

In addition, the user will be asked whether a sensitivity analysis should be conducted on the problem. This analysis produces *dual variables* for each node and *reduced costs* for each edge. Also, *tolerance* intervals are produced for each relevant edge cost, such that varying the individual edge cost within that range will not change the optimality of the current solution.

6.9 Traveling Salesman

This algorithm finds a near-optimal tour in a directed or undirected network. Each node in the network is to be visited exactly once along the tour. The "length" of an edge is the value specified by the COST data field. The length of a tour is the sum of the lengths of the edges on the tour. It is desired to find a tour of optimal length passing through all network nodes and using only existing network edges.

Five heuristics can be used to find an approximate solution to this traveling salesman problem. If the keyword MIN is specified then a tour of minimum length is sought, while if MAX is specified then a tour of maximum length is sought. The heuristic used is selected from: CHEAPEST insertion, FARTHEST insertion, NEAREST insertion, ORDINAL (arbitrary) insertion, or nearest NEIGHBOR. If none are specified, then the farthest insertion technique (which appears to work well in practice) is used. These heuristics tend to perform best for undirected networks, or directed networks with symmetric edge lengths.

Generally, each heuristic specifies an initial configuration, together with a method for selecting a next node and its placement within the current configuration. The *nearest neighbor* heuristic starts with a degenerate path (a node) and concatenates to the end of this path a node that is nearest (with minimum edge length) to the last node on the path. On the other hand, the *nearest insertion* procedure maintains a subtour, to which is added at each step a node closest to any node in the current subtour. This node is then inserted in the best possible position within the subtour, increasing the subtour length by as little as possible. In contrast, the *farthest insertion* procedure chooses a node that is farthest from any node in the current subtour, while the *ordinal (arbitrary) insertion* procedure selects the nodes in lexicographic order by node name. Finally, the *cheapest insertion* heuristic simultaneously selects the node as well as its position in the current subtour to achieve the smallest increase in subtour length.

For example, to find an approximate minimum length tour using the cheapest insertion technique, the command line

```
TSP CHEAPEST MIN
```

can be entered. The length of the shortest tour found is displayed at the terminal. If the terminal is ON, then the edges defining the tour and their lengths are also displayed.

7.0 External Data Files

Because the workbench used by NETSOLVE has space for only one network at a time, it is desirable to have a mechanism for transferring networks to and from the workbench. For example, the user may wish to store on a more permanent basis the present network, since upon exit from NETSOLVE this network is lost. Or, a user may wish to retrieve a previously stored network and place it onto the workbench. These functions, and others, are accomplished by reading and writing information on *external data files* – data files maintained externally to the NETSOLVE system, but accessible through the computer's operating system.

On the IBM PC/compatible series, all external data files associated with NETSOLVE are qualified by a user-specified *path* (in DOS 2.0 or later). The path normally assumed is "." and so the journal feature (by default) will write out information to the external data file NETSOLVE.JRN in the current drive and directory. The path associated with any external data file may be dynamically changed within a NETSOLVE session by means of the PATH command. By entering

    PATH \JSMITH

the system is instructed to qualify any external data files created or retrieved with the path "\JSMITH". More generally, any sequence of at most 16 characters can be specified as the path. **NOTE**: NETSOLVE does not check the validity of a path! The current path can be displayed at any time by entering the command line

    STATUS OUTPUT   .

As a concrete example, suppose that after creating, listing, and modifying the network JOBNET, we would like to store it permanently in an external data file. The command

    SAVE

will copy the current network to the external data file \JSMITH\JOBNET.NET (if JSMITH is the current path). At a later occasion, either during the same or another session, this network can be retrieved using the command

    GET JOBNET

(assuming the path is set to "\JSMITH"; note that the suffix .NET is automatically supplied by NETSOLVE). The network presently on the workbench will be destroyed and overwritten with JOBNET, however. For this reason, if a network is presently on the workbench and has not itself been saved, the system will give the user a second chance to save the present network before proceeding.

The disk provided with the NETSOLVE system contains two sample networks which may be read in using the GET command. The first network NET1 is a directed network having 4 nodes and 6 edges. (Try using MINFLOW on this example.) The second network NET2 is an undirected network having 6 nodes and 9 edges; this example can be analyzed using the MATCH, MST, SPATH, and TSP algorithms if desired. Figures 2 and 3 display these sample networks.

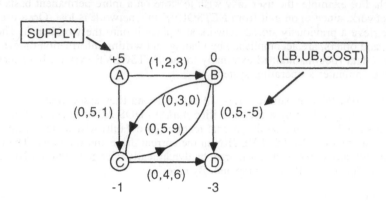

Figure 2. NET1

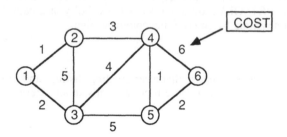

Figure 3. NET2

There are a few points worth bearing in mind about the SAVE and GET commands.

• The SAVE command, as illustrated here, writes the current network out in a condensed "internal" format. Thus the contents of the external data file will be unintelligible to humans, but will be perfectly acceptable when read back using the GET command.

• External data files are dynamically allocated, opened, and closed by the NETSOLVE system. The user need not do such things prior to entering the NETSOLVE system.

• The GET command provides an alternative to the CREATE command when the same network is used repeatedly. This feature can be especially useful for conducting sensitivity analyses on a given network.

• When a network is stored in internal format using SAVE, certain relevant network characteristics (e.g. node/edge formats) as well as certain system parameters (default values, current decimal setting) are written out. Upon issuing a GET, these network characteristics and system parameters are restored.

An alternative method of storing a network that writes out only network, but not system, parameters is described in Section 8.2.

8.0 Advanced Features

The previous sections have illustrated how to use the NETSOLVE command language to perform a variety of functions, such as creating, listing, modifying, and analyzing a network. In the present section, certain advanced uses of the NETSOLVE commands are described under the general topics of "Defaults," "Representation Types," and "Multiple Journals."

8.1 Defaults

When a network is constructed using the extended form of CREATE, the NF and EF specifications on the command line, or NF=(NAME) and EF=(FROM TO) if not explicitly stated, are used to direct the flow of input information. Thus, only those node/edge data fields so specified need actually be entered during the two phases of CREATE. However, the data fields not specified are in fact defined also, by assigning to them the current system *default* values for these fields. For instance, in defining a network via

```
CREATE HIGHWAY UNDIR EF=(FROM TO COST)
```

only the data fields NAME, FROM, TO, COST are expected by the system. We have previously seen that if only the FROM and TO fields are actually entered on a line, then the COST data field receives the current default value for the COST field. In the same spirit, the other unspecified node data field SUPPLY and the other unspecified edge data fields LOWER, UPPER are assigned values using the current default values for these fields. Normally, these will be the values given in Table 2, so that all nodes will have SUPPLY=0.0 and all edges will have LOWER=0.0 and UPPER=999999.0 in this example. Notice that the value 999999.0 given here represents the system's version of "infinity".

Table 2. Normal System Default Values

Data Field	Default Value
SUPPLY	0.0
LOWER	0.0
UPPER	999999.0
COST	1.0

A similar substitution of default values occurs when using the APPEND or ADD commands. The default values can, however, be easily changed by the use of the DEFAULT command. For example, the command line

```
DEFAULT LOWER=2.5 SUPPLY=-1
```

will reset the default values for LOWER and SUPPLY to those designated, and these new values will be used in subsequent processing. The current default values for all data fields can be displayed at any time by issuing

```
STATUS DEFAULT   .
```

Because the default values are considered to be *system* parameters, they are not reset by the CLEAR command.

The concept of default values can be used to streamline the effort required to input networks under certain circumstances. As an example, suppose one had a network with all (or many) input COST values of 1. Then instead of entering this same value repeatedly for every such edge, we could use the normal default value for COST (1.0) and the command line

```
CREATE HIGHWAY UNDIR
```

to enter only the specified information (FROM,TO) for each edge. The COST field will automatically be assigned the default value of 1.0 for every edge; any exceptions to this can be rectified by using the CHANGE command on such edges.

A similar strategy can be followed if the desired input value is not the normal default value by simply using the default command to set this value for the desired field prior to entering the CREATE command. If, for instance, half of the edges had a COST value of 2.0 and the rest had a COST value of 4.0 then the first group of edges could be entered using

```
DEFAULT COST=2.0
CREATE HIGHWAY UNDIR   .
```

The remaining group could be entered via the command lines

```
DEFAULT COST=4.0
APPEND   .
```

Other variations are possible as well. For example, if the second group of edges referred to above had a variety of COST values instead of a common value, this portion could be entered by using

```
APPEND EF=(FROM TO COST)   .
```

Thus the edge format is redefined to show COST explicitly, and the three fields FROM, TO, COST can be entered for those edges with varying COST values.

At any time, the NFORMAT and EFORMAT commands can be used to redefine the current node and edge formats. Subsequent listing commands can then be used to display the defaulted/entered values for any desired data field. This may be a desirable way to verify the values assigned to unspecified data fields after using CREATE or APPEND.

Another use of defaults occurs in conjunction with executing certain network algorithms. For example, the assignment algorithm ASSIGN requires that the data field SUPPLY be defined (1 for "supply" nodes, -1 for "demand" nodes). Rather than entering the values 1 and -1 repeatedly for each node, the DEFAULT command can be used (with SUPPLY=1) for entering only the supply nodes, and then used again (with SUPPLY=-1) for entering only the demand nodes. In addition, the ASSIGN command requires values for each of the edge data fields LOWER and UPPER. If a standard assignment problem (without required or excluded pairings) is being analyzed, then the normal default values for LOWER and UPPER will suffice. Namely, only the COST field need be entered for edges; the default values LOWER=0.0 and UPPER=999999.0 are appropriate for specifying such a problem, and thus will automatically be associated with every edge.

8.2 Representation Types

When networks are stored on or retrieved from external data files, two types of representations are possible. The first type, an INTERNAL representation, provides a compact way of encoding data. This representation is most appropriate for saving a network which is at a later time to be read back into NETSOLVE.

A second representation type, called the EDGELIST representation, is more appropriate for reading in an external data file not generated by NETSOLVE or for writing out the current network in a number of different formats (e.g. for use by other computer codes outside of NETSOLVE). The EDGELIST representation employs basically the same format as that used to enter network data within CREATE or APPEND. Namely, it consists of two *header* records: the first contains the single keyword EDGELIST, the second contains keywords giving in turn the network orientation, node format, and edge format. (Table 3 lists the possible keywords for these characteristics.) Following the header is the set of node data (arranged with respect to the node format) and then the set of edge data (arranged with respect to the edge format). Each node or edge has one associated record, and the node records are separated from the edge records by a blank record.

Table 3. Acceptable Keywords for Second Header Record

```
Orientation:    DIR     UNDIR

Node Format:    NAME    SUPPLY

Edge Format:    FROM    TO    LOWER    UPPER    COST
```

To output the current network so that node data are listed in the order SUPPLY, NAME and edge data are listed in the order LOWER, COST, FROM, TO the following command lines are used

```
NF SUPPLY NAME
```

```
EF LOWER COST FROM TO
SAVE SAMPLE1 EDGELIST   .
```

The network is saved in file \JSMITH\SAMPLE1.NET (assuming that \JSMITH is the current path) using the EDGELIST representation. By allowing the flexibility of selecting and arranging how the output data appear, this option can be useful in formatting, or reformatting, a given network for use in other codes external to NETSOLVE. The current setting for DECIMALS dictates the number of decimal digits appearing in numerical data that is output using the EDGELIST representation.

In an analogous way, a network constructed externally to NETSOLVE can be read into the system by using the command line

```
GET SAMPLE2   .
```

It is assumed here that the input network is to be found on \JSMITH\SAMPLE2.NET and that the network is arranged in EDGELIST form. Note that all EDGELIST representations have an initial record containing the word EDGELIST. In this way, the NETSOLVE system can detect the representation type and the keyword EDGELIST is not needed on the GET command line.

Unlike the case when saved in INTERNAL representation, a network saved in EDGELIST form does not record the current settings for system parameters (defaults, decimals). Accordingly, when the GET command is used to retrieve such a network, the present values of system parameters are not changed. This feature can be useful for redefining the values appearing in certain data fields. For example, to change all node supplies in the current network to -1, the following commands can be issued

```
NF NAME
SAVE SAMPLE3 EDGELIST
DEFAULT SUPPLY=-1
GET SAMPLE3   .
```

A final command can be used to merge a new network (maintained on an external data file) into the current network. The new network must have the same orientation (directed/undirected) as the current network and must be node disjoint from the current network (no duplication of node names). Moreover, the new network must be represented in EDGELIST form. (For example, it could have been previously saved using the EDGELIST keyword.) Upon entering the command line

```
MERGE NEWNET
```

the new network in \JSMITH\NEWNET.NET will be merged into the current network on the workbench. This option can be useful when individual pieces of a large network have been separately created and now need to be reassembled. Once the pieces have been merged together, edges can be added between the various pieces to "glue" them together.

8.3 Multiple Journals

Section 4.3 discussed how the journal can be turned on and off, in order to record in a *cumulative* way various output segments during a NETSOLVE session. There are occasions where it would be desirable to wipe clean the current journal and restart the recording of output, and other occasions when one would like to route output to separate external files. The present section discusses how this can be done.

It is important to realize that the standard (default) journal name is NETSOLVE, and journal output will be placed in the external file NETSOLVE.JRN, appropriately qualified by the current path name. Therefore, if the command

```
JOURNAL ON
```

is issued and the journal file NETSOLVE.JRN already exists from a previous session, the user will be asked whether or not to <u>append</u> new journal output to this file or simply to <u>overwrite</u> the old information. This provides a mechanism for keeping a cumulative record of successive sessions, as well as a cumulative record within a session.

In addition, the user can route output to a new journal (say, called SECOND) by using an expanded form of the JOURNAL command:

```
JOURNAL SECOND ON   .
```

This informs the system that subsequent output will be channeled to the external file called SECOND.JRN. As before, turning the journal off (using JOURNAL OFF) and on again (using JOURNAL ON) will selectively place output in this second journal file. In this way, a number of different journal files can be maintained for receiving selected portions of the output generated during a given session.

Appendix A. Commands and Command Syntax

A.1 Commands

The following commands provide information about NETSOLVE:

ALIAS	:	Lists aliases for NETSOLVE keywords
COMMANDS	:	Lists the available commands
HELP "COMMAND"	:	Provides information on specific commands
STATUS	:	Lists NETSOLVE system parameters

All descriptions of command syntax use
" ... " to denote a user-supplied expression
< ... > to denote an optional expression
< ... >*   when an expression can be repeated

A NETSOLVE session is ended by the QUIT command.

A.2 Command Syntax

```
ADD EDGE "NODENAME" "NODENAME" <"EPAR"="VALUE">*
ADD NODE "NODENAME" <SUPPLY="VALUE">
 "NODENAME" : Name of node (at most 8 characters)
 "EPAR"     : Selected from COST,LOWER,UPPER
 "VALUE"    : Real or integer quantity

ALIAS

APPEND <NF=(<"NFMT">*)> <EF=(<"EFMT">*)>
 "NFMT" : Selected from NAME,SUPPLY (must include NAME)
 "EFMT" : Selected from COST,FROM,LOWER,TO,UPPER (must
            include FROM,TO)

ASSIGN <"OPT">
 "OPT"    : Optimization type (MIN/MAX).  The default is MIN.

CHANGE EDGE "NODENAME" "NODENAME" <"EPAR"="VALUE">*
CHANGE NODE "NODENAME" <"NPAR"="VALUE">*
 "NODENAME" : Name of node (at most 8 characters)
 "EPAR"     : At least one selected from COST,LOWER,UPPER
 "NPAR"     : At least one selected from NAME,SUPPLY
 "VALUE"    : Real or integer quantity

CLEAR <NOSAVE>

COMMANDS
```

```
CREATE
CREATE "NETNAME" "TYPE" <NF=(<"NFMT">*)> <EF=(<"EFMT">*)>
 "NETNAME" : Name of new network (at most 8 characters)
 "TYPE"    : Selected from DIR,UNDIR
 "NFMT"    : Selected from NAME,SUPPLY (must include NAME)
 "EFMT"    : Selected from COST,FROM,LOWER,TO,UPPER (must
              include FROM,TO)

DECIMALS "DIGITS"
 "DIGITS" : Number of decimal digits (0-7)

DEFAULT <"PARAMETER"="VALUE">*
 "PARAMETER" : At least one taken from COST,LOWER,SUPPLY,UPPER
 "VALUE"     : Real or integer quantity

DELETE EDGE "NODENAME" "NODENAME"
DELETE NODE "NODENAME"
 "NODENAME" : Name of node (at most 8 characters)

EFORMAT <"EFMT">*
 "EFMT" : Selected from COST, FROM, LOWER, TO, UPPER
           (must include FROM, TO)

GET "NETNAME"
 "NETNAME" : Name of a network previously saved in an external
             data file (at most 8 characters)

HELP <"COMMAND">
 "COMMAND" : Any valid command name

JOURNAL "STATE"
JOURNAL "FILENAME" ON
    "STATE" : Either ON or OFF
 "FILENAME" : Name of external data file for journal (at most
              8 characters)

LIST NODE <"NODENAME">
LIST EDGE <"NODENAME" "NODENAME">
LIST FROM  "NODENAME"
LIST TO    "NODENAME"
LIST ADJACENT   "NODENAME"
 "NODENAME" : Name of node (at most 8 characters, if omitted
              then all nodes/edges are listed)

LPATH "NODENAME1" "NODENAME2"
 "NODENAME1" : Name of source node
 "NODENAME2" : Name of destination node
```

```
MATCH "OPT" <SLCOST="VALUE">
 "OPT"   : Optimization type (selected from MIN, MAX)
 "VALUE" : Common self-loop weight for matching any node to
           itself.  Generally should be a large positive
           (negative) value for a MIN (MAX) problem.

MAXFLOW "NODENAME1" "NODENAME2"
 "NODENAME1" : Name of source node
 "NODENAME2" : Name of destination node

MERGE "NETNAME"
 "NETNAME" : Name of new network (at most 8 characters)

MINFLOW <"OPT">
 "OPT"   : Optimization type (MIN/MAX).  The default is MIN.

MST <"OPT">
 "OPT"   : Optimization type (MIN/MAX).  The default is MIN.

NFORMAT <"NFMT">*
 "NFMT" : Selected from NAME, SUPPLY (must include NAME)

NOTE "MESSAGE"
 "MESSAGE" : Text line output to journal (at most 72
             characters)

PATH "QUALIFIER"
 "QUALIFIER" : A valid path name of no more than 12
               characters.

QUIT <NOSAVE>

SAVE
SAVE "NETNAME" <"REP">
 "NETNAME" : Name under which network is to be saved (at
             most 8 characters)
 "REP"     : Either EDGELIST (for edgelist representation)
             or INTERNAL (for internal representation).
             If not specified, INTERNAL is assumed.

SPATH "NODENAME1" "NODENAME2"
 "NODENAME1" : Name of source node or "*" indicating all nodes
 "NODENAME2" : Name of destination node or "*" indicating all
               nodes
```

STATUS <"QUERY">*
 "QUERY" : Indicates category of information (DEFAULT,
            OUTPUT, SIZE, NETWORK) or specific subentries
            in categories (COST, LOWER, SUPPLY, UPPER for
            DEFAULT and DECIMALS, EFORMAT, JOURNAL, NFORMAT,
            PATH, TERMINAL for OUTPUT). A maximum of four
            queries are allowed on any command line. If none
            are specified, then all categories are listed.

TERMINAL "STATE"
 "STATE" : Either ON or OFF

TRANS <"OPT">
 "OPT"   : Optimization type (MIN/MAX). The default is MIN.

TSP <"METHOD"> <"OPT">
TSP <"OPT"> <"METHOD">
 "METHOD": Heuristic used (CHEAPEST, FARTHEST, NEAREST,
            nearest NEIGHBOR, or ORDINAL insertion).
            The default is FARTHEST insertion.
 "OPT"   : Optimization type (MIN/MAX). The default is MIN.

Appendix B. NETSOLVE Keyword Aliases

```
ADD          A
ADJACENT     ADJ
APPEND       APP
CHANGE       C
CHEAPEST     CHEAP
CLEAR        ERASE
COMMANDS     COMMAND    COMMS       COMM
COST         VALUE      LENGTH      LEN         PROFIT      WEIGHT
CREATE       CR
DECIMALS     DECIMAL    DEC
DEFAULT      DEF
DELETE       DEL
DIR          D
EDGE         EDGES      E           ARC         ARCS
EFORMAT      EF
FARTHEST     FAR
HELP         H
JOURNAL      JOUR
LIST         L
LOWER        LB
MERGE        ME
NEAREST      NEAR
NEIGHBOR     NEIGH
NETWORK      NET        GRAPH
NFORMAT      NF
NODE         NODES      N           VERTEX      VERTICES  V
NOSAVE       NO
ORDINAL      ORD
PATH         DOSPATH
QUIT         END        STOP        BYE
SAVE         S
SIZE         STORAGE
SLCOST       LCST
STATUS       ST
SUPPLY       SUP        DISTANCE    DIST
TERMINAL     TERM
UNDIR        U
UPPER        UB
```

Index

Activity
  critical, 390, 403
  dummy, 392
  time, 405
Adjacency list, 28, 31
Adjacency matrix, 27–28
Algorithm
  assignment (see Algorithm, Hungarian)
  branch-and-bound, 336
  Christofides', for capacitated Chinese
      postman, 304
  Christofides', for traveling salesman, 343
  Clarke and Wright, 350
  Dantzig shortest path, 99–102
      generalized, 111
  defined, 5, 33
  Dijkstra shortest path, 83
  Dijkstra two-tree shortest path, 92–93
  directed postman, 292–293
  double-sweep, 104–109
  dynamic flow, 206–209
  earliest arrival flow, 218–219
  earliest event time, 399
  Euler tours, 280–282
  event-numbering, 394
  exact, 34
  flow-augmenting, 187

flow with gains, 157–161
Floyd shortest path, 94–99
    generalized, 111
Ford shortest path, 88–90
greedy, 50
heuristic, 34, 341
Hungarian, 254–255
k-shortest path, 102
label-correcting methods for shortest
    paths, 82
label-setting methods for shortest paths,
    82
latest event time, 401
maximum branching, 59–65
maximum cardinality matching, 246
maximum dynamic flow, 213–214
maximum flow, 188–195
maximum-weight matching, 260–262
minimum cost (maximum flow), 209–211
mixed postman, 298
negative cycle, 139–142
network simplex, 145–148
other shortest path, 112–113
partitioning, 90–92
path decomposition, 195
path-finding, 33–34
path-scanning, 308

[Algorithm]
  preflow-push, 201–206
  simplex, 20–22
  spanning tree, 49
  Steiner tree, 57–59
  successive shortest path, 142–144
  taxonomy, 35
Arborescence
  defined, 59
  root of, 61
  in flow-augmenting algorithm, 187
  in salesman problem, 333
  shortest path, 84
  spanning, 59
  weight of, 60
Arc
  backward, 185
  bottleneck, 197
  capacity, 10
  defined, 4
  forward, 185
  head of, 4
  increasable, 184
  length, 77
  reducible, 184
  tail of, 4
  weight, 60
Arc routing problem, 278
Aronson, J., 227
Assignment, 234
Assignment problem, 250

Balinski, M. L., 179
Barr, R. S., 124
Basic solution, 21, 137
Basis, 21, 137
Basketball conference scheduling, 237–239
Bazarra, M. S., 126, 150
Bellmore, M., 336
Beltrami, E., 300
Berge, C., 323
Bisection search, 44
Bodin, L., 300, 350
Bottleneck, 112
Bowers, M. R., 126
Bradley, G. H., 146

Branch, 9
Branch-and-bound technique, 336
Branching
  algorithm, 60
  defined, 59
  spanning, 59
  weight of, 60
Brown, G. G., 167

Capacitated arc routing problem, 310
Capacitated minimum spanning tree, 54
Capacity, 10, 198
Cardinality of a set, 5
Center
  absolute, 364, 372
  defined, 364, 372
  general, 364, 372
  general absolute, 364, 372
  methods for finding, 372–379
  multi-, 387
Chinese postman, see Postman problem
Chord, 9
Christofides, N., 303, 343
Chvátal, V., 327
Circuit board drilling, 318
Clarke, G., 350
Complementary slackness conditions
  in assignment problem, 252
  defined, 20
  in directed postman problem, 292
  in flow with gains algorithm, 153
  in maximum-weight matching algorithm,
    259–260
  in minimum-cost flow algorithm, 134
Component
  connected, 8
  strongly connected, 323
Computational complexity
  defined, 35–38
  of Dijkstra's algorithm, 88
  of double-sweep algorithm, 107
  exponential, 36
  of Floyd's algorithm, 101
  of Ford's algorithm, 90
  of generalized shortest-path algorithms,
    111

polynomial, 36
of preflow-push maximum-flow
    algorithm, 203
Computer wiring, 318
Conservation of flow, 10
Constraint set, 18
Convolution, 110
Cooper, R. B., 127
Cornuejols, G., 363
Crew scheduling, 234
Critical path method (CPM), 398
Cut
  capacity of, 180
  defined, 180
Cutset
  defined, 9
  proper, 10
  simple, 10
Cycle
  absorbing, 154
  defined, 8
  directed, 8
  fundamental set of, 23
  generating, 153
  Hamiltonian, 317, 321, 323
  odd, 241, 244–245
  salesman, 317, 321
  shrinking of, 247
  simple, 8

Dantzig, G. B., 92, 94, 99
Data structures, 27
Dearing, P. M., 387
Decision box, 406
Degree
  in-degree, 6, 287
  out-degree, 6, 287
Degree-constrained spanning tree, 54
Deo, N., 77, 82
Dijkstra, E., 83
Disconnecting set, 9
Distance
  point-arc, 369
  point-vertex, 367
  vertex-arc, 367
  vertex-vertex, 367

Distribution system design, 362
Divoky, J. J., see Hung, M.
Dual graph, 6
Dual linear program, 19, 152

Edge, 3
Edge list, 27, 30–31
Edmonds, J., 195, 282
Eisner, H., 406
Elmaghraby, S., 406, 411
Employment scheduling, 16–17
Equipment replacement, 15–16
Euler, L., 1
Euler graph, 279
Euler tour, 279
Event, algorithm for numbering, 394

f-point, 366
Facility layout, 14
Finite termination
  of double-sweep algorithm, 107
  of flow with gains algorithm, 163
  of maximum-flow algorithm, 195
  of maximum dynamic flow algorithm,
    216
  of minimum-cost (maximum-flow)
    algorithm, 213
Float
  free, 402
  independent, 403
  total, 402
Flow
  augmentation, 158
  canonical, 162
  defined, 10
  dynamic, 182, 207
  earliest arrival, 217
  earliest departure, 227
  feasible, 11
  latest arrival, 223
  latest departure, 224
  lexicographic, 217, 226
  maximum, 178, 189
  minimum-cost, 123, 209
  temporally repeated, 214
  with gains, 151

Floyd, R., 94
Ford, L., 88, 195
Fulkerson, D., 195
Forest, 9

Gain, of a path, 113
Gain factor, 151
Gallo, G., 77
Garfinkel, R. S., 336
GASNET3, 129
Generalized addition, 103
Generalized minimization, 102
Glover, F., 90, 92
Golden, B., 304, 305, 308, 350
Goldman, A. J., 387
Gouila-Houri, A., 324
Graph
  bipartite, 234
  complete, 6
  connected, 8
  defined, 3
  directed, 4
  inverse, 164, 220
  matrix of, 28–30
  mixed, 296
  planar, 6, 14, 200
  strongly connected, 323
  symmetric, 287
  time-expanded replica, 207
  undirected, 3
Graph theory, 1
Group technology, 48, 69

Hakimi, S., 375, 387
Halpern, J., 387
Hamilton, Sir W. R., 317
Handler, G., 387
Happ, W. W., see Pritsker, A. A. B.
Held, M., 335, 336
Heuristic
  construction, 38
  defined, 34
  improvement, 38
  insertion procedure, 342–343
  partitioning and decomposition, 39

  tour construction, 342
  tour improvement, 342, 347
Hung, M., 92

Incidence matrix, 17, 28–30
Independence, 405
Instance of a problem, 34
Interior point, 367

Jarvis, J. P., 39, 126
Johnson, E., 195, 282

k-change, 347
Karp, R., 195, 335, 336
Karzanov, A. V., 203
Kernighan, B. W., see S. Lin
Knapsack problem, 79–80
Konigsberg bridge problem, 1, 2, 279
Kruskal, J. B., 53

Larson, R. C., 126
Lenstra, J. K., 318
Lexicographic flow, 222
Lexicographic preference, 226
Lin, S., 347
Lin, Y., 291
Linear programming
  assignment problem as, 250
  basic concepts, 18–20
  directed postman problem as, 292
  dual, 19
  flow with gains as, 151
  maximum cardinality matching as, 241
  maximum-flow problem as, 189
  maximum-weight matching as, 259
  minimum-cost flow as, 130
  minimum-cost flow with gains, 151
  primal, 19
Locally optimal solution, 347
Location theory, 362
Lockbox location, 363
Loop, 5
Lower bound
  assignment-based, 329
  minimum spanning tree-based, 333

Magirou, V. F., 318
Maimon, O., *see* Halpern, J.
Matching
 defined, 234
 maximum cardinality, 239–250
 maximum-weight, 250
Matrix
 of a graph, 28–30
 lower triangular, 136–137
 shortest-path length, 95
Maximum-capacity route, 48
Maximum dynamic flow, 207
Maximum-flow problem, 178
Maxwell, W. L., 182
McBride, R. G., *see* Brown, G. G.
Median
 absolute, 364, 372
 defined, 364, 372
 general, 364, 372
 general absolute, 364, 372
 methods for finding, 379–383
 multi-, 387
 weighted absolute, 364
Mei-Ko, K., 278
Minieka, E., 376
Mirchandani, P., 387

Nemhauser, G., 336
NETSOLVE
 description of, 39–42
 for finding longest paths, 411–413
 for finding maximum flows, 231–233
 for finding minimum cost flows, 167–170
 for finding minimum spanning trees,
  67–71
 for finding optimal assignments, 268–269
 for finding optimal matchings, 267–268
 for finding shortest paths, 114–116
 for finding traveling salesman tours,
  353–356
 for solving the transportation problem,
  170–171
Network
 defined, 5
 generalized project, 405

project, 390
 time-expanded static, 182
Node
 active, 201
 defined, 4
 sink, 10
 source, 10
 transshipment, 10
Nonnegativity constraints, 18
NP-complete, 37
NP-hard, 38

Objective function, 18
Odoni, A. R., *see* Larson, R. C.
Optimum communication spanning tree, 55
Order picking in a warehouse, 319

Pallotino, S., *see* Gallo, G.
Pang, C., *see* Deo, N.
Path
 alternating, 242
 augmenting, 242, 258
 critical, 398
 defined, 6
 directed, 8, 186
 flow-augmenting, 159, 184, 210
 Hamiltonian, 336
 *k*th shortest path, 102
 length of, 77
 shortest, 77, 81–84
 simple, 8
 weighted augmenting, 257
Postman problem
 capacitated, 303
 defined, 80, 278
 for directed graphs, 287–289
 for mixed graphs, 296–301
 for undirected graphs, 283–284
Predecessor function, 87
Preflow, 201
Prim, R. C., 53
Pritsker, A. A. B., 406, 411
Production lot sizing, 78–79
Production scheduling, 126

Program evaluation and review technique
    (PERT), 404

Ratliff, H. D., 319
Reduced cost, 135, 146
Residual capacity, 184
Residual network, 139, 186
Rinnooy Kan, A. H. G., *see* Lenstra, J. K.
Root arc, 137
Root node, 137
Rosenthal, A. S., *see* Ratliff, H. D.

Salesman problem, 37, 317
Sequencing in job shops, 319–320
Shier, D. R., 39, 104
Shortest path
    classification of, 82
    definition, 77
Simplex algorithm, 20–22, 145, 166
Sink, 10
Site selection, 179
Slack variable, 18
Source, 10
Star, 28, 32–33
Steckhan, H., 336
Symmetric difference, 5
Subgraph
    generated by edges, 8
    generated by nodes, 8

Tansel, B. C., 364, 387
Transportation problem, 148–149, 290
Traveling salesman problem (*see* Salesman
    problem)
Tree
    alternating, 244
    defined, 9
    Hungarian, 245

spanning, 9, 12, 49, 51, 334
    Steiner, 47, 57–58
Triangle inequality, 321
Truemper, K., 155
Turner, J. S., *see* Barr, R. S.

Umpire assignment, 236–237

Variance, 405
Vehicle routing and scheduling, 320
Vehicle routing problem, 350
Vertex
    adjacent, 6
    defined, 3
    degree, 6
    exposed, 242
    incidence, 6
    initial, 6
    inner, 244
    matched, 242
    outer, 244
    penultimate, 98
    predecessor, 324
    strongly connected subset, 323
    successor, 324
    terminal, 6
    unordered, 3

Weight
    of an arborescence, 60
    of a branching, 60
    matching, 250, 257
    of a tree, 46
Whitehouse, G., *see* Pritsker, A. A. B.
Wilson, R. C., *see* Maxwell, W. L.
Wong, R., *see* Golden, B.
Wright, J. W., 350

Zhao, Y., *see* Lin, Y.

T - #0203 - 071024 - C0 - 229/179/26 - PB - 9780367402808 - Gloss Lamination